# BASIC STATISTICAL METHODS

# BASIC STATISTICAL METHODS

## Fourth Edition

## N. M. Downie
*Purdue University*

## R. W. Heath
*Formerly Stanford University*

**HARPER & ROW, PUBLISHERS**
New York   Evanston   San Francisco   London

Sponsoring Editor: George A. Middendorf
Project Editor: Cynthia Hausdorff
Designer: T. F. Funderburk
Production Supervisor: Robert A. Pirrung

BASIC STATISTICAL METHODS, Fourth Edition

Library of Congress Cataloging in Publication Data

Downie, Norville Morgan, 1910-
    Basic statistical methods.

    Bibliography:  p.
    1. Statistics.  I.  Heath, R. W., joint author.
II.  Title.
HA29.D68   1974                 519.5              73-13299
ISBN  0-06-042731-0

# Contents

# Preface

This edition, like its three predecessors, is written to meet the need of the beginning student in the social sciences for a short, clear, elementary statistics book. We believe that such a book should treat the computation, interpretation, and application of commonly used statistics. No extensive attempt has been made to derive formulas or to involve statistical theory, since the mathematical background of the typical user of this book precludes effective presentation of these topics.

Like the earlier editions, this book is essentially divided into three parts. The first nine chapters present descriptive statistics, and the next seven are an introduction to statistical inference. The third part consists of two unrelated chapters, first an introduction to test theory and construction, and second, a look at the more frequently used distribution-free statistical tests.

Many portions of the fourth edition have been rewritten, and new material has been added. Altogether, we have attempted to present an up-to-date elementary text for a rapidly developing field.

Problems designed to offer practice in the techniques discussed in each chapter appear throughout the book. The answers appear in Appendix P. At the request of many users of this textbook, the senior author has written a separate study guide for use with the text.

Appendixes A to N contain material that will become more and more useful to the student as he progresses through the book. Appendix A, after the student becomes familiar with it, will save hours of computational time. The other appendixes, B through N, are associated with the statistical concepts and tests introduced in the text.

Many sincere thanks are due to the various authors and publishers who gave us permission to use the material that appears in Appendixes A to N. Special acknowledgments appear at the bottom of each appendix.

We are especially indebted to the late Ronald A. Fisher of Cambridge, to Dr. Frank Yates of Rothamsted, and to Messrs. Oliver and Boyd, Ltd., of Edinburgh, for permission to reprint Appendixes C, D, and F from their book *Statistical Tables for Biological, Agricultural, and Medical Research*.

N. M. Downie
R. W. Heath

# BASIC STATISTICAL METHODS

# Introduction
## Chapter 1

Students often wonder what statistics is and why they should bother to study the subject. For hundreds of years people have been collecting statistics. An early tribal chief, for instance, had so many armed warriors, so many horses, took a certain number of the enemy. Today we have vast quantities of data associated with sports, the stock market, traffic, law enforcement, and hundreds of other human activities. From one point of view then statistics may be considered collections of data associated with human enterprises. In a more limited sense, each individual is a statistic. From the viewpoint of a life insurance company, each of us is a statistic.

Statistics may also be considered to be a method that can be used to analyze data—that is to organize and make sense out of a large amount of material. It is with this manipulation of data that we shall be concerned in this book. Statistical methodology may be looked upon as being of three types—descriptive, correlational, and inferential.

To illustrate the three types suppose that we use the entering freshman class of a large university. Each student's folder will contain the scores of the admissions tests used by the university, such as the college boards, a high school transcript, results of a physical examination, and, perhaps, scores based upon interest and personality inventories. Taken as a whole these folders present a mass of information about the freshmen. To learn about the freshmen all these data must be studied. Let us look at a few things that we can do with them. Suppose that we use only the scores on the verbal and mathematical parts of the Scholastic Aptitude Test. We might summarize all of these scores into two distributions. We might draw graphs that would show the differences in the scores of males and females, the differences among individuals in the several schools of the university, or the differences among students from varying backgrounds. Then we could find an average score on each of the two parts of the test for the whole group and for the different subgroups. If we are concerned with the relationship of individual scores to these averages, we can change these raw scores into another type of a more meaningful nature. Centiles or standard *scores* show how an individual in the group stands in reference to others. All of the foregoing operations are included in what we call *descriptive statistics* because they give us information, or describe, the sample we are studying.

We could also examine the scores on the verbal part of this test and see if they are related to scores on the mathematical part. This is called computing a correlation coefficient. We could also correlate these test scores with the first semester grades made by these freshmen, or with their high school ranks, or with the various scores on the interest and personality inventories. The results of correlational work are useful in making predictions of future behavior. If we know that a relationship exists between two variables, then scores on one may be used to predict scores on the other. In statistics this study of prediction is referred to as regression analysis. The results of correlational analysis are used to study the reliability and validity of educational and psychological tests. *Correlational analysis,* then, is a major part of statistical methodology.

Third we have *inferential statistics.* Usually when samples are studied, the investigator is interested in going beyond the sample and making an inference about the population from which the sample was drawn. Populations are frequently so large that the only way their characteristics will ever be known is through the study of samples drawn systematically from the population. It follows then that from measures of averages and variability based upon samples we make inferences about the size of the same traits in the population. The use of inferential statistics is basic to experimental research in all branches of science.

There are reasons for studying statistics other than knowing how to use the subject in a research task. A knowledge of statistics is basic to the intelligent reading of a research article or a modern text in science. Without the background one would get from a first course in statistics, these accounts of modern science are unintelligible. Statistics is also of use to the informed instructor in building and analyzing tests and in preparing grades. Statistics may also contribute to the general education of a consumer. Modern advertising makes all sorts of claims, often bolstering them with impressive statistics. The intelligent consumer looks critically at these claims and the statistics used to support them.

## A BRIEF HISTORY OF STATISTICS

Statistics has a long and venerable history. Perhaps the earliest use of statistics was when an ancient chief counted the number of effective warriors that he had or the number he would need to defeat his enemy, or when he figured how much might judiciously be collected in taxes. In later times, statistics were used to report death rates in the great London plague and in the study of natural resources. These uses of statistics, which encompass a broad field of activity referred to as "state arithmetic," are purely descriptive in nature.

In the seventeenth and eighteenth centuries mathematicians were asked by gamblers to develop principles that would improve the chances of winning at cards and dice. The two most noted mathematicians who became involved in this, the first major study of probability, were Bernoulli and DeMoivre. In the 1730s DeMoivre developed the equation for the normal curve. Important work on probability was conducted in the first two decades of the nineteenth century by two other mathematicians, LaPlace and Gauss. Their work was an application of probability principles to astronomy.

Through the eighteenth century statistics was mathematical, political, and governmental. In the early nineteenth century, a famous Belgian statistician, Quetelet, applied statistics to investigations of social and educational problems. Walker (1929)[1] credits Quetelet with developing statistical theory as a general method of research applicable to any observational science. Beyond any doubt, the individual who had the greatest effect upon the introduction and use of statistics in the social sciences was Francis Galton. In the course of his long life he made notable contributions in the fields of heredity and eugenics, psychology, anthropometry, and statistics. Our present understanding of correlation, the measure of agreement between two variables, is credited to him. The mathematician Pearson collaborated with Galton in later years and was instrumental in developing many of the correlation and regression formulas that are in use today. Among Galton's contributions was the development of centiles or percentiles.

The famous American psychologist James McKeen Cattell studied in Europe in the 1880s and contacted Galton and other European statisticians. On his return to the United States he and his students, including E. L. Thorndike, began to apply statistical methods to psychological and educational problems. The influence of these men was great; in a few years theoretical and applied statistics courses were commonly taught in American universities.

In the twentieth century new techniques and methods were applied to the study of small samples. The major contributions in small-sample theory were made by the late R. A. Fisher, an English statistician. Although most of his methods were developed in an agricultural or biological setting, it was not long before social scientists recognized the utility of Fisher's methods and made use of his ideas. Today statistics is the major methodological tool of the research worker in the social sciences. The student interested in the history of statistics is referred to a brief but thorough article written by Dudycha and Dudycha (1972).

[1] Complete references are given at back of book.

## A NOTE TO THE STUDENT

It is probable that some students using the text may have found mathematics unpleasant or difficult in the past. Since this book is concerned with numbers and formulas, they may fear that the subject is complicated and obscure. These students should set their minds at ease. One does not have to be a genius, or fond of mathematics, to learn the methods presented here. We believe that a grasp of the elements of seventh- and eighth-grade arithmetic will suffice. The only further prerequisites are the ability to take square roots and to subtract negative numbers. It is possible to brush up on these with a little practice.

A more important problem for most students is that of precision or accuracy. It has been our experience that most errors come not from using the wrong principle but from carelessness in the simple operations of addition and subtraction.

Statistics is one of those subjects that cumulates. One topic leads to another and the second is built upon the first. The work has to be kept up to date. If knowledge of statistics is built on an incomplete foundation, the whole structure will surely topple. The problems found at the end of each chapter will help to test the stability of the student's progress.

# A Review of Fundamentals
## Chapter 2

In this chapter we shall be concerned with two topics: a review of simple arithmetic and elementary algebra and a discussion of the fundamental nature of measurement. Many of the errors that occur in statistical computations are not caused by a lack of knowledge of statistics but by mistakes in very simple arithmetic. Hence we shall start with a rapid review of the rules that must be followed if computations are to be correct.

## REVIEW OF ARITHMETIC

### Decimals

ADDITION AND SUBTRACTION. When adding or subtracting decimals, align the numbers so that the decimal point of each number is directely below that of the number above.

To add 3.094, 235.67, and 45.7, we align the numbers like this:

$$
\begin{array}{r}
3.094 \\
235.67 \\
\underline{45.7} \\
284.464
\end{array}
$$

MULTIPLICATION. In multiplying decimals, the product has as many decimal places as there are in both multiplicand and the multiplier taken together, as shown below:

$$
\begin{array}{r}
1.072 \\
\times\ .02 \\
\hline
.02144
\end{array}
\qquad
\begin{array}{r}
.00007 \\
\times\ .2 \\
\hline
.000014
\end{array}
\qquad
\begin{array}{r}
1.2 \\
\times\ 1.2 \\
\hline
1.44
\end{array}
$$

DIVISION. When two decimals are divided, the number of decimal places in the quotient is equal to the number of decimal places in the dividend minus the number of places in the divisor when there is no remainder.

$$\frac{.012}{.3} = .04 \qquad \frac{2.0648}{.2} = 10.324 \qquad \frac{.008}{8} = .001$$

## Fractions

ADDITION AND SUBTRACTION. When two or more fractions are to be added or subtracted, they must first be reduced to fractions having the same or a common denominator, as below:

$$\frac{1}{2} + \frac{1}{3} = \frac{3}{6} + \frac{2}{6} = \frac{5}{6} \qquad 2\frac{3}{4} + \frac{1}{2} = \frac{11}{4} + \frac{2}{4} = \frac{13}{4} = 3\frac{1}{4}$$

$$\frac{3}{8} - \frac{3}{16} = \frac{6}{16} - \frac{3}{16} = \frac{3}{16} \qquad \frac{x}{y} + \frac{a}{b} = \frac{xb}{yb} + \frac{ya}{yb} = \frac{xb + ya}{yb}$$

MULTIPLICATION. To multiply fractions, multiply all the numerators and place this quantity over the product of all of the denominators, as shown:

$$\frac{1}{2} \times \frac{2}{3} = \frac{2}{6} = \frac{1}{3}$$

$$\left(\frac{3}{4}\right)\left(\frac{2}{3}\right)\left(\frac{4}{6}\right)\left(\frac{4}{5}\right) = \frac{96}{360} = \frac{4}{15}$$

When multiplying fractions, considerable time is saved if terms common to both the numerator and the denominator are canceled. This is equivalent to dividing the numerator and denominator of a fraction by the same number. The size of the product is not changed, Let us rework the last example:

$$\frac{\overset{1}{\cancel{3}}}{\underset{1}{\cancel{4}}} \times \frac{\overset{1}{\cancel{2}}}{\underset{1}{\cancel{3}}} \times \frac{\overset{1}{\cancel{4}}}{\underset{3}{\cancel{6}}} \times \frac{4}{5} = \frac{4}{15}$$

First we cancel the first 3 in the numerator with the 3 in the denominator. Next the first 4 in the numerator is canceled by the 4 in the denominator. The 2 in the numerator is divided into the 6 in the denominator, leaving a 3 in that position. Then we have $1 \times 1 \times 1 \times 4$ in the numerator and $1 \times 1 \times 3 \times 5$ in the denominator. The product of all terms in the numerator is equal to 4; the product of those in the denominator is 15. Our answer is again $\frac{4}{15}$.

DIVISION. To divide one fraction by another, invert the fraction which is the divisor and proceed as in multiplication.

$$\frac{3}{4} \div \frac{2}{3} = \frac{3}{4} \times \frac{3}{2} = \frac{9}{8} = 1\frac{1}{8}$$

$$2\frac{3}{4} \div \frac{11}{7} = \frac{11}{4} \times \frac{7}{11} = \frac{7}{4} = 1\frac{3}{4}$$

$$\frac{x}{y} \div \frac{a}{b} = \frac{x}{y} \times \frac{b}{a} = \frac{xb}{ya}$$

## Negative Numbers

ADDITION. To add numbers, all of which are negative, add the numbers in the usual fashion and place a minus sign in front of the sum.

```
 -6
 -8
-12
----
-26
```

When the signs are mixed and there are only two numbers, that is, one negative and one positive, subtract the smaller from the larger and give the remainder the sign of the larger.

| −6 | −22 | 56 | 19 |
|----|-----|----|-----|
| 8 | 28 | −72 | −30 |
| 2 | 6 | −16 | −11 |

When adding a series of numbers with different signs, add all the positive numbers and then the negative ones and combine the results as above.

```
 -4
 -7
  8
 13
-12
 -5
----
 21
-28
----
 -7
```

SUBTRACTION. To subtract a negative number from another number, change its sign and proceed as in addition.

| 12 | −22 | −4.48 |
|----|-----|-------|
| −(−8) | −(−8) | −(−8.24) |
| 20 | −14 | 3.76 |

MULTIPLICATION. When two numbers have the same sign, either positive or negative, the product of the two numbers is positive. When the two numbers have different signs, one positive and one negative, the product of the two numbers is negative.

$$\begin{array}{cccc} 6 & -6 & 6 & -6 \\ \times\,2 & \times\,(-2) & \times\,(-2) & \times\,(2) \\ \hline 12 & 12 & -12 & -12 \end{array}$$

DIVISION. As in multiplication, when a positive or negative number is divided by a number of the same sign, the quotient is always positive. When the dividend and divisor are of unlike signs, the quotient is always negative.

$$\frac{6}{2} = 3 \qquad \frac{-6}{-2} = 3 \qquad \frac{6}{-2} = -3 \qquad \frac{-6}{2} = -3$$

## Use of Zero
The chief rule to remember when using zero is: When any number is multiplied by zero, the product is zero.

$2 \times 0 = 0$

$(.5)(3.55)(0)(4976) = 0$

## Exponents
We shall use exponents only in a limited way in elementary statistics, but the student should know what an exponent is and what it means. For example, in $2^3$ the 3 is the exponent, and it means to multiply $2 \times 2 \times 2$ or to raise 2 to the third power.

$3^2 = 3 \times 3 = 9$

$4^3 = 4 \times 4 \times 4 = 64$

$x^4 = (x)(x)(x)(x)$

## Removing Parentheses and Simplifying
Sometimes it is necessary to simplify a complex term. The general rule is to start by performing operations so that the parentheses located on the inside can be removed.

$[(12 + 4)4] - [(3 + 10) + (6 \times -12)]$

$= [(16)4] - [(13) + (-72)]$

$= 64 - (-59)$

$= 64 + 59$

$= 123$

## Proportions and Percentages

A proportion, the symbol for which is p, is defined as a part of a whole. If a pie is cut into six equal parts, each slice is a proportion and we can write that $p = \frac{1}{6}$ or .167.

To use another example, suppose that in a given class of 400 students, 40 receive A's as their final grade; 100, B's; 150, C's; 70, D's; and 40, F's. What proportion received each letter grade?

| | N | p | P |
|---|---|---|---|
| A | 40 | $\frac{40}{400} = .10$ | 10 |
| B | 100 | $\frac{100}{400} = .25$ | 25 |
| C | 150 | $\frac{150}{400} = .375$ | 37.5 |
| D | 70 | $\frac{70}{400} = .175$ | 17.5 |
| F | 40 | $\frac{40}{400} = .10$ | 10 |
| $N = 400$ | | 1.000 | 100.0 |

It should be noted that the sum of the proportions for a given example is always 1 and the maximum value of any single proportion is 1.

A percentage is obtained by multiplying a proportion by 100. The symbol for a percentage is *P*. For our example, the corresponding percentages are shown in the column at the right. It will be noted that the percentages for our data add up to 100.

A word of warning should be given about percentages and proportions. When the number of cases is small, percentages are unstable. That is, a change in one case can cause a relatively large change in the percentage. For example, when there are ten cases, a change in one case causes a change of 10 in terms of percent. It might be desirable to follow a rule that when the number of cases is less than 100 the use of percentages should be avoided. In fairness to the reader of the results of a study, the number of cases on which a percentage is computed should always be reported with the percentage.

A recent article stated that there was an increase of 132 percent in the number of new teachers of Russian between one year and the next, whereas there was only a 16.5 percent increase in the number of new high school teachers of English. It should have been noted in the article that there were 11,966 new teachers of English, but only 65 new teachers of Russian.

## Rounding Numbers

In rounding numbers to the nearest whole number or to the nearest decimal place, we proceed as follows:

| | |
|---|---|
| To the nearest whole number | 7.2 = 7 |
| | 7.8 = 8 |
| To the nearest tenth | 7.17 = 7.2 |
| | 7.11 = 7.1 |
| | .09 = .1 |
| To the nearest hundredth | 7.177 = 7.18 |
| | .674 = .67 |
| | 1.098 = 1.10 |

The general rule is that if the last digit is less than 5, it is dropped; if the last digit is more than 5, the preceding digit is raised to the next higher digit. The only complication arises when numbers end in 5. There is a general rule for this case. When the digit preceding the 5 is an odd number, this digit is raised to the next higher one; when it is an even number, the 5 is dropped. The following examples illustrate this rule:

$$8.875 = 8.88$$
$$8.05 \ = 8.0$$
$$5.25 \ = 5.2$$
$$66.975 = 66.98$$

Situations like the following arise:

$$\frac{37}{52}(3)$$

There are two ways of simplifying this. The first is to divide 37 by 52 and then multiply the quotient by 3. Or the numerator, 37, could be multiplied by 3 and the product then divided by 52 or multiplied by the reciprocal of 52. The second method is preferred because only one rounding operation is necessary.

SIGNIFICANT DIGITS. One question that frequently arises in recording numbers is How many digits should we have in our answers? As a general rule the answer should have only one digit more than exists in the raw data. For example, if we have a series of test scores, each of which contains two digits, then ordinarily we would have no more than three digits in the average or mean which we compute from these data. There is nothing to be gained in computing these averages to five or six decimal places. No meaningful accuracy is obtained from these large decimals. As a matter of fact, such large decimals mean nothing when computed on the basis of two-place numbers. A good rule is to have one more significant digit in the answer than was present in the original numbers. Here are some examples of the number of significant digits in a series of numbers.

| | |
|---|---|
| 78 | two |
| 786 | three |
| 78.2 | three |
| 1008 | four |
| 1976.09 | six |
| .0025 | two (the two zeroes merely point off the number in this case) |

Sometimes students get into trouble as a result of rounding numbers in their problems too freely. Suppose that we have an operation which consists of six distinct steps. At the conclusion of the computations for each step, the student rounds his results. A series of a half dozen such roundings in the course of the solution of a problem causes inaccuracies to enter the work. If we are going to express our answer to the nearest tenth, a good rule is to carry all operations through in terms of hundredths and round to the nearest tenth in the last step.

### Use of Appendix A
The student who learns to use Appendix A will save much time in his computational work. Suppose that one wants the following:

$$\frac{4}{39} + \frac{13}{37}$$

Of course, one could get the least common denominator and then combine the two fractions. But there is an easier way. The reciprocal of each denominator is obtained from the $1/n$ column of Appendix A and each is multiplied by its respective numerator.

$$\frac{1}{39} = .025641$$

$$\frac{1}{37} = .027027$$

$$4(.025641) + 13(.027027) = .102564 + .351351$$
$$= .453915$$
$$= .454$$

If one wished to divide a number by the square root of another number, this could be done most easily as follows:

$$\frac{13}{\sqrt{37}}$$

Using the $1/\sqrt{n}$ column of Appendix A we proceed:

$$13 \left( \frac{1}{\sqrt{37}} \right) = 13(.1644)$$
$$= 2.1372$$
$$= 2.14$$

## Square Root

In statistical operations, the student will find that there are many occasions when it is necessary to extract a square root. Appendix A contains the square roots of whole numbers from 1 to 1000. We shall use the Appendix to obtain an approximation of the square root of numbers. However, before Appendix A is used, the digits making up a number must be paired off, just as when square roots are determined in arithmetic. This is done by starting at the decimal point and working to the left and to the right of the decimal point as illustrated below.

|           |            | Square root |
| --------- | ---------- | ----------- |
| 45678.9   | 4 56 78. 90 | 214         |
| 4567.89   | 45 67. 89  | 67.6        |
| 4.56789   | 4. 56 79   | 2.14        |
| 45.6789   | 45. 67 89  | 6.76        |
| .004568   | . 00 45 68 | .0676       |

To obtain the square root of 45678.9, we look down column $n^2$ until we come to a number as close to 45679 as possible. The number in Appendix A is 45796. Going to the column to the left of 45796 we find 214, which is the approximate square root of 45679. Next let us look up the square root of 4567.89. The number in column $n^2$ closest to 456789 is 456976, the square root of which is 676. This, when pointed off, gives 67.6 as the square root of 4567.89. With the third of the above numbers, 4.56789, we proceed in the same manner as we did for the first, going down the $n^2$ column until we come to a number as close to 45679 as possible. Again, this gives us 214. Since our pairing is different, we have to point off differently, and in this case the square root becomes 2.14. Similarly we find the square roots of 45.6789 and .004568 to be 6.76 and .0676, respectively.

When there is an incomplete pair preceding the decimal, as in the first and third examples above, the $n^2$ column made up of five digits is used. Similarly, when the first pair is complete as in the second number, the six-digit part of the $n^2$ column is used. We use the five- and six-digit part of the table to get as much accuracy as possible from it. If we used the one-to four-digit part of the table, we would have fewer digits in the answers. Thus with a number like .004568 we use the six-digit part of the table instead of the four-digit part. Doing this takes us to 456976. To the left of this appears 676. The pair of zeros in the decimal results in one zero in the square root. Hence the square root of .004568 is .0676. Square roots extracted from a table are usually accurate to the tens digit.

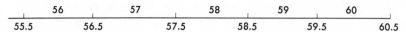

**Figure 2.1**    The upper and lower limits of continuous data.

## HOW VARIABLES ARE CONSIDERED IN STATISTICS

### Types of Measurement
We can classify data into two types: continuous and discrete, or discontinuous. Feet, pounds, minutes, and meters are examples of continuous data. With these we can make measurements of varying degrees of precision. For example, we can break meters into centimeters, centimeters into millimeters, and with intricate devices we can make measurements which are more and more precise. Such data can be considered as points on a line. The size and accuracy of the measurements that we can make along this line depend on the way that the measurements are made.

To illustrate, suppose that we measure a boy and we say that he is 57 inches tall. Does this mean that he is exactly 57 inches tall? Probably not. In reading our scale, we merely read that number of inches to which the boy's height was closest. This 57 inches includes then a segment of our line; that is, the segment extends from 56.5 to 57.4999 inches. We can round the latter and then say that 57 includes everything from 56.5 to 57.5 inches. Similarly a reading of 58 extends from 57.5 to 58.5. Each reading which represents continuous data has a lower limit and an upper limit, as seen in Figure 2.1.

Discrete or discontinuous data, on the other hand, are based on measurements that can be expressed only in whole units. The counting of people, for example, can occur only in whole units in contrast to measurements of length, which can be divided into smaller and smaller units. Other examples of discrete units are the number of words spelled correctly, the number of objects assembled, and the number of cars passing a point during a certain period of time. The student will note, however, that in statistical work most data tend to be treated as continuous, so we make such statements as: the typical graduate of college A has 2.8 children. The student should become accustomed to thinking of every number as having an upper and a lower limit.

### Measurement Scales
Stevens (1968) has written extensively on the types of measurement that are used in science. Although not all statisticians agree with Stevens on the types of statistics that should be used with the various measures in his classification, Stevens has devised a system that makes a logical approach to measurement. As he pointed out, if there were no measurement, there would be no statistics, and if measurements were accurate in all situations, there would be a much lessened demand for statistics.

Stevens recognizes four types of scale: *nominal, ordinal, interval,* and *ratio. Nominal* scales are used as measures of identity. Numbers may serve as labels to identify items or classes. In its simplest form, the numbers carried on the backs of athletes represent a nominal scale. Other examples of such scales are the classification of individuals into categories. For example, a sample of people being studied may be sorted into the following categories on the basis of religious preference: (1) Protestant; (2) Catholic; (3) Jewish; (4) other; and (5) none. Or they might be classified on the basis of sex, eye color, political party membership, urban-rural, and the like. Simple statistics are used with nominal data. For example, the number, proportion, or percentage of cases in each class or category may be determined.

When an *ordinal* scale is used in measurement, numbers reflect the rank order of the individuals or objects. Ordinal measures are arranged from the highest to the lowest or vice versa. The classical example of such a scale is the one used in evaluating the hardness of minerals. Hardness is defined as the degree of resistance to abrasion or scratching. On this scale 1 is characterized as being very soft and easily scratched, such as is the case with talc. At 10, the opposite end of the scale, is the diamond, which scratches all others and is itself scratched by none. Similarly a group of men may be arranged by physical or mental traits. Ordinal measures reveal, for instance, which person or object is larger or smaller, brighter or duller, harder or softer than the other. But such measures do not tell how much taller or how much heavier one is from the other. Statements such as James is taller than John, who is taller than William, who is taller than Paul, can be made. Statistically not much can be done with ordinal measures except to determine the median and centiles and to compute rank correlation coefficients.

The third type of scale, the *interval* scale, provides numbers that reflect differences among items. With interval scales the measurement units are equal. Examples of such scales are the Fahrenheit and centigrade thermometers, time as reckoned on our calendar, and scores on intelligence tests. In the last case we assume equal units of measurement. Many statistics are used with interval scales: arithmetic mean, standard deviation, and the product-moment correlation coefficient. Also our most widely used statistical tests of significance, the *t* test and the *F* test, may be used with such data. Interval scales show that a person or item is so many units larger or smaller, heavier or lighter, brighter or duller, etc., than another.

The final and highest type of scale is the *ratio* scale. The basic difference between this and the preceding type is that ratio scales have an absolute zero. It is true that interval scales (e.g., Fahrenheit and centigrade) also have zero points, but such points are arbitrarily chosen. Common ratio scales are measures of length, width, weight, capacity, loudness, and so on. In measuring temperature, the Kelvin scale, which has a zero point at $-273°C$ where there is a complete absence of heat, is of this type. When a ratio scale is used numbers reflect ratios among items, and data obtained with such scales may be subjected to the highest types of statistical treatments.

When data are in terms of feet, we can say that one length is twice or half that of another. When our measurements are on an interval scale, we cannot do this and make sense. For example, suppose that the maximum temperature today is 60°; the same day last year it was 30°. In this case we cannot state that it is twice as warm today as it was on the same date last year. What is the difference between these two conditions? When we were dealing with feet, we were using a measuring scale that was based on an absolute zero; in the second case we were using a scale which started 32 degrees below the freezing point of water. When measurements are on a ratio scale, meaningful comparisons can be made. As a matter of fact, when data are of this type, all of the usual mathematical and statistical manipulations may be made. However, in actual practice many of our measurements are based on interval scales and we apply practically all of our statistical techniques to these measurements.

What can we say about the measurements that we make in education, sociology, and psychology? First, we frequently assume that they have equal units of measurement. An inspection of some of these, such as intelligence quotients, reveals that this assumption is not likely to be true. Furthermore, our scales do not possess an absolute zero. The physicist can describe absolute zero on his heat scale. It is not difficult to visualize zero inches, pounds, or meters. But what does zero IQ mean? Or what does it mean when a boy gets a score of zero on a geography test? Actually we do not know what these scores of zero mean. Then it follows that we have no basis for stating that a child with an IQ twice the size of the IQ of another child is twice as bright. Neither can we say that the child whose score on an arithmetic test is double that of another child knows twice as much arithmetic as the first child.

## SAMPLES AND POPULATIONS

It is important to distinguish between the terms *samples* and *populations*. Let us use an illustration to do this. Suppose that we are interested in the mental ability of children in the second grade. One way to investigate this is to give intelligence tests to second graders. We begin by obtaining permission to administer the test to one group of second graders. We compute the average, or mean, score for the group; this mean score is a statistic and gives us the average test score of our sample. Since there are so many second-grade pupils, we could continue this process for a long time by drawing sample after sample. If each of these samples is a *random sample* (random sampling procedures will be discussed in Chapter 11), we can combine all the sample averages, or means, to obtain a grand mean. This grand mean will be our best estimate of the average intelligence of all the second graders. That is, the average of all of the means of our samples is used to tell us something about the population value. All second graders in the United States make up the population, or universe, from which the various samples are drawn. Values which refer to populations are called *parameters;* values which refer to samples are called *statistics*. Populations, as the term is used in statistics, are arbitrarily defined groups. They need not be as large as the one used here as an illustration. We could define the 552 seniors in a certain school system as our population, and from this we could draw samples. One of the major aspects of statistical research is making inferences about population characteristics on the basis of one or more samples that have been studied.

### Statistical Symbols

The student will soon find that there are a number of symbols used in statistics. However, there is no absolute conformity in the use of these symbols, and notational usage will vary somewhat from author to author. A common and simple statistical equation is

$$\Sigma f = N$$

This equation is read, "summation of, or the sum of, the frequencies (*f*) is equal to *N* (the number of cases)." The Greek capital sigma, $\Sigma$, is one of the most widely used statistical symbols, and is customarily used as an abbreviation for sum.

If we are dealing with one *variable,* a characteristic that manifests differences in magnitude or quantity, it is customary to designate it as *x*. One measurement or observation of a variable is usually symbolized by the corresponding capital letter; hence *X* is one measure for variable *x* and *Y* is one measure for variable *y*. Different individual measures on the same variable are denoted by numeric subscripts: $X_1, X_2, X_3, \ldots, X_N$.

The student will find it useful to be familiar with the following expressions:

| | |
|---|---|
| $X = Y$ | *X* equals *Y* |
| $X \neq Y$ | *X* does not equal *Y* |
| $X > Y$ | *X* is greater than *Y* |
| $X < Y$ | *X* is less than *Y* |
| $X \geq Y$ | *X* is equal to or greater than *Y* |
| $X \leq Y$ | *X* is equal to or less than *Y* |

As mentioned earlier in this chapter, the characteristics of a population are called *parameters,* whereas the characteristics of a sample are called *statistics.* In this book, as in most statistical writing, different symbols are used for *parameters* and *statistics.*

| Characteristic | Parameter | Statistic |
|---|---|---|
| Mean | $m$ or $\mu$ | $X$ |
| Standard Deviation | $\sigma$ | $s$ |
| Variance | $\sigma^2$ | $s^2$ |
| Proportion | $p$ | $p$ |
| Pearson correlation coefficient | $R$ | $r$ |
| Number of cases | $n$ | $N$ |

**EXERCISES**

**1.** Complete the following operations:

a. $87.6 - (-22.4)$
b. $-87.6 - (-12.2)$
c. $(.0007) (.07)$
d. $(-.7) (.08)$

e. $(-.87)(-3)$
f. $14 \div .7$
g. $-.99 \div .9$
h. $49 \div .007$

**2.** Complete the following operations as indicated. Where possible report answers to two decimal places.

a. $1/2 + 3/4 + 5/6$
b. $1/2 + 3/4 - 5/8$
c. $1/2 + 3/8 - (-3/4)$
d. $(5/4)(3/8)$
e. $(2/7)(-3/4)$
f. $(7/9) \div (2/3)$
g. $(2/3)(4/9) \div (5/9)$
h. $10[256 - (12)^2][170 - (13)^2]$

i. $(5/6)(2/3)(0)$
j. $5^4$
k. $(4)^3 - (5)^3$
l. $x(xy)$
m. $(3/4)^2$
n. $(x + y)^2$
o. $(fx)(x)$
p. $(x)(xy)(x^2y)$

**3.** Using Appendix A find the square root of each of the following:

a. 7724
b. 77240
c. 7.7240
d. 148824
e. 1488.2

f. .00007
g. .00016
h. 7.0009
i. $(1/8)$
j. 76453.678

**4.** Round each of the following to the nearest tenth:

a. 13.46
b. 42.23
c. 17.897

d. .089
e. 29.25
f. 29.75

**5.** How many significant digits in each of the following:

a. 432.76
b. 60.76
c. .77

d. .077
e. 7.0007

**6.** Using Appendix A simplify the following, expressing your answer to the nearest thousandth:

a. $1/31 + 1/69$
b. $11/32 + 4/127$
c. $1/\sqrt{15} + 1/\sqrt{26}$
d. $11/45 + 13/131 - 7/18$

e. $3/\sqrt{14} + 11/\sqrt{23} - 10/\sqrt{52}$
f. $(2/31)(1/33)$
g. $(3/\sqrt{69}) - 1/23$

# Frequency Distributions, Graphs, and Centiles
## Chapter 3

Often to make our data more interpretable and convenient, we set up a frequency distribution and draw graphs of various kinds to represent the data.

### THE FREQUENCY DISTRIBUTION

Suppose that we have given a geography test yielding the scores shown in Table 3.1. The first step is to determine the range. This is defined as the highest score minus the lowest score plus one. In Table 3.1 the highest score is 82 and the lowest is 28. In statistics we let a capital $X$ stand for any raw score or any measurement. If we were working with two sets of data at the same time, we could use a capital $Y$ for any measurement in the other set. The range then is

$$(X_H - X_L) + 1 \quad \text{or} \quad (82 - 28) + 1 = 55$$

The second step is to decide how large each of the intervals in the frequency distribution is going to be. A widely accepted practice is to have between 10 and 20 intervals in the frequency table. When there are less than 10 intervals, the coarseness of the grouping may cause inaccuracies; when there are more than 20, the work becomes laborious. The size of the intervals can be determined in a trial and error fashion.

**Table 3.1**  Scores on a Geography Test

| | | | | | | | |
|----|----|----|----|----|----|----|----|
| 56 | 78 | 62 | 37 | 54 | 39 | 62 | 60 |
| 28 | 82 | 38 | 72 | 62 | 44 | 54 | 42 |
| 42 | 55 | 57 | 65 | 68 | 47 | 42 | 56 |
| 56 | 56 | 55 | 66 | 42 | 52 | 48 | 48 |
| 47 | 41 | 50 | 52 | 47 | 48 | 53 | 68 |

If we let each interval cover 10 units, then the range will be included by 6 of these intervals, $55 \div 10 = 5.5$ or 6. Thus 10 is too large, because this results in only 6 intervals. Suppose that this time we try 5. Since 5 goes into 55 exactly 11 times, 5 would be acceptable as the size of our interval. If we try 3, we find that 3 goes into 55 about 18 times; so 3 could be used for the size of the interval. Let us consider one more, 2. Two is contained in 55 about 28 times. We would reject this as the size of the interval, because it results in too many intervals. Since we need between 10 and 20 intervals in our frequency distribution, we can take the average of 10 and 20, which is 15, and divide this into the range. In this case the range of 55 divided by 15 results in a value between 3 and 4, and either of these could accordingly be used as the interval size ($i$).

If you observe frequency distributions made by others, you will note that odd numbers are frequently used for the size of the intervals. The advantage of this is that the midpoints of each of the intervals will be whole numbers. Such numbers as 3 and 5 are commonly used for interval size ($i$). An exception to this practice is the use of multiples of 10, which often make convenient intervals.

Suppose for the data in Table 3.1 we decide to use an interval with a size of 3. We would note this by writing $i = 3$. We start building the frequency distribution that is shown in Table 3.2. The next problem is to decide where to start. A common practice is to let the bottom interval begin with a number which is a multiple of the interval size. In this case our lowest score is 28, the size of our interval is to be 3, so the bottom interval would begin at 27 and end at 29. These are the *integral limits*. You will remember from the last chapter that in statistics we deal with upper and lower limits (or exact limits), and so this bottom interval actually begins at 26.5 and ends at 29.5. If we subtract the first of these from the latter, we obtain a value of 3, which is the size of the interval. This interval size is not apparent unless we are aware of the exact limits of the interval. This bottom interval also has a midpoint that is one-half of the distance between the lower and the upper limits. The midpoint for this interval then is $26.5 + 1.5$, which equals 28. As noted above, when the size of an interval is an odd number, the midpoint will be a whole number.

After deciding upon the limits of the bottom interval, we determine the rest of the intervals by increasing each integral limit by 3. We stop when we reach the interval 81–83, which contains the highest score in the distribution. The usual practice is to set up frequency distributions with the lowest scores at the bottom.

The next task is to tally the scores. We take the scores one at a time and record each to the right of its appropriate interval by making a tally mark. When this is finished, we combine the tallies in the column to the right. This column is headed with $f$, which stands for frequencies. This column is summed, and we write the total at the bottom. We can write a capital $N$ in

**Table 3.2**  Setting Up a
Frequency Distribution

|         |           | $f$ |
|---------|-----------|-----|
| 81–83   | /         | 1   |
| 78–80   | /         | 1   |
| 75–77   |           | 0   |
| 72–74   | /         | 1   |
| 69–71   |           | 0   |
| 66–68   | ///       | 3   |
| 63–65   | /         | 1   |
| 60–62   | ////      | 4   |
| 57–59   | /         | 1   |
| 54–56   | 7HL ///   | 8   |
| 51–53   | ///       | 3   |
| 48–50   | ////      | 4   |
| 45–47   | ///       | 3   |
| 42–44   | 7HL       | 5   |
| 39–41   | //        | 2   |
| 36–38   | //        | 2   |
| 33–35   |           | 0   |
| 30–32   |           | 0   |
| 27–29   | /         | 1   |

$$N = 40$$

front of this sum to denote that the number of cases is 40, or we can write $\Sigma f = 40$. As noted previously, the symbol $\Sigma$ is read as "summation of" or "sum of." Note that for intervals where there were no tallies, we entered a zero in the $f$ column. Data in such a frequency distribution are said to be *grouped,* and formulas used with grouped data are applicable to this type of data only. In contrast, we also have ungrouped data; the formulas applied to them are often referred to as raw score formulas.

Suppose that we organize the same geography test scores in a different frequency table. This time the size of the interval will be 5. Since our lowest score is 28, the lowest interval will be 25–29. This interval has exact limits of 24.5 and 29.5 and a midpoint of 27. This second table appears as Table 3.3. This time we have only 12 intervals instead of 19 which resulted when we used an interval size of 3. In the actual analysis of these data, the interval of 5 would be preferred.

At this point let us summarize the steps in organizing a frequency distribution or frequency table.

**1.** Determine the range.

**2.** Divide this by 15 to estimate the approximate size of the interval.

**3.** List the intervals, beginning at the bottom. Let the lowest interval begin with a number which is a multiple of the interval size.

**Table 3.3**   Setting Up a
Frequency Distribution

|  |  | $f$ |
|---|---|---|
| 80–84 | / | 1 |
| 75–79 | / | 1 |
| 70–74 | / | 1 |
| 65–69 | //// | 4 |
| 60–64 | //// | 4 |
| 55–59 | ///// // | 7 |
| 50–54 | ///// / | 6 |
| 45–49 | ///// / | 6 |
| 40–44 | ///// / | 6 |
| 35–39 | /// | 3 |
| 30–34 |  | 0 |
| 25–29 | / | 1 |
|  | $\Sigma f = 40$ | |

**4.** Tally the frequencies.

**5.** Summarize these under a column labeled $f$.

**6.** Total this column and record the number of cases at the bottom. It might be mentioned here that if this number, obtained by adding the frequencies, is the same as the known number of cases, it does not follow that no mistake has been made. To check the work, the scores should be retallied.

## GRAPHS

### The Frequency Polygon
Of all the graphic devices used to illustrate statistical distributions, probably the most frequently encountered is the frequency polygon. This is because the frequency polygon is very easy to construct, and it is basically very simple to interpret. Several distributions, using a different type of line for each, may be portrayed on the same axes.

CONSTRUCTION OF THE FREQUENCY POLYGON. In all work with graphs, two axes are used. The vertical axis is always labeled the $y$ axis, and values taken along this axis are called *ordinate* values. The other axis, the $x$ axis, is called the *abscissa*. It is horizontal and meets the $y$ axis at right angles at a point called the origin (0). In constructing graphs of the frequency polygon type, the $x$ axis is longer than the $y$ axis. Usually a ratio of 3 to 2 or 4 to 3 will result in a good graph. For example, if the $x$ axis is 6 inches in length, the $y$ axis should be about 4 inches. Or if the $x$ axis is 8 inches, the $y$ axis should be about 6 inches.

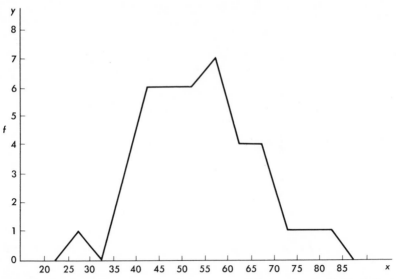

**Figure 3.1** Frequency polygon for the geography scores in Table 3.1.

To make graphs that can be easily and accurately read, a good quality graph paper should be used. The type that has ten squares to the inch is very good for statistical work. It is easier to build graphs of the desirable proportions if the $x$ axis is placed on the wider side of the graph paper. In building a frequency polygon, the frequency values are always placed on the $y$ axis. A lowercase $f$ is placed along the $y$ axis. The scores are placed on the $x$ axis. To illustrate this, suppose that we construct a frequency polygon for the data in Table 3.3. The highest frequency is 7, so we need no more than 7 units on the $y$ axis of Figure 3.1. On the $x$ axis we have placed score values.

The next step is to plot the frequencies. For any one of the intervals in Table 3.3 we can make an assumption that if all the scores in that interval were averaged, this average would be equal to the midpoint of the interval. It follows then that the midpoint is the value which best represents any given interval. Frequency values are plotted above the corresponding midpoints of each of the class intervals. Let us take the first or bottom interval of our table, 25–29. This interval has a midpoint of 27 and a frequency of 1. We go along the $x$ axis until we come to 27, and then we go up one unit above this point and place a point. The midpoint of the next interval is 32, and this interval has a frequency of 0. So a mark is placed at 32 on the $x$ axis. Next, the midpoint of the third interval is 37 with a frequency of 3. So at a point 3 units above 37 on the $x$ axis we place our next point. We proceed like this until all the frequencies have been plotted. Then with a ruler we connect these points with straight lines.

Rather than leave our graph suspended in space, we assume that there is

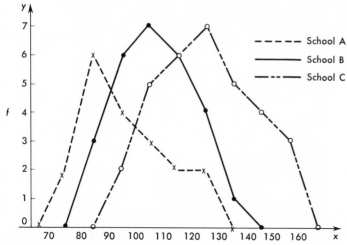

**Figure 3.2** Distribution of Stanford-Binet IQs for groups of kindergarten children in three schools of a city.

another interval above and below those shown in our table, and that each of these intervals has a frequency of zero. So at 22, the midpoint of the lower of these terminal intervals, we place a mark on the $x$ axis. A similiar mark is placed at 87 to provide our top. The graph is now anchored at both ends. Some find it much more convenient to construct frequency polygons with the numerical values of the midpoints of the intervals on the $x$ axis rather than the score values shown in Figure 3.1. The resulting picture is the same, no matter which method is used in the construction.

As mentioned earlier, the frequency polygon is very useful in portraying two or more distributions at once. Figure 3.2 shows the distributions of the intelligence quotients of three groups of kindergarten children in three different schools of a city. In such a graph similarities and differences are very apparent.

CONSTRUCTING FREQUENCY POLYGONS WHEN FREQUENCIES DIFFER. In many cases, as in Table 3.4, we find that the numbers in the different groups we wish to plot differ considerably. In this table we have the distribution of intelligence test scores for 180 children in one school ($f_1$) and for 500 children in another school ($f_2$). If we attempt to plot both these distributions on the same axes as they are now organized, we shall have difficulty. Because of widely different frequencies in corresponding intervals, it may be impractical to plot both distributions on the same graph. One line will be so far above the other that direct comparison is impossible.

In cases like these, a common practice is to convert each frequency into a percentage or proportion and to plot these. We have converted the data in

**Table 3.4**   Intelligence Test Scores of Children in Two Schools

| (1) Scores | (2) School A $f_1$ | (3) School B $f_2$ | (4) $P_1$ | (5) $P_2$ |
|---|---|---|---|---|
| 150–159 | 1 | 10 | .6% | 2% |
| 140–149 | 4 | 20 | 2.2 | 4 |
| 130–139 | 8 | 40 | 4.4 | 8 |
| 120–129 | 12 | 170 | 6.7 | 34 |
| 110–119 | 31 | 180 | 17.2 | 36 |
| 100–109 | 69 | 50 | 38.3 | 10 |
| 90–99 | 32 | 10 | 17.8 | 2 |
| 80–89 | 18 | 10 | 10.0 | 2 |
| 70–79 | 4 | 5 | 2.2 | 1 |
| 60–69 | 1 | 5 | .6 | 1 |
| | $\Sigma f_1 = 180$ | $\Sigma f_2 = 500$ | 100.0 | 100 |

Table 3.4 to percentages. Notice that we have two more columns, $P_1$ and $P_2$, the percentages of each group in the various intervals. We compute these percentages by dividing the frequency in each interval by the number in the total group and multiplying this result by 100. If we take the bottom frequency in the $f_1$ column, we have $(1/180)(100)$, which results in .6 percent. In this manner we compute all the percentages.

An easier way to change a series of frequencies to percentages is to find a constant multiplier and then to multiply each frequency in the distribution by this constant. Let us first consider the $f_1$ distribution in Table 3.4. There are 180 cases. It follows then that each case is equal to $(1/180)(100)$ percent, which in this case is .555. This .555 then is our constant multiplier; multiplying each frequency in column 2 by this constant and rounding results in the values shown in column 4. A check on the work is that the sum of column 4 is 100 percent. Sometimes there are minor variations in this sum caused by rounding. The $f_2$ data in column 3 are taken next. This time the constant multiplier is $(1/500)(100)$, which equals .2. So we multiply each of the $f_2$ frequencies by .2 and find the results shown in column 5.

The next step is to construct the frequency polygon as described earlier. This time, though, we place our percentages, instead of the frequencies, on the $y$ axis. Since 38.3, which is the largest percentage in columns 4 and 5, is smaller than 40, the largest value needed on our $y$ axis is 40. Along the $x$ axis we have the test scores plotted in the usual manner (see Figure 3.3). We plot our percentages above the midpoints of the appropriate intervals on the $x$ axis and connect these points with straight lines. These lines should be constructed using different types of line or different colors. Since we have two graphs on the same axis, we include a legend showing which polygon represents which set of data.

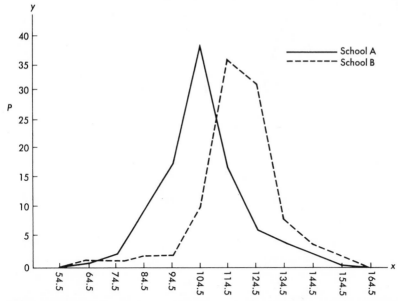

**Figure 3.3** Distribution of scores on an intelligence test in two schools, with the frequencies converted to percentages.

## Types of Curve

After making a few frequency polygons, we soon notice that the curves tend to have shapes that can be classified into types. Sometimes the frequencies tend to pile up on the left-hand side of the graph with a tail extending to the right, as shown in curve A of Figure 3.4. Such a curve is said to be *skewed*. If the tail goes to the right, we label this type of skew as being *positive*. (This conforms with the mathematicians' practice of calling the right end of a line positive and the left end negative). Curve B shows the tail extending to the left. This condition is called *negative skewness*. The student should remember that it is the tail of the distribution which determines the sign of the skewness and not the location of the pile-up of scores.

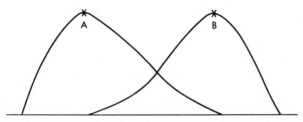

**Figure 3.4** Types of skewness: curve A is positively skewed; curve B is negatively skewed.

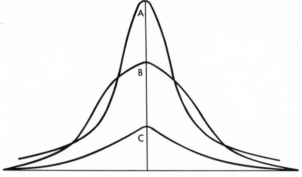

**Figure 3.5** Types of kurtosis: curve A is leptokurtic; curve B is mesokurtic; curve C is platykurtic.

Curves can also be classified on the basis of their peakedness, or *kurtosis*. Figure 3.5 shows the three types of kurtosis. Curve A is peaked and the tails are more elevated above the base line. Such a curve is said to be *leptokurtic*. Curve B is described as being *mesokurtic* and curve C, decidedly flattened, is said to be *platykurtic*. Measures of skewness and kurtosis will be presented in a later chapter.

Another type of curve that is by no means rare is the bimodal curve seen in Figure 3.6. This curve has two peaks or modes. It is also possible for a curve to have more than two modes.

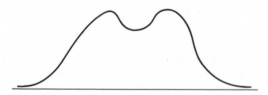

**Figure 3.6** A bimodal curve.

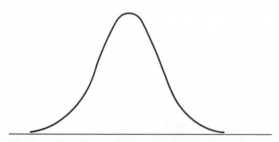

**Figure 3.7** The normal or bell-shaped curve.

The last type of curve that we shall consider is the one referred to as the normal or bell-shaped curve (Figure 3.7). This is the graph of the so-called normal distribution. Many statistical operations discussed in this book are based on the assumption that the data are normally distributed. Since this type of curve is of major importance, considerable time and space will be devoted to it later. At that time we shall discuss its properties and uses.

## Other Types of Graph

THE HISTOGRAM. The histogram (Figure 3.8) is very similar in construction to the frequency polygon. Everything is the same up to the point of plotting the frequencies. This histogram in Figure 3.8 is also based on the data in Table 3.3. Here is how this graph is made. The bottom interval 25–29 has a frequency of 1. Starting at the lower limit of this interval, 24.5, the interval is marked off, and since the interval has a frequency of 1, this interval has a height of 1 on the y axis. The next interval has a frequency of zero; hence there is a gap in our histogram at this point. At the interval beginning with 34.5, we mark off a bar with a height of 3, since there are three scores in this interval. We continue this process until we have entered all the frequencies into the graph. If the midpoints of the top line of each of the columns comprising the histogram were connected, we would have the same frequency polygon previously constructed.

The histogram is more time consuming to construct than the frequency polygon. Also, only one histogram can be placed clearly on one set of axes

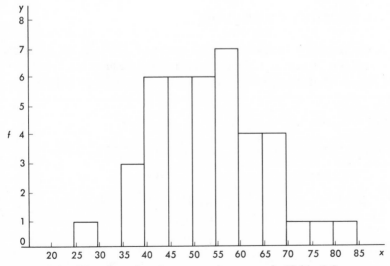

**Figure 3.8**  Histogram for the geography scores in Table 3.3.

**Table 3.5**   Seniors in the Five Colleges of Two Universities, A and B

| College | $f$ A | B | $p$ A | B | $P$ A | B |
|---|---|---|---|---|---|---|
| Engineering | 440 | 640 | .48 | .58 | 48% | 58% |
| Liberal Arts | 220 | 220 | .24 | .20 | 24 | 20 |
| Agriculture | 120 | 120 | .13 | .11 | 13 | 11 |
| Home Economics | 80 | 80 | .09 | .07 | 9 | 7 |
| Fine Arts | 60 | 40 | .06 | .04 | 6 | 4 |
| | 920 | 1100 | 1.00 | 1.00 | 100 | 100 |

unless a second is reversed and plotted below the $x$ axis. We saw how several frequency polygons were placed on the same axes to make comparisons and illustrate differences. This cannot be done as easily and clearly with histograms. Their use is thus more limited than that of some other types of graph.

GRAPHS FOR FREQUENCIES, PROPORTIONS, AND PERCENTAGES. Table 3.5 presents figures showing the number of seniors in five colleges of two universities. The frequencies have been converted into both proportions and percentages, $p$'s and $P$'s, respectively. Many types of graph can be used to display these data pictorially. One of the commonest of these is the bar graph (Figure 3.9). Another frequently encountered is the pie diagram seen in Figure 3.10. The weekly newsmagazines, the business sections of the daily newspapers, and our own professional journals include many illustrations of the various types of these graphs.

## CUMULATIVE FREQUENCY DISTRIBUTIONS

In Table 3.6 a frequency distribution of the scores of 376 boys on a test of mechanical ability is presented. Column 2 contains the frequencies. In column 3 we have the cumulative frequencies, $cf$. These are obtained as follows. We

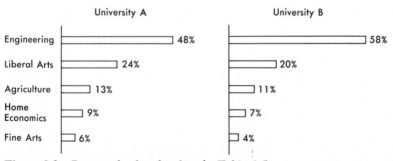

**Figure 3.9**   Bar graphs for the data in Table 3.5.

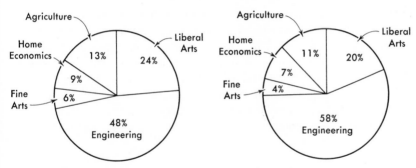

**Figure 3.10** Pie diagrams for the data in Table 3.5.

start at the bottom and note how many scores are below the upper limit of the bottom interval. The number is 4. So a 4 appears as the bottom entry in column 3. Next we ask how many scores fall below the upper limit of the next interval, that is, how many scores are below 14.5. The answer is 19, the frequency of all scores below this point. The cumulative frequency for the next interval is $23 + 19 = 42$. The process is continued by adding the frequency of each interval to the cumulative frequency of all the intervals below the given interval. If the work is correct, the $cf$ of the top interval will equal the number of cases.

### Obtaining the Cumulative Proportions and Percentages

As pointed out previously, the easiest way to convert a set of frequencies to proportions is to obtain a constant multiplier by dividing 1 by the number

**Table 3.6** Obtaining Cumulative Frequencies, Proportions, and Percentages

| (1) | (2) $f$ | (3) $cf$ | (4) $cp$ | (5) $cP$ |
|---|---|---|---|---|
| 60–64 | 2 | 376 | 1.000 | 100.0 |
| 55–59 | 12 | 374 | .995 | 99.5 |
| 50–54 | 20 | 362 | .963 | 96.3 |
| 45–49 | 32 | 342 | .907 | 90.7 |
| 40–44 | 46 | 310 | .824 | 82.4 |
| 35–39 | 58 | 264 | .702 | 70.2 |
| 30–34 | 64 | 206 | .548 | 54.8 |
| 25–29 | 58 | 142 | .377 | 37.7 |
| 20–24 | 42 | 84 | .223 | 22.3 |
| 15–19 | 23 | 42 | .112 | 11.2 |
| 10–14 | 15 | 19 | .050 | 5.0 |
| 5–9 | 4 | 4 | .011 | 1.1 |

$N = 376$

of cases and then multiplying each frequency by this constant to obtain the proportion. The quotient that results from dividing 1 by any number is called the *reciprocal* of that number. The reciprocals of the numbers from 1 to 1000 are presented in the $1/n$ column of Appendix A. For these data, we want to multiply each $cf$ by the reciprocal of 376. In Appendix A, we find that $1/376 = .00266$. The products of .00266 and each $cf$ are recorded in column 4, headed $cp$, cumulative proportions. To change these cumulative proportions to cumulative percentages, $cP$, each $cp$ is multiplied by 100. These values appear in column 5.

### Constructing the Cumulative Percentage or Ogive Curve

In setting up the cumulative proportion, cumulative percentage, or ogive curve as it may be called, we follow the general rules of building graphs which we discussed earlier. We attempt to have the ratio of the $y$ axis to the $x$ axis as 2 is to 3. The cumulative proportions or cumulative percentages are always placed along the vertical, or $y$, axis. As in the frequency polygon, the scores are entered on the horizontal, or $x$, axis. The values on the $y$ axis will range between 0 and 1, or 0 and 100, depending on whether we are using proportions or percentages. In this case we are going to plot the cumulative percentages. It may be recalled that when we were plotting a frequency polygon we placed the plotting points above the midpoint of each interval. *In plotting the cumulative percentage graph, we use the upper limits of each of the intervals and place our points above these.* This is consistent with the concept of cumulative frequencies or cumulative percentages. In finding these from our frequency table, we asked ourselves how many or what percentage

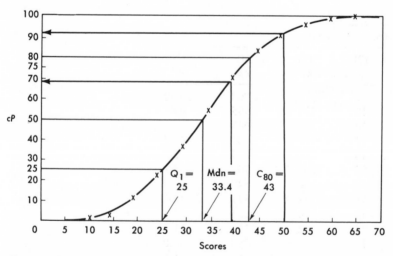

**Figure 3.11**   Cumulative percentage, or ogive curve, for the ability test scores in Table 3.6.

of the cases fell below the upper limit of each interval. Figure 3.11 is a construction of one of these curves.

The upper limit of the bottom interval is 9.5. Hence above 9.5 on the $x$ axis we go up 1.1 points and place our point. Above 14.5, the upper limit of the next interval, we count off 5 units on the $y$ axis and again place a mark. This process is continued until all $cP$ values are plotted. Then a smoothed curve is drawn. Some of the plotted points will not be on the curve, some appearing on one side and some on the other. Smoothed curves of cumulative distributions often take the shape seen in Figure 3.11 and are frequently referred to as S-shaped curves. The curve is brought to the base line by extending it to the next lower interval and giving that interval a $cP$ of zero.

## CENTILES

### Reading the Centile Points from the Ogive Curve

A centile or centile point is defined as a specific point in a distribution which has a given percentage of the cases below it. For example, the 88th centile ($C_{88}$) is that point in a distribution which has 88 percent of the cases below it.

If the cumulative percentage curve has been accurately constructed on a large piece of graph paper, centile points may be read from it with a high degree of accuracy. To read the points a ruler is placed on the $cP$ column at the centile point desired and a straight line is drawn over to the curve. From this point on the curve another line is drawn down to make a right angle with the $x$ axis. The point on the $x$ axis where this vertical line meets the $x$ axis indicates the desired centile point. In Figure 3.11 lines are drawn showing the values for $C_{80}$, the median $C_{50}$, and $C_{25}$. These values are approximately 43, 33, and 25, respectively. These values are very close to the computed values for the same statistics, which are 43.5, 33.1, and 25.4.

### Computation of Centiles

To illustrate the computation of centiles, the data in Table 3.6 will be used. First we shall compute $C_{50}$. By definition, this centile point will have 50 percent of the cases above and below it. That is, it is the midpoint of the distribution and is known as the *median*. Dividing the $N$ of 376 by 2 or taking 50 percent of it gives 188 cases. Hence we are interested in finding that point in the distribution with 188 cases above and below it.

We start by counting up from the bottom until we come as close to 188 cases as possible, but not exceeding it. This brings us to the point at the top of the interval 25–29 or at the bottom of the interval 30–34, this point being 29.5. There are 142 cases below this point. We need 46 more cases, the difference between 188 and 142. We must interpolate, for we need 46 of the 64 cases in the next interval. In other words, we need to go 46/64 of the distance through the length of the interval, which in this case is 5. We can write this as follows:

$$C_{50} = 29.5 + \frac{46}{64}(5)$$

$$= 29.5 + \frac{230}{64}$$

$$= 29.5 + 3.59$$

$$= 33.1$$

This can be checked by coming down from the top:

$$C_{50} = 34.5 - \frac{18}{64}(5)$$

$$= 34.5 - \frac{90}{64}$$

$$= 34.5 - 1.4$$

$$= 33.1$$

Suppose we wish to find $C_{12}$ for the same data. First we take 12 percent of 376, which is 45.12 cases. We can count 42 cases from the bottom, which brings us to the top of the interval, 19.5. At this point we must interpolate:

$$C_{12} = 19.5 + \frac{45.12 - 42}{42}(5)$$

$$= 19.5 + \frac{3.12}{42}(5)$$

$$= 19.5 + \frac{15.60}{42}$$

$$= 19.5 + .37$$

$$= 19.9$$

Next we shall compute $C_{88}$. We could start by taking 88 percent of $N$ and counting up from the bottom as we did for $C_{12}$. However, we can make our work much easier by taking 12 percent of $N$ and coming down from the top. For example, 12 percent of $N = 45.12$; then

$$C_{88} = 49.5 - \frac{45.12 - 34}{32}(5)$$

$$= 49.5 - \frac{11.12}{32}(5)$$

$$= 49.5 - \frac{55.60}{32}$$

$$= 49.5 - 1.74$$

$$= 47.8$$

The operation of obtaining any centile point may be summarized in the following formula:

$$\text{Any centile point} = ll + \frac{Np - cf}{f_i}(i) \qquad (3.1)$$

where $ll$ = lower exact limit of interval in which we are interpolating

$N$ = number of cases

$p$ = proportion corresponding to desired centile

$cf$ = cumulative frequency of cases below interval in which we are interpolating

$f_i$ = frequency of the interval in which we are interpolating

$i$ = size of the class interval

Applying this equation to finding $C_{12}$ to the data in Table 3.6 we have:

$$C_{12} = 19.5 + \frac{(376)(.12) - 42}{42}(5)$$

$$= 19.5 + \frac{45.12 - 42}{42}(5)$$

$$= 19.5 + \frac{3.12}{42}(5)$$

$$= 19.5 + .37$$

$$= 19.9$$

Several of the centile points have special names, such as *median* for $C_{50}$. $C_{25}$ is known as the first quartile and is written $Q_1$. Similarly $C_{75}$, the third quartile, is represented by the symbol $Q_3$. In addition to these there are 9 decile points that divide the distribution into ten equal parts. For example, $C_{10}$ equals $D_1$ and $C_{20}$ equals $D_2$.

When the data are ungrouped, the individual measures can be arranged from low to high and the centile points obtained by counting. This can be illustrated by finding the median in the following examples.

Suppose that we have the following 11 scores:

20, 19, 18, 17, 16, 15, 14, 13, 12, 11, 10

In this case the score of 15 has 5 scores on each side of it and hence is the median.

Here is another series:

20, 19, 18, 17, 16, 15, 14, 13, 12, 11

In this series of 10 scores we count up 5 cases (or down 5). This brings us to the point 15.5, which is the median, halfway between scores of 15 and 16.

The preceding cases are easy, since the median can be obtained by inspection. Suppose that our series is as follows:

21, 20, 19, 18, 17, 17, 17, 16, 15, 14

Again we have 10 measures and again we wish to find the point that has 5 scores on each side of it. We start from the bottom and count cases 14, 15, and 16, which occupy the first 3 positions. However, the three 17's make it impossible to finish this problem as we did the first two. We continue up to the bottom of the number 17 or to the top of 16. This point is 16.5. Here we have to interpolate. We have used 3 cases and need 2 more. These three 17's are assumed to be spread equally through the interval 16.5–17.5, and since we want only two of these 17's, we shall take a point that is two-thirds of the way through this interval as the median. In summary, $C_{50} = 16.5 + \frac{2}{3} = 16.5 + .67 = 17.2$.

## Use of Centiles

Centiles are widely used in educational circles in reporting the results of standardized tests. In their favor, it can be said that they are very easy to understand. Even if one does not know that a person who has a centile score of 77 is at a point above 77 percent of those upon whom the test was standardized, at least the 77th centile looks like 77 percent, and even the most uninformed teacher can understand what that means. Since centiles appear to be like percents, there is little difficulty in understanding their meaning. Also they give an adequate indication of an individual's rank in a group.

Centiles do, however, have serious limitations, and many makers and users of tests no longer bother with them. If you examine a set of centile norms or a profile sheet based on such norms, you will note that the centile norms are piled up at the middle of the distribution. In Figure 3.12, for example, a raw score of 33 is equivalent to $C_{50}$, a raw score of 36, to a centile point of 60, and a raw score of 30, to a centile point of 40. A change in 6 raw score units is equivalent to a change in 20 centile units. There is, then, a piling up

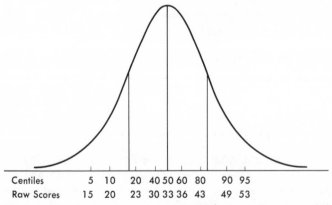

| Centiles | 5 | 10 | 20 | 40 | 50 | 60 | 80 | 90 | 95 |
| --- | --- | --- | --- | --- | --- | --- | --- | --- | --- |
| Raw Scores | 15 | 20 | 23 | 30 | 33 | 36 | 43 | 49 | 53 |

**Figure 3.12** Centile and raw-score equivalents for a set of data.

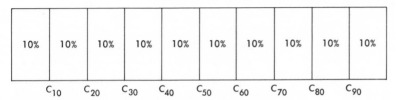

**Figure 3.13**   The rectangular distribution of centiles.

of centile points at the center of the distribution, and differences between them at this part of the curve have little meaning.

To state that an individual who is at $C_{47}$ on a certain test differs from the individual who is at $C_{51}$ on the same test is making much ado about nothing. At the center of distribution, the use of centile scores tends to exaggerate differences that are not actually in existence. Centiles are unequal units of measurement and cannot be treated arithmetically. That is, there is no justification for averaging them, combining them, or treating them in any mathematical fashion. As statistics go, they are dead ends. Nothing should be done with them. If we wish to manipulate data which have been reduced to centiles, we should convert the data back to raw scores and work with these. Since centiles are unequal units of measurement, some statistical workers believe that we would be better off without them. With standardized tests, centiles are becoming less frequently used as a method of reporting norms. However, in certain circles they will be with us for a long time to come.

Let us see why centiles pile up at the center of the distribution. If we go back to the original definition of a centile point as a point in a distribution with a certain percentage of the distribution of the cases below it, it follows that $C_{10}$ is that point in the distribution with 10 percent of the cases below it. To say that 10 percent of the area of a curve falls below $C_{10}$ would be another way of stating this (see Figure 3.13). If we take $C_{20}$ next, it is by definition that point with 20 percent of the cases in the distribution below it. An additional 10 percent of the area is added below $C_{20}$ in our figure. This may be continued until all the area or all the cases of a distribution are included. The resulting distribution looks like Figure 3.13, which, instead of being a normal curve, is a rectangle. The distribution of centiles is described as being rectangular. Measurements made in social and biological sciences tend to take the form of the normal curve. Distortions come about when such data are converted to centiles which have a distribution of another shape.

Sometimes confusion arises in the interpretation of tables of centiles. For example, note that Table 3.6 gives a centile value for each interval (*cP*). However, these centile points correspond to the *upper limit* of each interval. Sometimes tables are prepared which indicate the centile point equivalent to the *midpoint* of each interval. This practice is common in reporting norms for standardized tests.

## CENTILE RANKS

Some test makers use centile ranks as indicators of positions in a group. A centile rank of a score denotes the percentage of the cases in a distribution that scored lower than the score mentioned. If we say that a score of 69 has a centile rank of 20, we mean that 20 percent of the group scored below this score of 69. Centile ranks may be read from the same graphs as centile points, but this time the reading process is reversed. For example, suppose that one wanted the centile rank of a score of 50 for the data plotted in Figure 3.11. Taking the point 50 on the abcissa, a perpendicular is erected and extended to the ogive curve. From this point where the perpendicular intersects the ogive, a line is drawn to the y axis. This point on the y axis corresponds to the centile rank. From Figure 3.11, it is seen that a score of 50 has a centile rank of 93. Similarly a score of 40 has a centile rank of 68.

### Computation of Centile Ranks

If we wish to find just one centile rank and do not wish to construct an ogive curve, we would proceed as follows. Suppose that we wish to find the centile rank of a score of 52 in the distribution making up Table 3.6. In this table it is seen that the score of 52 falls in the interval 50–54. Below the bottom of this interval there are 342 cases and 20 cases fall within the interval itself. Now we must ascertain how many cases there are below the score, or point, of 52. The point 52 is halfway through the interval 50–54, 49.5 being the lower limit plus 2½ of the entire interval of 5. The 20 cases in the interval are assumed to be equally spread throughout the interval. Then we need one-half of these 20 cases, or 10 cases. Adding these 10 to the 342 cases below the interval, we find that 352 cases are below a score of 52. We then change this number of cases to a percentage of the total number of cases. The result is the centile rank.

$$\frac{352}{376}(100) = 93.6 = 94$$

This is approximately the same value for the centile rank of a score of 52 as shown in Figure 3.11.

## EXERCISES

**1.** For each of the following determine the exact limits, the midpoint, and the interval size:

a. 3–5

b. 25–29

c. 20–29

d. (−5)–(+5)

e. (−9)–(−5)

f. .5–1.4

g. .25–.49

**2.** Select an appropriate interval size for each of the following:

| | Low Score | High Score |
|---|---|---|
| **a.** | 1 | 12 |
| **b.** | 40 | 90 |
| **c.** | 0 | 140 |
| **d.** | 3 | 25 |
| **e.** | −2 | +2 |

**3.** The following are the scores of 60 students on a 100-item spelling test:

```
82  74  83  46  80  57  59  94  76  72
52  77  48  48  61  65  86  65  73  54
74  64  60  63  68  41  66  55  46  75
76  64  68  67  68  27  67  53  68  78
59  72  71  67  68  62  58  69  54  62
64  72  61  67  39  57  57  75  69  61
```

    **a.** Using an interval of 5 set up a frequency distribution for these scores.
    **b.** Construct a frequency polygon for this distribution.
    **c.** On the same axes construct an histogram for the data.
    **d.** What terms should be used to describe this distribution?

**4.** Below are the scores of two groups of boys on a test of spatial relations.

| Scores | Distribution A $f$ | Distribution B $f$ |
|---|---|---|
| 60–63 | 3 | 1 |
| 56–59 | 11 | 0 |
| 52–55 | 17 | 2 |
| 48–51 | 27 | 4 |
| 44–47 | 26 | 6 |
| 40–43 | 20 | 10 |
| 36–39 | 15 | 8 |
| 32–35 | 10 | 7 |
| 28–31 | 8 | 5 |
| 24–27 | 6 | 5 |
| 20–23 | 4 | 0 |
| 16–19 | 2 | 1 |
| 12–15 | 0 | 1 |
| 8–11 | 0 | 0 |
| 4–7 | 1 | 0 |
| | 150 | 50 |

    **a.** Make a frequency polygon for each of these distributions on the same axes.
    **b.** What terms may be used to describe these distributions?

**5.** For distribution A of problem 4 find the following centile points:

    **a.** $C_{10}$     **c.** $C_{20}$     **e.** Median     **g.** $Q_3$
    **b.** $C_{90}$     **d.** $C_{80}$     **f.** $Q_1$     **h.** $D_7$

**6.** The following scores were made on a simple test of psychomotor skills. Find the median of each.

   **a.** 9, 8, 7, 6, 5, 2, 2
   **b.** 10, 9, 7, 6, 5, 4, 2, 2
   **c.** 10, 9, 8, 8, 8, 8, 7, 6, 2
   **d.** 0, 1, 1, 4, 4, 4, 5, 7, 9, 10, 10
   **e.** 0, 2, 3, 4, 8, 9, 10, 11

**7.** Construct an ogive curve for the data in problem 4, distribution A. From the curve read the same centile points that you determined in problem 5 above. Compare the results.

**8.** Find the centile rank of each of the following scores in distribution A of problem 4:

   **a.** 20      **c.** 40
   **b.** 30      **d.** 60

**9.** From the ogive constructed in problem 7 above read the centile ranks for the scores in problem 8 above. Compare the results with those obtained in problem 8.

# Averages
## Chapter 4

One of the important ways of describing a group of measurements or scores is by the use of averages. Usually three averages are considered: the mean, the median, and the mode. Of these three measures of central tendency, the mean is the most commonly encountered and, as we shall see later, it is the one that is basic to many other statistical computations. In each of these averages we have a single numerical value that represents a group of individuals.

### THE ARITHMETIC MEAN

The mean is another term for arithmetic average. Everyone who has ever computed an "average" has computed a mean. Very simply, the mean is the sum of the scores divided by the number of cases. Suppose that we have five scores: $X_1 = 10$, $X_2 = 12$, $X_3 = 15$, $X_4 = 18$, and $X_5 = 20$. Then the mean $(X)$, read as "$X$ bar," would be

$$X = \frac{X_1 + X_2 + X_3 + X_4 + X_5}{N}$$

$$= \frac{10 + 12 + 15 + 18 + 20}{5}$$

$$= \frac{75}{5} = 15$$

Instead of writing the equation for the mean as above, we shorten it to

$$X = \frac{\Sigma X}{N} \tag{4.1}$$

where $X$ = mean

$\Sigma X$ = the sum of the scores

$N$ = the number of cases

## The Mean for Grouped Data

Suppose that we have the grades of 50 students in a course in elementary statistics listed in Table 4.1. In this school, course grades are quantified by making an A equal to 6, a B equal to 5, etc., as shown in column 1 of the table. Our task is to find the average final grade for the course. In column 2 are the frequencies for each of the letter grades. The values in column 3 are the products of the row values in each of the two previous columns, that is $6 \times 12 = 72$. Columns 2 and 3 are summed and the sums are used in the following formula:

$$X = \frac{\Sigma fX}{N} \tag{4.2}$$

$$= \frac{233}{50}$$

$$= 4.66$$

$$= 4.7$$

What we have done here is to shorten the work of adding twelve 6's, eighteen 5's, thirteen 4's, five 3's, and two 2's. Grouping often saves time and work.

When the interval size is greater than 1, the method used in Table 4.2 may be applied. In this table the data of Table 3.3 are again used. As we have already mentioned, the point that best represents any interval is the midpoint of that interval. Accordingly, the midpoint of each interval is entered in column 2, and these are designated as $X$. Each of these $X$ values is then multiplied by its corresponding $f$ and the product is placed in column 3. The values in this column are added and formula 4.2 is used:

**Table 4.1**  Computation of the Mean for Grouped Data When the Interval Size is 1

|   | (1) $X$ | (2) $f$ | (3) $fX$ |
|---|---|---|---|
| A | 6 | 12 | 72 |
| B | 5 | 18 | 90 |
| C | 4 | 13 | 52 |
| D | 3 | 5 | 15 |
| F | 2 | 2 | 4 |
|   |   | $N = 50$ | $\Sigma fX = 233$ |

$$X = \frac{\Sigma fX}{N}$$

$$= \frac{2130}{40}$$

$$= 53.2$$

This procedure may be shortened by using a method that starts with an arbitrary origin or reference point, as in Table 4.3. The steps followed in obtaining the mean by this method are outlined below:

**1.** We take the midpoint of one of the intervals as the arbitrary reference point. As far as the result is concerned, it makes no difference which interval midpoint is used. In the example illustrated, the midpoint of the interval 50–54, 52, is taken as the arbitrary reference point.

**2.** A second column is set up. This column is labeled $x'$ and can be read as "$x$ prime" and is defined as deviations from the arbitrary reference point. Since the interval 50–54 was taken as the interval containing our reference point, there is no deviation, and a zero is placed in column 2 for this interval. The interval of 55–59 deviates 1 interval from the arbitrary reference point, and a 1 is entered for this interval in column 2. This is continued upward until each interval has a value. We do the same thing for the intervals below the reference point. However, this time we place a minus sign in front of each of our deviations.

**3.** Next we multiply each $f$ by its $x'$ and enter the product in the column labeled $fx'$.

**4.** Sum this column.

**Table 4.2** Computation of the Mean for Grouped Data

|  | (1) | (2) Midpoint | (3) |
|---|---|---|---|
|  | $f$ | $X$ | $fX$ |
| 80–84 | 1 | 82 | 82 |
| 75–79 | 1 | 77 | 77 |
| 70–74 | 1 | 72 | 72 |
| 65–69 | 4 | 67 | 268 |
| 60–64 | 4 | 62 | 248 |
| 55–59 | 7 | 57 | 399 |
| 50–54 | 6 | 52 | 312 |
| 45–49 | 6 | 47 | 282 |
| 40–44 | 6 | 42 | 252 |
| 35–39 | 3 | 37 | 111 |
| 30–34 | 0 | 32 | 0 |
| 25–29 | 1 | 27 | 27 |
|  | $N = 40$ |  | $\Sigma = 2130$ |

**Table 4.3**  Computation of the Mean for Grouped Data, Short Method

|        | (1) $f$ | (2) $x'$ | (3) $fx'$ |
|--------|:---:|:---:|:---:|
| 80–84  | 1 | 6 | 6 |
| 75–79  | 1 | 5 | 5 |
| 70–74  | 1 | 4 | 4 |
| 65–69  | 4 | 3 | 12 |
| 60–64  | 4 | 2 | 8 |
| 55–59  | 7 | 1 | 7 |
| 50–54  | 6 | 0 | 0 |
| 45–49  | 6 | −1 | −6 |
| 40–44  | 6 | −2 | −12 |
| 35–39  | 3 | −3 | −9 |
| 30–34  | 0 | −4 | 0 |
| 25–29  | 1 | −5 | −5 |
|        | $N = 40$ | | $\Sigma fx' = 10$ |

**5.** The mean is then computed by substituting in this equation:

$$X = M' + \frac{\Sigma fx'}{N}(i) \qquad\qquad (4.3)$$

where $M'$ = arbitrary reference point
  $i$ = the size of the class interval
    and the other symbols are as previously defined

Then for our problem:

$$X = 52 + \frac{10}{40}(5)$$

$$= 52 + \frac{50}{40}$$

$$= 52 + 1.25$$

$$= 53.2$$

which is the same result obtained above.

In using this method it is customary to select as the reference point an interval near the center of the distribution, since this results in smaller numbers for the computational work. However, some prefer to begin with the bottom interval, which leads to all positive values. Regardless of the interval chosen, the results will always be the same.

It might be noted that the mean obtained here is not identical to that obtained by adding all the scores and dividing by $N$. Such differences between the means produced by the two methods are usually very small and of no practical significance. This difference is brought about by the error of grouping. The true mean here obtained by adding the raw scores and dividing by 40 is 53.3. Such slight discrepancies are brought about because we do not always meet the assumption that the average of the scores included in any interval equals the midpoint of that interval. In some intervals the average is higher than the midpoint, in others, lower. In the long run, these deviations tend to cancel each other and any chance error that remains tends to be small.

## A Definition of the Mean

We previously noted that the mean is the arithmetic average. This tells us actually very little about what the mean is. We shall now arrive at a more precise definition of the mean. Suppose that we have the following very simple distribution:

| (1) $X$ | (2) $x$ | (3) $x^2$ |
|---|---|---|
| 10 | 4 | 16 |
| 8 | 2 | 4 |
| 6 | 0 | 0 |
| 4 | −2 | 4 |
| 2 | −4 | 16 |
| $\Sigma X = 30$ | $\Sigma x = 0$ | $\Sigma x^2 = 40$ |
| $\bar{X} = 6$ | | |

Here we have five measures with a sum of 30 and a mean of 6. Column 2 is headed by a lowercase $x$. This is a new statistic to us and is called a *deviation score* or measure. It is defined as $X - \bar{X}$. Lowercase $x$ is the actual deviation of a score from the mean. The first score, 10, is 4 units above the mean. The second score, 8, deviates by 2 units, and so it continues for the other measures.

A value of zero is obtained by summing column 2. This then is the definition of the mean: that point about which the sum of the deviations is zero. Deviations are sometimes referred to as *moments,* and the analogy is made between the mean of a distribution and the fulcrum of a seesaw when the seesaw is in a state of equilibrium. The sum of the moments on one side of the fulcrum is equal to the sum of the moments on the opposite side.

That the sum of the deviations about the mean equals zero may be demonstrated algebraically as follows:

$$\Sigma(X - \bar{X}) = \Sigma X - \Sigma \bar{X}$$
$$= \bar{X}N - \bar{X}N$$
$$= 0$$

Since $\bar{X} = \Sigma X/N$, it follows that $\Sigma X = \bar{X}N$. This is substituted for $\Sigma X$ in the first equation. Summing the mean over all scores equals multiplying the mean by $N$. Thus $\Sigma X = XN$. This is substituted for $\Sigma X$ in the original equation. The values of these two terms have been shown to be the same, $XN$; hence their difference is zero.

In column 3 each deviation has been squared, and the sum of this column taken and found to be 40. This leads to another description of the mean as being that point in a distribution about which the sum of the squares of the deviations is at a minimum. For these data the sum of the squared deviations about any point other than the mean would be greater than 40.

## AVERAGING MEANS

Often we are given the means of two or more samples and we wish to find the mean of all the measures combined into one group. This is done by computing the weighted mean. Suppose that a test is given to three groups with the following results:

$$X_1 = 60 \qquad N_1 = 10$$
$$X_2 = 50 \qquad N_2 = 60$$
$$X_3 = 40 \qquad N_3 = 30$$

where $X$ and $N$ stand for the mean and the number of individuals in groups 1, 2, and 3, respectively. We wish to find the mean of the three groups combined, $X_T$.

Previously we learned that the mean is equal to the sum of the measures divided by the number of cases:

$$X = \frac{\Sigma X}{N}$$

If we have the mean and the number of cases, we can obtain the sum of the measures by solving the equation

$$\Sigma X = X(N)$$

This is exactly the procedure followed in obtaining the mean for the total group. We get the sum of the measures for each group and add these, then divide this obtained sum by the total number of cases as follows:

| $X$ | $N$ | $\Sigma X$ |
|---|---|---|
| 60 | 10 | 600 |
| 50 | 60 | 3000 |
| 40 | 30 | 1200 |
| | $\Sigma N = 100$ | $\Sigma X = 4800$ |

$$X_T = \frac{4800}{100} = 48$$

It should be noted that this mean cannot be obtained by averaging the three sample means. In this case the average of the three means would be 50. Only when the number in each sample is identical can the means of the samples be averaged directly to obtain the mean of the total group.

### Effect on the Mean of Adding a Constant to, or Subtracting One from, Each Measure in a Distribution

Often we are plagued with decimals, very large numbers, or negative numbers. Our computational work can be made much easier if we are aware that there are operations that can be performed on measures or scores which do not affect the basic results. Suppose that we have a group of scores, some of which are negative. Let us add a constant to each of the scores so that all of the scores will be positive. Or suppose that we have scores which range from 400 to 780. This time we reduce these scores by subtracting 400 from each. The question now is how has the mean of each group been affected. Such operations are known as coding and are illustrated in Table 4.4.

In this table we have 11 scores from each of which a constant of 500 has been subtracted. Both columns have been summed and the mean of each found. Note that to the mean of the second column (the coded mean), the constant 500 must be added to obtain the mean of the distribution. The student must remember that whenever data are coded, the statistics computed are *for the coded data* and such values must frequently be decoded. From this we see that when a constant is subtracted from each score in a distribution, the same constant is subtracted from the mean. Similarly, the mean

**Table 4.4**  Coding Data by Subtraction

| $X$ | $X - 500$ |
|---|---|
| 750 | 250 |
| 710 | 210 |
| 690 | 190 |
| 680 | 180 |
| 660 | 160 |
| 650 | 150 |
| 620 | 120 |
| 580 | 80 |
| 570 | 70 |
| 550 | 50 |
| 510 | 10 |

$\Sigma X = 6970$  $\qquad$ $\Sigma(X - 500) = 1470$

$\overline{X} = 6970/11$  $\qquad$ $\overline{X} - 500 = 1470/11$

$\quad = 633.6$  $\qquad\quad$ $\overline{X} - 500 = 133.6$

$\qquad\qquad\qquad\qquad\quad$ $\overline{X} = 133.6 + 500$

$\qquad\qquad\qquad\qquad\qquad\quad = 633.6$

is increased by the constant when every score in a distribution has a constant added to it. The same logic applies to coding by a constant multiplier or divisor. The mean must be decoded by multiplication if each score has been divided by a common number.

## THE MEDIAN

In the previous chapter a description was given of the methods of computing the median for grouped and ungrouped data. The median was defined as the point in a distribution with an equal number of cases on each side of it. The application of the method discussed in Chapter 3 to the data in Table 4.3 results in a median of 52.8 as summarized here:

$$\text{Median} = 49.5 + \frac{4}{6}(5)$$

$$= 49.5 + \frac{20}{6}$$

$$= 49.5 + 3.3$$

$$= 52.8$$

## THE MODE

A third average which we shall mention very briefly here is the mode, the symbol of which is Mo. For ungrouped data, the mode is defined as that datum value which occurs most frequently. When the data have been arranged into a frequency table, the mode is defined as the midpoint of the interval containing the largest number of cases. For example, the mode of the data in Table 4.3 is 57, the midpoint of the interval 55–59.

## USE OF THE DIFFERENT AVERAGES

Of the three measures of central tendency, the mean is the most frequently encountered. However, there are cases when the use of the mean is not justified. If we examine the incomes of a sample of male citizens between the ages of 20 and 60 of a state in the union, we will find that these incomes do not make a normal distribution. The frequency polygon for these data is positively skewed. The lowest value in the distribution is zero and the curve has its frequencies massed at the lower end with a mode in the $5000–6000 region, and there the tail extends to the right as some incomes may reach into the millions of dollars. A mean for these data would be pulled toward the high salaries, that is, in the direction of the skew.

Suppose that ten individuals make the following contributions to an organization:

$10.00
     .10
     .10
     .10
     .10
     .05
     .05
     .05
     .05
     .05
──────────
Σ = $10.65

The mean for these data is $1.06 and the median is $.075 or $.08. This illustrates that every single measure in a distribution affects the size of the mean. This is not so of the median. The top measure could have been a thousand or a million dollars, and the median would still have been 8 cents.

For the foregoing data, the median gives a much more accurate picture of the typical contribution than does the mean. It then follows that when a distribution is positively or negatively skewed, the best average to be used is the median. In a positively skewed distribution the mean will be higher than the median, and the median higher than the mode. The order of magnitude of the three averages is reversed in a negatively skewed distribution. In a normal distribution the mean, median, and mode are identical.

The mode is seldom used. It is very easy to compute, but it suffers from the fact that it is very unstable. Let us refer to Table 4.3 again. We noted that the mode for this distribution was 57, the midpoint of the interval 55–59. Now suppose one of these frequencies is removed from this interval and placed in the interval 45–49. This interval would now have a frequency of 7, and the mode of the distribution would now be 47. Then a change in the location of just one case would bring about a change in the mode of 10. A statistic that fluctuates so much is too unstable for any but the crudest of uses.

When data are collected using a nominal scale, the mode would be the appropriate statistic to use as a measure of popularity. The median is associated with ordinal data. The mean may be used with interval or ratio data, provided the distribution of the data approximates a normal curve.

When more computational work is to be done later, the mean should be used as an average. It will be seen that many of our other statistics are determined from the mean. The median is usually limited to descriptive use in statistical work. Of these two measures, the mean is more reliable, since it varies less from sample to sample. This will be discussed later in a chapter on sampling statistics.

## OTHER AVERAGES

The student in psychology will occasionally be confronted with data that should have a geometric or harmonic mean used as the average.

The geometric mean of two measures is the square root of their product; of three measures, the cube root of their product; and of $n$ measures, the $n$th root of their product. Thus the geometric mean of 2 and 8 is 4:

$$\sqrt{2(8)} = \sqrt{16} = 4$$

In general the geometric mean (GM) is equal to:

$$GM = \sqrt[n]{(X_1)(X_2)(X_3) \cdots (X)_n} \tag{4.4}$$

From this formula it can be seen that this mean cannot be used when the value of any measure is zero or has a negative sign. This mean is used in psychophysics or in other data concerned with measures of rates of change. We shall refer to it later when we discuss correlation.

A second infrequently encountered mean is known as the harmonic mean and is defined as the reciprocal of the arithmetic mean of the reciprocals of the measures.

$$HM = \frac{1}{1/N(1/X_1 + 1/X_2 + 1/X_3 + \cdots + 1/X_N)} \tag{4.5}$$

This mean is used in averaging rates when the time factor is variable and the act being performed is constant. See problem 9.

## EXERCISES

**1.** Group the following test scores with an interval of 1 and find the mean using formula 4.2.

| | | | | | | |
|---|---|---|---|---|---|---|
| 13 | 11 | 10 | 9 | 8 | 6 | 4 |
| 12 | 11 | 10 | 9 | 7 | 6 | 3 |
| 12 | 11 | 10 | 8 | 7 | 5 | 3 |
| 12 | 10 | 9 | 8 | 7 | 5 | 2 |
| 11 | 10 | 9 | 8 | 6 | 4 | 1 |

**2.** Find the three averages for the following data, which represent the scores of a group of high school seniors on a short intelligence test.

| Score | $f$ | Score | $f$ |
|---|---|---|---|
| 69–71 | 5 | 48–50 | 3 |
| 66–68 | 0 | 45–47 | 1 |
| 63–65 | 3 | 42–44 | 0 |
| 60–62 | 2 | 39–41 | 2 |
| 57–59 | 4 | 36–38 | 0 |
| 54–56 | 6 | 33–35 | 1 |
| 51–53 | 4 | 30–32 | 1 |

3. The following are scores on a short skills test. Find the three averages for each.

    **a.** 10, 12, 13, 14, 15, 16, 17, 18, 20
    **b.** 8, 8, 9, 10, 11, 16, 18
    **c.** 5, 6, 7, 7, 7, 8, 10, 12
    **d.** 15, 18, 18, 18, 18, 20, 20

4. The following scores were obtained on the mechanical scale of an interest inventory.

| Score | $f$ |
|-------|-----|
| 30 | 2 |
| 29 | 1 |
| 28 | 3 |
| 27 | 8 |
| 26 | 17 |
| 25 | 22 |
| 24 | 32 |
| 23 | 26 |
| 22 | 12 |
| 21 | 8 |
| 20 | 6 |
| 19 | 4 |
| 18 | 0 |
| 17 | 1 |

$$N = 142$$

    **a.** Find the three averages.
    **b.** What inference may be drawn about the shape of this distribution?

5. In an alumni survey the graduates of a certain class of an engineering school reported the number of their children as follows:

| Number of Children | Families |
|--------------------|----------|
| 9 | 1 |
| 8 | 0 |
| 7 | 3 |
| 6 | 4 |
| 5 | 12 |
| 4 | 45 |
| 3 | 62 |
| 2 | 86 |
| 1 | 107 |
| 0 | 50 |

    **a.** What is the average number of children per family?
    **b.** Find $Q_1$ and $Q_3$ for this distribution.

6. Following are the salaries of a group of high school teachers:

| Salary | f |
|---|---|
| 18,000 and up | 5 |
| 15,000–17,999 | 15 |
| 14,000–14,999 | 80 |
| 13,000–13,999 | 40 |
| 12,000–12,999 | 25 |
| 11,000–11,999 | 30 |
| 10,000–10,999 | 18 |
| 9,000–9,999 | 10 |
| 8,000–8,999 | 6 |
| 7,000–7,999 | 6 |
| Below 7,000 | 4 |

**a.** Obtain an appropriate average for these data.
**b.** Find $Q_3$ and $Q_1$.

**7.** For each of the following compute the weighted mean.

**a.** $X_1 = 42$, $N_1 = 60$; $X_2 = 48$, $N_2 = 80$; $X_3 = 40$, $N_3 = 42$.
**b.** $X_1 = 60$, $N_1 = 10$; $X_2 = 50$, $N_2 = 100$.

**8.** On a trip a family averaged 40 mph for the first hour, 50 mph for the second hour, and 58 mph for the third hour. What was their average speed?

**9.** Suppose that the family mentioned above covered the first 100 miles at 40 mph, the next 100 miles at 50 mph, and the last 100 miles at 58 mph. What was their average speed?

# Variability
## Chapter 5

In the previous chapter we were concerned with the computation, meaning, and use of averages. A little thought will soon point up the fact that averages in themselves do not adequately describe a distribution. Averages locate the center of a distribution but tell us nothing about how the scores or measurements are arranged in relation to the center. Let us take two illustrations. Suppose that we have two distributions of scores on the same test, each with a mean of 67. In the first of these the highest score is 72 and the lowest is 62. The second distribution has a high score of 107 and a low score of 25. The range of the first distribution is 11 and that of the second distribution is 83. Thus to give a better picture of a distribution we need both a measure of central tendency and one of variability or dispersion. In statistical work we would say that the first of our illustrations is a homogeneous group. The individuals within it are very similar in reference to the trait measured. The other group is described as heterogeneous, the variability being great. Both of these terms are widely used in educational, sociological, and psychological work.

## MEASURES OF VARIABILITY

### The Range
The range has already been defined as the high score minus the low score plus one. We used it in the first step in setting up frequency distributions. We will spend little more time on it here, other than to state that of all the measures of variability, range is the most unstable. By this we mean that from sample to sample, the range varies more than any of the other measures. An illustration will show why this is so. Suppose that we have a distribution of scores, the lowest of which is 30 and the highest is 103. The next score below

103 happens to be 90. By the use of our formula, the range is found to be 74; but 13 of the points making up this range are the result of the high score of 103. The chances are good that the next sample will not contain this high deviate score, and hence the range will be much smaller. The range, like the mode, is a very unstable statistic, since it may vary considerably from sample to sample.

The range can be used justifiably when we want a hasty measure of variability and do not have time to compute one of the others. Each of the statistics discussed later in this chapter is a better measure. Of course, in dealing with a population instead of with a sample, the range becomes more useful.

### The Average Deviation

The average deviation, also referred to as the mean deviation, is no longer widely used in statistical work, having been replaced by the standard deviation. A brief consideration of it here, however, may make the following material on the standard deviation easier. To begin, we shall again define the symbol $x$ as a deviation of any score from the mean of its distribution. In symbols we can write this as follows:

$$x = X - \bar{X} \tag{5.1}$$

where $x$ = the deviation of a score from the mean

$X$ = a raw score

$\bar{X}$ = the mean

In any distribution the sum of these deviations from the mean is equal to zero. As previously noted, this is an important fact about the mean. About no other point in any distribution is this true. Our definition of the mean is that it is that point about which the sum of the deviations is equal to zero.

Since the sum of the deviations about the mean is equal to zero, it follows that we can obtain no average deviation unless we change our procedure. In actual practice, we sum the deviations disregarding the signs. The equation for the average deviation is written

$$AD = \frac{\Sigma|x|}{N} \tag{5.2}$$

where $|x|$ = the absolute deviations of the scores from the mean, that is, without regard to sign

$N$ = the number of cases

The calculation of the average deviation is shown in Table 5.1. For this distribution, then, the scores are, on an average, 6 units from the mean.

**Table 5.1**   Computation of the Average Deviation $(N = 10)$

| $X$ | $x$ |
|---|---|
| 26 | 10 |
| 24 | 8 |
| 22 | 6 |
| 20 | 4 |
| 18 | 2 |
| 16 | 0 |
| 14 | −2 |
| 10 | −6 |
| 6 | −10 |
| 4 | −12 |

$\Sigma X = 160$    $\Sigma x = 0$
$\overline{X} = 16$    $\Sigma|x| = 60$

$$AD = \frac{\Sigma|x|}{N}$$

$$= \frac{60}{10}$$

$$= 6$$

## The Standard Deviation

THE STANDARD DEVIATION FOR UNGROUPED DATA. Of all the measures of variability, the standard deviation is by far the most widely encountered, mainly because it is used in so many other statistical operations. We shall begin by showing how it is computed for both ungrouped and grouped data. With ungrouped data, the process starts in the same fashion as does that for the average deviation. That is, we first compute the mean. Then we compute the deviation of each score from this mean (see Table 5.2). Then we square each of these deviations and add this column. A check on our work would be that the sum of the $x$ column or the sum of the deviations about the mean should be zero. We find that this is true for our problem. To find the standard deviation, we use the following formula and substitute in it as shown:

$$s = \sqrt{\frac{\Sigma x^2}{N}} \tag{5.3}$$

$$= \sqrt{\frac{198.40}{10}}$$

$$= \sqrt{19.840}$$

$$= 4.5$$

**Table 5.2**   Computation of the Standard
Deviation for Ungrouped Data

| $X$ | $x$ | $x^2$ |
|---|---|---|
| 20 | 7.6 | 57.76 |
| 18 | 5.6 | 31.36 |
| 16 | 3.6 | 12.96 |
| 14 | 1.6 | 2.56 |
| 13 | .6 | .36 |
| | $\Sigma = +19$ | |
| 11 | −1.4 | 1.96 |
| 10 | −2.4 | 5.76 |
| 9 | −3.4 | 11.56 |
| 8 | −4.4 | 19.36 |
| 5 | −7.4 | 54.76 |
| | $\Sigma = -19$ | |
| $\Sigma X = 124$ | $\Sigma x = 0.0$ | $\Sigma x^2 = 198.40$ |
| $X = 12.4$ | | |

In some texts the formula for the standard deviation is given as

$$s = \sqrt{\frac{\Sigma x^2}{N - 1}}$$

When using this formula, one is actually correcting the sample standard deviation for bias. For reasons that will not be discussed here, the sample standard deviation is a biased estimate of the parameter standard deviation, and this bias is a downward one. That is, the simple standard deviation tends to be smaller than the population standard deviation. The equation for the population standard deviation is

$$\sigma = \sqrt{\frac{\Sigma x^2}{N}} \tag{5.4}$$

It can be argued that one always studies samples to make inferences about populations and hence the sample standard deviation should always be corrected for bias. Actually, if the sample is of any size at all, whether $N$ or $N - 1$ is used in the formula for the sample standard deviation is of little importance. For example, if $N = 9$, whether one divides by 8 or 9 does make a difference in the size of $s$. But if $N = 200$, the difference obtained by dividing by 199 instead of 200 is trivial. Very often we are interested only in how a group of scores varies about the sample mean. This is the case when we have a classroom test. Until we get to the use of statistics in making statistical inferences (Chapter 11) we shall use formula 5.3 in obtaining the standard deviation.

**Table 5.3**    Computation of
the Standard
Deviation
Directly from
Raw Scores

| X | $X^2$ |
|---|---|
| 20 | 400 |
| 18 | 324 |
| 16 | 256 |
| 14 | 196 |
| 13 | 169 |
| 11 | 121 |
| 10 | 100 |
| 9 | 81 |
| 8 | 64 |
| 5 | 25 |
| $\Sigma X = 124$ | $\Sigma X^2 = 1736$ |

If a calculating machine is available, it is much easier to use the so-called raw-score formula for getting the sum of the deviations squared. The sum is usually referred to as the sum of the squares and is one of our most useful statistics. This technique is illustrated in Table 5.3. In this table we copy the scores and label the first column X. The second column is merely the square of each X. (Refer to Appendix A for the square of any number from 1 to 1000.) Both columns are then summed. It should be pointed out that when a calculating machine is used, it is not necessary to copy the original scores or their squares as shown in Table 5.3. These are entered into the machine one at a time, and as the process goes along, the sum of each is accumulated. At the end, both values may be read directly from the machine.

For this method we obtain the sum of the squares by the following equation:

$$\Sigma x^2 = \Sigma X^2 - \frac{(\Sigma X)^2}{N} \tag{5.5}$$

$$= 1736 - \frac{(124)^2}{10}$$

$$= 1736 - \frac{15376}{10}$$

$$= 1736 - 1537.6$$

$$= 198.4$$

This value for the sum of the squares is the same as obtained by totaling the deviations of each score from the mean. Even without a calculating machine students may find this method easier than the first. After we obtain the sum of the squares, we next substitute our values into equation 5.3 and solve for the standard deviation. Very often the term sum of the squares is abbreviated *SS*.

Formulas 5.3 and 5.5 may be combined to give a formula that may be used in calculating the standard deviation from raw scores as follows:

Formula (5.3)
$$s = \sqrt{\frac{\Sigma x^2}{N}}$$

Formula (5.5)
$$\Sigma x^2 = \Sigma X^2 - \frac{(\Sigma X)^2}{N}$$

Substituting equation 5.5 in 5.3 we have

$$s = \sqrt{\frac{\Sigma X^2 - \frac{(\Sigma X)^2}{N}}{N}}$$

Rearranging terms

$$s = \sqrt{\frac{N\Sigma X^2 - (\Sigma X)^2}{N}\left(\frac{1}{N}\right)}$$

$$s = \frac{1}{N}\sqrt{N\Sigma X^2 - (\Sigma X)^2} \tag{5.6}$$

STANDARD DEVIATION FOR GROUPED DATA. Just as there are two methods for finding the mean of grouped data, there are two corresponding techniques for finding the standard deviation. The first uses the midpoints of the class intervals; this will be illustrated by using the data from Table 4.2, which have been reproduced in Table 5.4. The mean of this distribution is 53.25.

**Table 5.4**  Computation of Standard Deviation Using Midpoints

| | (1) $f$ | (2) Midpoint $X$ | (3) $fX$ | (4) $fX^2$ |
|---|---|---|---|---|
| 80–84 | 1 | 82 | 82 | 6724 |
| 75–79 | 1 | 77 | 77 | 5929 |
| 70–74 | 1 | 72 | 72 | 5184 |
| 65–69 | 4 | 67 | 268 | 17956 |
| 60–64 | 4 | 62 | 248 | 15376 |
| 55–59 | 7 | 57 | 399 | 22743 |
| 50–54 | 6 | 52 | 312 | 16224 |
| 45–49 | 6 | 47 | 282 | 13254 |
| 40–44 | 6 | 42 | 252 | 10584 |
| 35–39 | 3 | 37 | 111 | 4107 |
| 30–34 | 0 | 32 | 0 | 0 |
| 25–29 | 1 | 27 | 27 | 729 |
| | $N = 40$ | | $\Sigma fx = 2130$ | $\Sigma fx^2 = 118810$ |

The steps are as follows:

**1.** Add a new column, 4, headed $fX^2$.
**2.** Multiply each value in column 3 by the corresponding value in column 2 to obtain the $fX^2$ for each row.
**3.** Sum the $fX^2$ column.
**4.** Then substitute values in the following equation:

$$s = \sqrt{\frac{\Sigma fX^2}{N} - \bar{X}^2} \qquad\qquad (5.7)$$

$$= \sqrt{\frac{118810}{40} - 53.25^2}$$

$$= \sqrt{2970.25 - 2835.56}$$

$$= \sqrt{134.69}$$

$$= 11.6$$

The second method is associated with the method of finding the mean using an arbitrary origin. This technique is illustrated using the data from Table 4.3, which are reproduced in Table 5.5. Inspection of this table shows that the process is identical up to and including column 4. We start by selecting an arbitrary reference point. In this problem we chose the midpoint of the interval 50–54. Then we gave this interval a deviation value of zero and set up our $x'$ column (column 3) as shown. We obtain the value in column 4, $fx'$, by multiplying the values in the two previous columns. We obtain the

**Table 5.5**  Computation of Standard Deviation for 40 Scores on a Statistics Test

| (1) | (2) $f$ | (3) $x'$ | (4) $fx'$ | (5) $fx'^2$ | (6) $f(x' + 1)^2$ |
|---|---|---|---|---|---|
| 80–84 | 1 | 6 | 6 | 36 | 49 |
| 75–79 | 1 | 5 | 5 | 25 | 36 |
| 70–74 | 1 | 4 | 4 | 16 | 25 |
| 65–69 | 4 | 3 | 12 | 36 | 64 |
| 60–64 | 4 | 2 | 8 | 16 | 36 |
| 55–59 | 7 | 1 | 7 | 7 | 28 |
|  |  |  | $\Sigma = +42$ |  |  |
| 50–54 | 6 | 0 | 0 | 0 | 6 |
| 45–49 | 6 | −1 | −6 | 6 | 0 |
| 40–44 | 6 | −2 | −12 | 24 | 6 |
| 35–39 | 3 | −3 | −9 | 27 | 12 |
| 30–34 | 0 | −4 | 0 | 0 | 0 |
| 25–29 | 1 | −5 | −5 | 25 | 16 |
|  |  |  | $\Sigma = -32$ |  |  |
|  | $N = 40$ |  | $\Sigma fx' = 10$ | $\Sigma fx'^2 = 218$ | $\Sigma f(x' + 1)^2 = 278$ |

$fx'^2$ values in column 5 by multiplying the values in the two previous columns. Then we sum the various columns as shown.

To find the standard deviation we again have to find the sum of the squares. For grouped data this is done as follows:

$$\Sigma x^2 = i^2 \left[ \Sigma fx'^2 - \frac{(\Sigma fx')^2}{N} \right] \qquad (5.8)$$

$$= 5^2 \left( 218 - \frac{10^2}{40} \right)$$

$$= 25 \left( 218 - \frac{100}{40} \right)$$

$$= 25(218 - 2.5)$$

$$= 25(215.5)$$

$$= 5387.5$$

We obtain the standard deviation by solving in the usual fashion:

$$s = \sqrt{\frac{\Sigma x^2}{N}}$$

$$= \sqrt{\frac{5387.5}{40}}$$

$$= \sqrt{134.69}$$

$$= 11.6$$

We can check the accuracy of the work in several ways. The first of these, called Charlier's check, is a check on the accuracy of the sums in our $fx'$ and $fx'^2$ columns. To carry out this check, we set up column 6 as shown in Table 5.5. The heading here tells us to take each $x'$ value, add 1 to it, square it, and then multiply it by the frequency in the interval. For the top interval this becomes 6 plus 1 which is 7; 7 squared is 49, and this multiplied by the frequency, 1, is still 49. When each interval has been so treated, the values in column 6 are summed. Charlier's check is made by substituting in the following equation:

$$\Sigma f(x' + 1)^2 = \Sigma fx'^2 + 2\Sigma fx' + \Sigma f \qquad (5.9)$$

$$278 = 218 + 2(10) + 40$$

$$= 218 + 20 + 40$$

$$= 278$$

We shall present a second rough check on our computed standard deviation, after we discuss the meaning of a standard deviation.

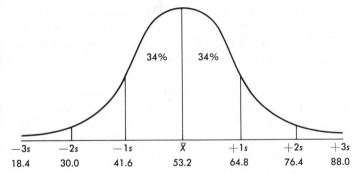

| | | | | | | |
|---|---|---|---|---|---|---|
| −3s | −2s | −1s | X̄ | +1s | +2s | +3s |
| 18.4 | 30.0 | 41.6 | 53.2 | 64.8 | 76.4 | 88.0 |

**Figure 5.1**   Standard deviation units and the normal curve.

INTERPRETATION OF STANDARD DEVIATION. A standard deviation helps describe the normal curve. When our data take the shape of the normal curve, standard deviation units measured off along the base line, starting from the mean, always cut off certain proportions of the area under the curve. In the problem illustrated in Figure 5.1, the mean 53.2 is recorded at the center of the curve. Then 1 standard deviation unit is added to this mean, 53.2 + 11.6, resulting in a score of 64.8. Similarly, 1 standard deviation is measured off on the other side of the mean. In a normal curve, these two standard deviation units taken together include approximately 68 percent of the area. We shall in future chapters refer to this as two-thirds the area. If we measure off 2 standard deviation units on each side of the mean, we include between these two points approximately 95 percent of the area. And when we take 3 standard deviation units, over 99 percent of the area of the curve is included. Actually, about 13 cases in 10,000 are left over on each side of the mean out beyond the +3 and −3 standard deviation points. In a later chapter on the normal curve we shall see where these values come from.

Let us return to Table 5.5 and see how our computed standard deviation checks when we count the number of cases included by 1 standard deviation unit on each side of the mean. As previously noted, the points corresponding to +1 and −1 standard deviation are 64.8 and 41.6, respectively. Let us locate 64.8 on Table 5.5. For all practical purposes, we might say that it is the point at the top of the interval 60–64. So we shall start counting down the number of cases between this and 41.6. We have to determine how many of the 6 cases in the interval 40–44 are above 41.6 and are to be included in our sum. The top of this interval is 44.5. The value 41.6 is 2.9 or approximately 3 units below this. Hence we want ⅗ of the 6 units or approximately 4 of them. So starting from the top (64.8), we add 4 + 7 + 6 + 6 + 4, which equals 27. The number of the cases in this problem is 40. A standard deviation taken on each side of the mean would then include 27/40 of the distribution, which in terms of percents reduces to between 67 and 68 percent. This check may not always come as close to the values of the normal curve; but if the computational work is correct and if the obtained distribution is more or less normal, the percentage areas of the normal curve should be approximated.

RELATION OF THE RANGE TO THE STANDARD DEVIATION. We noted above that approximately 6 standard deviations cover the range. This is true only when the number of cases is large. As $N$ decreases, the number of standard deviations needed to include all of the cases decreases. Here is how it looks.[1]

| $N$ | Number of Standard Deviations Included in the Range |
|---|---|
| 5 | 2.3 |
| 10 | 3.1 |
| 25 | 3.9 |
| 30 | 4.1 |
| 50 | 4.5 |
| 100 | 5.0 |
| 500 | 6.1 |
| 1000 | 6.5 |

EFFECT OF ADDING OR SUBTRACTING A CONSTANT TO EACH MEASURE ON A STANDARD DEVIATION. Previously we learned that adding or subtracting a constant to or from each measure in a distribution increased or decreased the mean by the same constant. When measures are so treated the standard deviation remains the same, since the range of scores is not affected by the addition or subtraction of a constant. This is illustrated below:

| $X$ | $x$ | $x^2$ | $X + 10$ | $x$ | $x^2$ |
|---|---|---|---|---|---|
| 20 | 4 | 16 | 30 | 4 | 16 |
| 18 | 2 | 4 | 28 | 2 | 4 |
| 16 | 0 | 0 | 26 | 0 | 0 |
| 14 | −2 | 4 | 24 | −2 | 4 |
| 12 | −4 | 16 | 22 | −4 | 16 |
| $\Sigma X = 80$ | | $\Sigma x^2 = 40$ | $\Sigma(X + 10) = 130$ | | $\Sigma x^2 = 40$ |
| $X = 16$ | | | $X = 26$ | | |

Note that the sum of the squares is similar for both the coded and uncoded data, hence in each case the standard deviation $= \sqrt{40/5} = \sqrt{8} = 2.8$.

When each measure in a distribution is multiplied or divided by a constant, the standard deviation and the mean are multiplied or divided by the same constant, as follows:

[1] Adapted from L. H. C. Tippett. "On the Extreme Individuals and the Range of Samples from a Normal Population," *Biometrika,* 1925, *17,* 386.

| X | x | $x^2$ | 2X | x | $x^2$ |
|---|---|---|---|---|---|
| 20 | 4 | 16 | 40 | 8 | 64 |
| 18 | 2 | 4 | 36 | 4 | 16 |
| 16 | 0 | 0 | 32 | 0 | 0 |
| 14 | −2 | 4 | 28 | −4 | 16 |
| 12 | −4 | 16 | 24 | −8 | 64 |

$$\Sigma X = 80 \qquad \Sigma x^2 = 40 \qquad \Sigma 2X = 160 \qquad \Sigma x^2 = 160$$
$$\overline{X} = 16 \qquad\qquad\qquad \overline{X} = \;\;32$$

$$s = \sqrt{40/5} \qquad\qquad s = \sqrt{160/5}$$
$$\;\;= \sqrt{8} \qquad\qquad\qquad = \sqrt{32}$$
$$\;\;= 2.83 \qquad\qquad\qquad = 5.66$$

The standard deviation of the coded data is exactly twice that of the uncoded measures.

Much time and effort can be saved by coding data. Large numbers are made smaller, negative numbers may become positive, and decimals may be converted to whole numbers. We must remember, however, to uncode our final answers so that our statistic will pertain to the original data, not to the coded material.

THE VARIANCE. Much of our statistical work is handled by using another statistic to describe variability. This statistic is called the variance and is simply the standard deviation squared.

$$s^2 = \frac{\Sigma x^2}{N} \tag{5.10}$$

The variance ($s^2$) and other symbols are as previously described. In this textbook we shall have limited use for the variance, but a good share of modern statistics is based upon the manipulation of variances. The variance is frequently referred to as the mean-square (MS).

AVERAGING STANDARD DEVIATIONS. In Chapter 4 we learned that if we wished to average two or more means, we could do it by the so-called weighting method. Standard deviations cannot be treated in the same manner. When two or more standard deviations are to be averaged, the following formula is to be used:

$$s_T = \sqrt{\frac{N_A(\overline{X}_A{}^2 + s_A{}^2) + N_B(\overline{X}_B{}^2 + s_B{}^2)}{N_A + N_B} - \overline{X}_T{}^2} \tag{5.11}$$

where   $s_T$ = standard deviation of combined groups
$N_A, N_B$ = number of individuals in each of the two groups
$\overline{X}_A, \overline{X}_B$ = means of the two groups
$\overline{X}_T$ = weighted mean of the two groups combined
$s_A, s_B$ = standard deviations of the two groups being combined

If more than two groups are being combined, an additional element is placed in both the numerator and denominator for each additional group.

### The Quartile Deviation

The quartile deviation, symbol $Q$, is frequently called the semi-interquartile range. In a previous chapter two quartile points were mentioned, $Q_1$, the equivalent of the 25th centile, and $Q_3$, the equivalent of the 75th centile. The quartile deviation or the semi-interquartile range is half the distance between these two quartile points. In symbols we would write this as follows:

$$Q = \frac{Q_3 - Q_1}{2} \qquad (5.12)$$

This statistic is easy to calculate. To illustrate we have again set up in Table 5.6 the distribution of geography scores previously used for illustration. We must first calculate $Q_1$ and $Q_3$. By definition, $Q_1$ is the point in this distribution which has 25 percent of the scores below it. Twenty-five percent of 40 cases is 10 cases. We start counting cases from the bottom, and we find that below the lower limit of the interval 45–49 we have exactly 10 cases. No interpolation is necessary in this case, and we record that $Q_1 = 44.5$. To find $Q_3$ we repeat this same process except that instead of going up 75 percent of the cases, we come down from the top 25 percent of the cases. By counting down we find that when we get to 64.5 (the lower limit of the interval 65–69), we have included 7 cases. We need 3 more. Since there are 4 cases in this interval

**Table 5.6**  Computation of the Quartile Deviation

| | $f$ | |
|---|---|---|
| 80–84 | 1 | |
| 75–79 | 1 | |
| 70–74 | 1 | |
| 65–69 | 4 | |
| | | 7 |
| 60–64 | 4 | |
| 55–59 | 7 | |
| 50–54 | 6 | |
| 45–49 | 6 | |
| | | 10 |
| 40–44 | 6 | |
| 35–39 | 3 | |
| 30–34 | 0 | |
| 25–29 | 1 | |
| | $N = 40$ | |

we need to go three-fourths of the way through the interval. We can show the operation as follows:

$$Q_3 = 64.5 - \frac{3}{4}(5)$$

$$= 64.5 - \frac{15}{4}$$

$$= 64.5 - 3.75$$

$$= 60.75$$

To find $Q$ we proceed as follows:

$$Q = \frac{Q_3 - Q_1}{2}$$

$$= \frac{60.75 - 44.5}{2}$$

$$= \frac{16.25}{2}$$

$$= 8.125$$

$$= 8.12$$

INTERPRETATION OF $Q$. In the previous chapter we found that the median for the data in Table 5.6 was 52.8. In a normal distribution, if we take the median and add and subtract one quartile deviation on each side of it, we will cut off approximately 50 percent of the cases. That is, for the data at hand we would expect to find 50 percent of the cases to fall between 60.9 (the median plus one $Q$) and 44.7 (the median minus one $Q$). This is shown graphically in Figure 5.2.

It also happens that if we measure off four quartile deviations on each side of the median, we will include practically all the cases. We can state this briefly by saying that eight $Q$'s approximately cover the range.

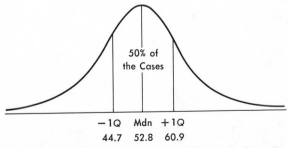

**Figure 5.2**  Relationship of $Q$ to the normal curve.

WHEN TO USE THE QUARTILE DEVIATION. Since the quartile deviation is asso-ciated with the median, it follows that whenever the median is used as a measure of central tendency, the quartile deviation is an appropriate measure of variability. It may be recalled that the median is the statistic to be used as a measure of central tendency when we have a skewed distribution. Even when distributions are skewed, the check using the middle 50 percent of the cases will work.

## Summary of the Measures of Variability

At this point we shall briefly summarize the four measures of variability.

**1.** *The Range.* This is least stable of all four measures. It is of limited use, except when speed is an issue or in simple situations such as in setting up a frequency distribution.

**2.** *The Quartile Deviation.* This statistic is always associated with the median, and it is used whenever the median is used as a measure of central tendency. This is usually the case in skewed distributions. It is noted here that in a normal distribution the quartile deviation and the standard deviation have a constant relationship. $Q = .6745s$.

**3.** *The Average Deviation.* This could be used with the mean in a normal distribution. In the past it was a widely used statistic, but today it is almost completely replaced by the standard deviation.

**4.** *The Standard Deviation.* This is the most reliable of all four measures and the one most frequently encountered. It is associated with the mean, and we use it when we are planning to make interpretations associated with the normal curve. As we shall see, this statistic has many uses in modern statistics and is one of our most important tools.

## MEASURES OF SKEWNESS AND KURTOSIS

### Moments

Both of the major statistics already discussed, the average and the standard deviation, are related to a group of statistics known as moments. The first four moments are as follows:

$$m_1 = \frac{\Sigma(X - \bar{X})}{N} = \frac{\Sigma x}{N} = 0$$

$$m_2 = \frac{\Sigma(X - \bar{X})^2}{N} = \frac{\Sigma x^2}{N} = s^2$$

$$m_3 = \frac{\Sigma(X - \bar{X})^3}{N} = \frac{\Sigma x^3}{N}$$

$$m_4 = \frac{\Sigma(X - \bar{X})^4}{N} = \frac{\Sigma x^4}{N}$$

And the *n*th moment about the mean would be

$$m_n = \frac{\Sigma(X - \bar{X})^n}{N} = \frac{\Sigma x^n}{N}$$

## Skewness

The skewness of a distribution, $Sk$ or $g_1$, as it may be called, is obtained as follows:

$$Sk(g_1) = \frac{m_3}{m_2\sqrt{m_2}} = \frac{\Sigma x^3/N}{(\sqrt{\Sigma x^2/N})^3} \tag{5.13}$$

or, the ratio of the mean of the cube of the deviations about the mean to the cube of the standard deviation.

If a distribution of measures is normal in shape, the sum of the cubes of the deviations above the mean will equal the sum of the cubes of deviations below the mean, and the total sum of the cubes of the deviations will be zero and $Sk$ will be zero. If the distribution is positively skewed, the sum of the cubes of the deviations above the mean will be greater than the sum of the deviations below the mean, and $Sk$ will be positive. Similarly, when conditions are reversed, $Sk$ will be negative in sign. The larger the value of $Sk$, the greater the amount of skewness.

## Kurtosis

Kurtosis ($Ku$ or $g_2$) of a distribution is determined by the use of the following formula:

$$Ku(g_2) = \frac{m_4}{m_2{}^2} - 3 = \frac{\Sigma x^4/N}{(\Sigma x^2/N)^2} - 3 \tag{5.14}$$

or, the ratio of the mean of the fourth power of the deviation about the means to the variance ($m_2{}^2$) squared minus 3.

When the value of $Ku$ is zero, the shape of the distribution is mesokurtic, the shape of the normal curve. When $Ku$ is negative, the curve is platykurtic and, when $Ku$ is positive, the curve is leptokurtic.

## EXERCISES

**1.** Calculate the standard deviation for the data in problems 1, 2, 3, and 4 of Chapter 4.

**2.** The administration of a short statistics test resulted in the following scores. Calculate the mean and standard deviation (use formula 5.6) of the scores.

| | | | | |
|---|---|---|---|---|
| 44 | 37 | 30 | 28 | 22 |
| 42 | 36 | 30 | 27 | 20 |
| 40 | 35 | 30 | 27 | 18 |
| 38 | 34 | 29 | 26 | 16 |
| 38 | 32 | 28 | 24 | 10 |

**3.** A group of seniors majoring in psychology made the following scores on the verbal test of the Graduate Record Examination. Calculate the mean and standard deviation of the above scores. Code your data and use formula 5.6.

| | | | | | | | |
|---|---|---|---|---|---|---|---|
| 750 | 640 | 600 | 570 | 540 | 490 | 450 | 400 |
| 700 | 630 | 590 | 570 | 540 | 490 | 440 | 380 |
| 680 | 630 | 590 | 560 | 530 | 480 | 440 | 360 |
| 660 | 610 | 580 | 560 | 500 | 470 | 430 | 350 |
| 650 | 600 | 570 | 540 | 490 | 470 | 420 | 320 |

**4.** The same statistics test was administered to three classes with the following results. Compute the mean and standard deviation of the three groups combined.

| | $X$ | $s$ | $N$ |
|---|---|---|---|
| Class A | 82 | 10 | 25 |
| Class B | 86 | 12 | 35 |
| Class C | 88 | 16 | 20 |

**5.** Find the quartile deviation for problems 5 and 6 of Chapter 4.

**6.** On a short attitude scale on drug usage a group of university seniors scored as follows:

| | | | | | | |
|---|---|---|---|---|---|---|
| 9 | 7 | 8 | 8 | 7 | 8 | 9 |
| 7 | 2 | 8 | 7 | 5 | 6 | 5 |
| 6 | 5 | 6 | 4 | 5 | 3 | 4 |
| 4 | 3 | 1 | 3 | 3 | 2 | 2 |
| 7 | 9 | 6 | 1 | 1 | 1 | 1 |

  **a.** Find $Q$.

  **b.** Is this statistic used correctely here?

**7.** Compute and interpret $g_1$ and $g_2$ for the following scores:
10, 6, 4, 2, 1

# Standard Scores and the Normal Curve
## Chapter 6

## STANDARD SCORES

The formula for the standard score is

$$z = \frac{X - \overline{X}}{s} = \frac{x}{s} \tag{6.1}$$

where $X$ = any raw score or unit of measurement
$\overline{X}, s$ = mean and standard deviation of the distribution of scores

To illustrate the computation and nature of standard scores, let us take the following scores, which are a part of a distribution with a mean of 60 and a standard deviation of 10:

| $X$ | $x$ | $z$ |
|-----|-----|-----|
| 70 | 10 | 1.00 |
| 60 | 0 | .00 |
| 50 | −10 | −1.00 |
| 54 | −6 | −.60 |
| 46 | −14 | −1.40 |

In the first column we have the raw scores ($X$). The mean is subtracted from each of these, and then this deviation from the mean, or $x$, is divided by the standard deviation to change the deviation values into standard score values. The raw score of 60 is at the mean. There is no deviation; hence the standard score is zero. A raw score of 70 is 1 standard deviation above the mean. This results in a $z$ score of 1. When we change raw scores to standard scores, we are expressing them in standard deviation units. These standard scores tell us how many standard deviation units any given raw score deviates from the mean.

Since 3 standard deviations on either side of the mean include practically

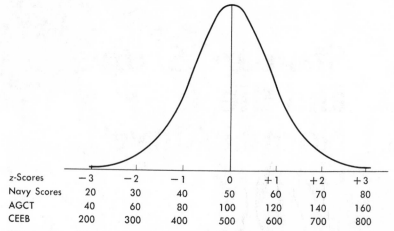

| z-Scores | −3 | −2 | −1 | 0 | +1 | +2 | +3 |
|---|---|---|---|---|---|---|---|
| Navy Scores | 20 | 30 | 40 | 50 | 60 | 70 | 80 |
| AGCT | 40 | 60 | 80 | 100 | 120 | 140 | 160 |
| CEEB | 200 | 300 | 400 | 500 | 600 | 700 | 800 |

**Figure 6.1**  Distribution of the various types of standard scores.

all of the cases, it follows that the highest $z$ score usually encountered is +3 and the lowest is −3. We can describe the distribution of $z$ scores by saying that they have a mean of zero and a standard deviation of 1. This is shown in Figure 6.1. Thus any time we see a standard score, we should be able to place exactly where an individual falls in a distribution. A student with a $z$ score of 2.5 is 2.5 standard deviations above the mean on that test distribution and has a very good score. These standard scores are equal units of measurement and hence can be manipulated mathematically. It should be noted here also that changing a distribution of scores to $z$ scores does not change the shape of the original distribution of scores. If the distribution was positively skewed to begin with, $z$ scores made from such a distribution would be positively skewed.

Since $z$ scores are expressed in decimals and since about half of them are negative, they are rather cumbersome to handle. Many times so-called linear transformations are made. Such transformations consist of making the scale larger, so that negative scores are eliminated, and of using a larger standard deviation, so that decimals are done away with. Transformed scores can be obtained from the following equation:

Standard score = $z$(new standard deviation) + the new mean

A common form for these transformations is based upon a mean of 50 and a standard deviation of 10. In equation form this becomes

Standard score = $z(10) + 50$

or, starting with the raw scores, we have

Standard score = $\dfrac{(X - \overline{X})}{s} (10) + 50$

This system with a mean of 50 and a standard deviation of 10 is very popular. It has been widely used by branches of the Armed Forces for many years. With this system we have a mean of 50 and a range of between 20 and 80. All scores are positive and all can be rounded to two-place numbers. Figure 6.1 shows this distribution of scores and also several others. The third row in Figure 6.1 shows the scores used on the American College Test (ACT), a test battery used to select college freshmen. Here the mean is 15 and the standard deviation is 5. The last row of these figures shows the type used by the College Entrance Examination Board and the Graduate Record Examination. Here we note a mean of 500 and a standard deviation of 100. Any system can be set up, but those just noted are most frequently encountered.

Some standard scores have been normalized. By this we mean that the distributions of these scores has been made to conform to that of the normal curve. We shall consider the normalizing of scores later. Here we shall briefly note some of these normalized standard scores. A very common standard score is the $T$ score, which has a mean of 50 and a standard deviation of 10. $T$ scores are frequently normalized. Then there are stanines (standard nines), which have a mean of 5 and a standard deviation of approximately 2. Many of the standard scores used in nationwide testing programs have been normalized.

## Uses of Standard Scores

Since standard scores are equal units of measurement and since their size is the same from distribution to distribution, they become a very useful tool both in the reporting of test scores and in doing research using test results. When the results of different tests taken by the same individuals are to be compared, this is best done by the use of standard scores. This process is illustrated in Table 6.1.

In part A of Table 6.1 the scores of students on three elementary school tests are presented. Only the scores of three students are listed. At the bottom are shown the mean and standard deviation of each test. Looking at the data in part A of Table 6.1, it becomes apparent that, as they stand, they convey little meaning. Which student had the best overall performance? On which test did the different students do best? or worst? None of these questions can be answered as the scores are shown.

Suppose that we now change these scores to standard scores. For this we shall use the system with a mean of 50 and a standard deviation of 10. We shall start with the geography test, and for all of these we shall use this transformation equation:

$$\text{Standard score} = \frac{X - \bar{X}}{s}(10) + 50$$

The standard score in geography for student A in this test is

$$SS = \frac{60 - 60}{10}(10) + 50$$

$$= 0 + 50$$

$$= 50$$

The geography score for student B becomes

$$SS = \frac{72 - 60}{10}(10) + 50$$

$$= \frac{120}{10} + 50$$

$$= 12 + 50$$

$$= 62$$

This is continued until all scores are transformed. The results of transforming the scores in part A of Table 6.1 are shown in part B. This may seem to be a laborious and time-consuming process. However, if there are many scores to be transformed, it is most convenient to arrange the scores from high to low and find the deviation of each from the mean (since these are in order, the rest of the process becomes very easy). Scores with a negative deviation are the same distance below the new mean as the positive scores with the same deviations are above it.

**Table 6.1**   Comparing and Combining Scores Made on Different Tests by the Use of Standard Scores

| Student | Part A:  Raw Scores Geography | Spelling | Arithmetic |
|---------|-----------|----------|------------|
| A | 60 | 140 | 40 |
| B | 72 | 100 | 36 |
| C | 46 | 110 | 24 |
| etc. | | | |
| Mean | 60 | 100 | 22 |
| Standard Deviation | 10 | 20 | 6 |

| Student | Part B:  Standard Scores Geography | Spelling | Arithmetic | Average |
|---------|-----------|----------|------------|---------|
| A | 50 | 70 | 80 | 67 |
| B | 62 | 50 | 73 | 62 |
| C | 36 | 55 | 53 | 48 |
| etc. | | | | |

Let us now examine part B of Table 6.1. Note that student A is at the mean in geography, 2 standard deviations above the mean in spelling, and 3 standard deviations above the mean in arithmetic. His average performance on these three tests was 67, 1.7 standard deviations above the mean. For student A we can then say that his performance is average in geography, excellent in spelling, and superior in arithmetic. In this manner we can consider the achievement of each of the students on each of the three tests, and we can get an average measure of his performance on all three tests. It should be noted that the only justifiable manner to compare scores is to first change the scores to standard scores, and then do the comparing. Standard scores change the raw scores to equal and comparable units.

Another use of standard scores is in determining final grades for a course. Let us take a course which has three examinations of 1 hour each during the semester and a final examination 2 hours long. At the end of the semester the instructor averages the three 1-hour examinations and then combines this average with the final examination in some manner or other. This method is not correct. Suppose that the means and standard deviations of the three 1-hour examinations were as follows:

|   | First Test | Second Test | Third Test | Final Exam |
|---|---|---|---|---|
| $\bar{X}$ | 52 | 87 | 62 | 124 |
| $s$ | 8 | 17 | 11 | 21 |

If a student's scores on these three 1-hour examinations are averaged, each will not contribute equally to the average. The second test with the largest standard deviation would contribute more to the final average than the other two, and the first examination would contribute least. The proposed method tends to equalize contributions of 1-hour examinations to final scores.

In grading situations, if the teacher wants each of a series of tests to contribute equally to the final grade, it follows that the scores of individuals on each test should be changed to standard scores and that these should be averaged. Then each 1-hour test is contributing more equally to the final grade. Suppose now that the instructor wants the final examination to have twice as much weight as each of the other tests in the final grade. These final grades should also be changed to standard scores. Then for any student his overall average is determined by taking his standard score on each of the first three tests, adding these, then adding to this sum 2 times his standard score on the final examination and dividing the total by 5. In symbols:

$$\frac{z_1 + z_2 + z_3 + z_f(2)}{5}$$

where $z_1,\ z_2,\ z_3 =$ the standard scores on the three 1-hour examinations

$zf =$ standard score on the final examination

$2 =$ weight final examination is to have

This method gives each examination the weight that the instructor desires it to have. Some teachers convert all scores to standard scores before entering them into their class books. This is strongly recommended.

Science teachers and some others have special problems with laboratory grades, project grades, and the like. It is desired that the laboratory grades, for example, contribute a certain amount to the final grade. This can only be done with any accuracy by changing these to standard scores and giving them the desired weight in the final average as shown above for the combination of 1-hour examinations and finals.

## THE NORMAL CURVE

### General Nature of the Normal Curve

For some time we have been talking about the normal curve. Now we shall examine it in detail, discuss its characteristics, its properties, and some of the basic ways in which it can be used. In the eighteenth century gamblers were interested in the chances of beating various gambling games and they asked mathematicians to help them out. DeMoivre (1733) was the first to develop the mathematical equation of the normal curve. In the early nineteenth century Gauss and LaPlace further developed the concept of the curve and probability. It was at about the same time that errors of observation made by astronomers were represented by a curve of this type. Today the normal curve is referred to as the curve of error, the bell-shaped curve, the Gaussian curve, or DeMoivre's curve.

By now the shape of this curve is familiar to you. Its maximum height is at the mean. In the language of the curve we say that the maximum ordinate (ordinates are given the symbol $y$) is at the mean. All other ordinates are shorter than this one. The normal curve is also said to be asymptotic. By this we mean that theoretically the tails never touch the base line but extend to infinity in either direction. In actual practice, however, 3 standard deviations on either side of the curve will include practically all of the cases. As mentioned previously, the skewness of the normal curve is zero and its peakedness is described as being mesokurtic.

In our educational and psychological work, we assume that certain traits are normally distributed. In actuality, probably no distribution ever takes on the absolute form of the normal distribution. Many of our frequency distributions are very close to the normal one, and we assume that they have a normal distribution. To the extent that our distributions differ from normal, error enters into our work. The normal curve is important not primarily because *scores* are assumed to be normally distributed, but because the *sampling* distributions of various statistics are known or assumed to be normal. Hence the normal curve's importance is primarily in sampling statistics. (This will be discussed in Chapters 11 and 12.)

Appendix B is based upon a normal curve with a mean of zero, a standard deviation of 1, and an area reduced to unity. It is referred to as the unity normal curve or the normal curve in standard score units. In solving problems we assume that our cases are spread evenly over this area. Suppose that we have a problem with 500 cases. In solving the problem we find that .262 of the area of the curve falls above a certain point. To determine the number of cases above this point, we merely multiply our proportion by the number of cases, in this example .262(500). We also use this area of the curve in talking about probability. Suppose that as an illustration we take a raw score which has an equivalent standard score of +1.88. By referring to the normal probability table, Appendix B, we note that approximately .03 of the area of the curve falls above this point. We then say that the probability of obtaining a score equal to or higher than this one of +1.88 is .03 or 3 in 100.

### Areas Under the Normal Curve
In the chapter in which we discussed standard deviations, we mentioned some of the relationships between standard deviation units and the normal curve. The most important of these relationships are again repeated in Figure 6.2. To summarize, 1 standard deviation taken on each side of the mean includes a total area of 68.26 percent of the curve, or approximately two-thirds of the cases. In terms of probability we can state that the chances are about two out of three of a score in any normally distributed sample falling within the area of 1 standard deviation on each side of the mean. A second standard deviation measured beyond the first cuts off 13.59 percent of the area. By adding all of the area included by 2 standard deviation units on both sides of the mean, we have accounted for more than 95 percent of the area or cases. If we continue and measure off another, or third, standard deviation on each side of the mean, we cut off another piece equal to 2.15 percent of the area. The sum of all of the areas included by these 6 standard deviation units is equal to 99.74 percent

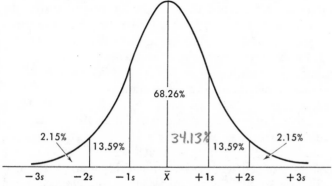

**Figure 6.2** Percentages under the normal curve at various standard deviation units from the mean.

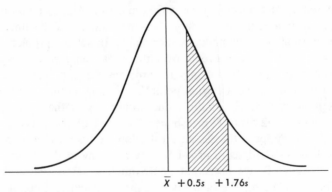

$\bar{X}$    $+0.5s$    $+1.76s$

**Figure 6.3** Percentage of the area of the normal curve between two points on the same side of the mean.

of the total. From this it follows that .26 percent of the cases are beyond three standard deviation units from the mean. This means 26 in 10,000. Dividing this by 2 to distribute these equally on both sides of the mean, we see that on each side we can expect 13 cases in 10,000 to fall beyond 3 standard deviation units from the mean.

AREAS CUT OFF BETWEEN DIFFERENT POINTS. Suppose, to begin with, we take the problem illustrated in Figure 6.3. We wish to find the proportion of the area or of the number of cases included between the two points $+.5s$ and $+1.76s$ units above the mean. The problems are all worked using the normal probability table, Appendix B. We begin by going to the table and finding the area of the curve cut off between the mean and a point equivalent to a standard score of .5 above the mean. This value appears in column 2 of the table and is found to be .1915. Next we continue down column 1 of the table until we come to a standard score of 1.76. By looking in column 2, we find that .4608 of the area is included between the mean and this point. Then the area of the curve between these two points is the difference between the two points, .4608 − .1915, which equals .2693. We can then state that approximately 27 percent of the cases fall between these two points, or that the probability of a score's falling between these two points is .27.

In the next illustration, we shall take two points on different sides of the mean. This time we wish to determine what proportion of the normal curve falls between a standard score of −.48 and one of +1.5 (Figure 6.4). There are no values for negative standard scores in Appendix B. As far as areas are concerned, equal standard scores, whether positive or negative, include equal areas when taken from the mean. From the table we find that a standard score of −.48 cuts off an area of .1844 between it and the mean. A standard score of +1.5 likewise includes .4332 of the area of the curve between it and the mean. The area included between both points is then equal to the sum of

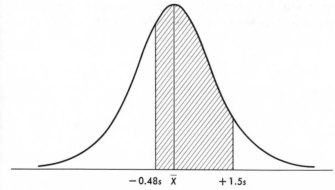

$-0.48s$    $\bar{X}$     $+1.5s$

**Figure 6.4** Percentage of the area of the normal curve between two points on different sides of the mean.

these two areas, $.1844 + .4332$, which is equal to $.6176$ or approximately 62 percent of the area.

Before we go any further, we might examine the other columns in Appendix B. Column 3 is labeled *Area in Larger Portion,* and column 4, *Area in Smaller Portion.* Any time we take a point on the base line of a curve and erect a perpendicular at this point, we divide the curve into two areas, a larger and a smaller area. For any given standard score, the sum of these two areas is equal to unity. Column 5 of the table, *Ordinate,* gives the size of the ordinates for the various standard scores.

Here is another type of problem that can be solved using these tables. Suppose that we have a distribution of test scores for which the following statistics have been computed: $\bar{X} = 80$, $s = 16$, $N = 510$. We wish to know what percentage of the scores in this distribution fall above a raw score of 110, assuming a normal distribution (Figure 6.5). This can of course be solved for

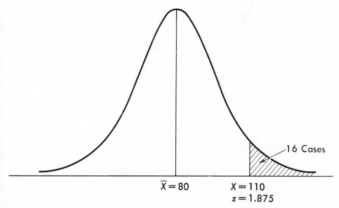

$\bar{X} = 80$      $X = 110$
$z = 1.875$

16 Cases

**Figure 6.5** Percentage of cases in a normal distribution falling above a certain score.

any score in this distribution. Since the table of the normal curve is in standard units, the first step is to change the raw score of 110 to a standard score as follows:

$$z = \frac{X - \bar{X}}{s}$$

$$= \frac{110 - 80}{16}$$

$$= \frac{30}{16}$$

$$= 1.875$$

Inasmuch as this score is halfway between 1.87 and 1.88, we must interpolate. This time we are interested in the area in the smaller portion of the curve at a given standard score point. The value from the table is half the distance between .0307 and .0301, which equals .0304. Next we multiply our $N$ by this proportion to find the number of cases above this raw score of 110. We find that 510(.0304) = 15.504 cases. This in round numbers is approximately 16 cases.

POINTS ABOVE OR BELOW WHICH CERTAIN PROPORTIONS OF THE AREA FALL. Now we are going to reverse the process we have been using. Suppose that we wish to find that point in the distribution which has 10 percent of the cases below it and 90 percent of the cases above it (Figure 6.6). This point would be the tenth centile. Again we enter the normal probability table and using column 4, the area in the smaller portion, we locate the value closest to .10.

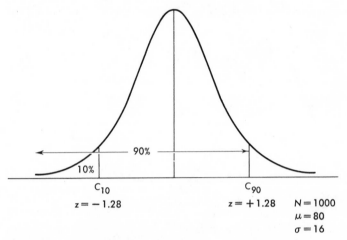

**Figure 6.6** Points in the normal curve above and below which different percentages of the curve lie.

In the table we find this to be .1003, and reading to the left we find that a standard score of 1.28 divides the area of the curve into the two proportions as desired. Since this standard score is to the left of the mean, it has to have a minus sign in front of it and is correctly written as −1.28. We could have solved this problem just as well by using column 3 of the table, the area in the larger portion. We would have gone down this column until we came as close to .90 as possible. This value is .8997, which puts us in the same row as before. The procedure from this point on is identical. If we wanted to find the point in the distribution with 10 percent of the cases above it ($C_{90}$), the work would be identical to that just completed, except that in the end the value of the standard score would be +1.28 because this time our point is to the right of the mean.

Suppose that we now want to know above which score in the distribution 90 percent of the cases will fall. In the last paragraph we found that a $z$ of 1.28 is the point below which 90 percent of the cases fall. To find the raw score corresponding to this $z$ score of 1.28 we solve the equation of $z$ for a different unknown.

$$z = \frac{X - \mu}{\sigma}$$

$$1.28 = \frac{X - 80}{16}$$

$$X - 80 = 16(1.28)$$

$$X = 80 + 20.48$$

$$X = 100.5$$

Similarly, the raw score equivalent of the point below which 10 percent of the cases fall is

$$-1.28 = \frac{X - 80}{16}$$

$$X - 80 = -20.48$$

$$X = 80 - 20.48$$

$$X = 59.5$$

## NORMALIZING A DISTRIBUTION OF SCORES

In the paragraphs that follow, the process of normalizing a distribution of scores will be demonstrated. From the results, the normal curve for the data will be plotted. This normal curve for any set of data is referred to as the *curve of best fit* for that set of data. The best-fitting curve for any set of data

**Table 6.2**   Normalizing a Distribution of Scores

| (1) | (2) | (3) Upper | (4) | (5) | (6) Proportion | (7) | (8) | (9) |
| --- | --- | --- | --- | --- | --- | --- | --- | --- |
| | $f_0$ | Limit | $x$ | $z$ | Below | Within | $f_e$ | $f_e$ |
| 90–94 | 1 | 94.5 | 30.6 | 2.51 | .9940 | .0119 | 1.785 | 1.8 |
| 85–89 | 3 | 89.5 | 25.6 | 2.10 | .9821 | .0276 | 4.140 | 4.1 |
| 80–84 | 8 | 84.5 | 20.6 | 1.69 | .9545 | .0548 | 8.220 | 8.2 |
| 75–79 | 12 | 79.5 | 15.6 | 1.28 | .8997 | .0919 | 13.875 | 13.8 |
| 70–74 | 28 | 74.5 | 10.6 | .87 | .8078 | .1306 | 19.590 | 19.6 |
| 65–69 | 36 | 69.5 | 5.6 | .46 | .6772 | .1573 | 23.595 | 23.6 |
| 60–64 | 12 | 64.5 | .6 | .05 | .5199 | .1605 | 24.075 | 24.1 |
| 55–59 | 18 | 59.5 | −4.4 | −.36 | .3594 | .1388 | 20.820 | 20.8 |
| 50–54 | 10 | 54.5 | −9.4 | −.77 | .2206 | .1016 | 15.240 | 15.2 |
| 45–49 | 8 | 49.5 | −14.4 | −1.18 | .1190 | .0631 | 9.465 | 9.5 |
| 40–44 | 8 | 44.5 | −19.4 | −1.59 | .0559 | .0331 | 4.965 | 5.0 |
| 35–39 | 5 | 39.5 | −24.4 | −2.00 | .0228 | .0148 | 2.220 | 2.2 |
| 30–34 | 1 | 34.5 | −29.4 | −2.41 | .0080 | .0080 | 1.200 | 1.2 |

$N = 150$
$\bar{X} = 63.9$
$s = 12.2$

$\Sigma = .9940$          $\Sigma f_e = 149.1$

has the same mean and standard deviation and is based upon the same number of cases as the original data.

This process will be demonstrated by using the data in Table 6.2. These data are based on 150 cases, with a mean of 63.9 and a standard deviation of 12.2. Column 1 lists the intervals and column 2 shows the observed frequencies, which are labeled this time $f_0$. After setting up these two columns, we proceed as follows:

**1.** Determine the upper limit of each interval and record these in column 3.

**2.** Determine the $x$ values of column 4 by subtracting the mean, 63.9, from each of the upper limits in column 3.

**3.** Change each of these $x$ values of column 4 to a standard score, $z$, by dividing by the standard deviation, 12.2.

**4.** From the normal probability table, Appendix B, determine the proportion of the area in the normal curve below each of these standard scores. For the top score, 2.51, we see in the table that .9940 of the area is below this point. These proportions are recorded in column 6.

**5.** The values in column 7 are determined as follows. The bottom value in column 7 has the same value as that at the bottom of column 6, for the proportion within and below this interval is the same as the proportion below the upper limit of the interval 30–34. The proportion within the second interval is obtained by subtracting .0080, the proportion within the bottom interval, from .0228, the proportion below the upper limit of the second interval. This results in a value of .0148. For the proportion within the

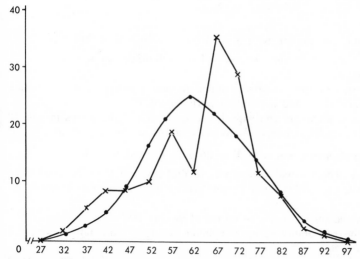

**Figure 6.7**  Frequency polygon and normalized curve for the same data.

third interval, we subtract .0228 from .0559. This process is continued until all the values in column 7 are determined. The sum of this column should be close to 1, but it is usually actually a little less than this, for there are always cases in the extremes of the tails of the normal curve that are not taken into account in this process.

**6.** The values in column 8 are obtained by multiplying the number of cases, 150, by each of the proportions in column 7. Again these add up to slightly less than 150.

**7.** In column 9 these expected frequencies ($f_e$) are rounded to the nearest tenth.

In Figure 6.7, the axes have been set up in the usual manner for constructing a frequency polygon. First the $f_o$'s are plotted and these points connected with a rule as in the usual method for plotting a frequency polygon. Then the values in column 9 are plotted and these are connected by means of a smooth curve. In Figure 6.7 we have the curve of best fit for these data superimposed upon the frequency polygon for the original data.

The student may now be wondering when he should normalize a curve or even if there is any justification for performing the act. Later, in Chapter 14 on chi-square, we shall see that one of the uses of chi-square is to see if a distribution of measures departs from normal. To make a chi-square test, we must be able to set up the expected frequencies for any set of data. Also there are times when a distribution of a sample of scores is not normal; yet the research worker has a hunch or knows that the distribution of the trait with which he is concerned is normally distributed in the population. Since he can justify this, he may find it useful to normalize the data in his sample.

EXERCISES

1. Given a normal distribution with $\bar{X} = 30$, $s = 6$, and $N = 500$.
   a. Find the $z$-score equivalent of each of the following raw scores: 45, 36, 30, 22, 18.
   b. Transfer each of the $z$ scores obtained in part a above to a distribution with a mean of 50 and a standard deviation of 10.
   c. How many cases would you expect to exceed each of the scores in part a?
   d. What centile point corresponds to each of the scores in part a?

2. Given a normal distribution with $\bar{X} = 80$, $s = 12$, and $N = 800$.
   a. How many scores would you expect to find above a score of 100?
   b. What is the probability of a person getting a score of 100 or more on this test?
   c. How many scores fall between scores of 90 and 100?
   d. What is the probability of an individual having a score fall between 90 and 100?
   e. How many scores would you expect to find between scores of 60 and 100?

3. A group of tests was administered with the following results. On the right are the scores of one member of the class, Betty.

| Test | $\bar{X}$ | $s$ | Betty's Score |
|---|---|---|---|
| Algebra | 48 | 12 | 44 |
| Physics | 40 | 8 | 30 |
| English | 80 | 15 | 108 |

   a. Using a mean of 50 and a standard deviation of 10, find Betty's standard score on each test.
   b. What percentage of students taking the test did she surpass in English? In physics? In algebra?
   c. What inferences may be drawn from these scores?

4. Suppose that there is a course in elementary psychology with 1200 students enrolled and that these students are graded on the curve. They have been given a test resulting in a mean of 50 and a standard deviation of 10. It has been determined that 5 percent will get A's, 20 percent B's, 50 percent C's, 20 percent D's, and 5 percent F's. Find the score values for each of these grades.

5. Given a large sample, which of the following would you expect to approach a normal distribution?
   a. The heights of adult males in centimeters.
   b. The incomes of 40-year-old males in any state.
   c. The heights of trees in an adult forest stand.
   d. The life of auto tires guaranteed for 40,000 miles.
   e. Scores on a test of musical aptitude.
   f. Means of an infinite number of samples ($N = 20$) drawn randomly from an infinitely large sample.
   g. Errors made in reading the fine scale on the dial of a piece of laboratory apparatus.

**6.** Below are the scores of a sample of 130 students on the Minnesota Paper-Form Board test.

| | |
|---|---|
| 60–62 | 2 |
| 57–59 | 4 |
| 54–56 | 8 |
| 51–53 | 10 |
| 48–50 | 18 |
| 45–47 | 16 |
| 42–44 | 14 |
| 39–41 | 14 |
| 36–38 | 10 |
| 33–35 | 18 |
| 30–32 | 14 |
| 27–29 | 2 |

$$N = 130$$
$$X = \phantom{0}42.8$$
$$s = \phantom{00}8.2$$

**a.** Assume a normal population distribution and determine what percent of cases you would expect to find between the mean and the following scores in similar samples: 60, 38, 28.

**b.** Find the percentage and number of cases expected to fall between the following pairs of scores: 35 and 45, 50 and 55, 56 and 60.

**c.** How many cases would you expect to find above a raw score of 50? Below a score of 35?

**7.** Apply the normalizing process to the data of problem 6 and plot the best-fitting curve for these data. On the same axes, plot the frequency polygon for the original data.

# Correlation—
# The Pearson *r*
## Chapter 7

Correlation is basically a measure of relationship between two variables. If we notice some of the common things we measure, we see that high grades in English tend to be associated with high grades in foreign languages. Both of these tend to be associated with high scores on intelligence tests. In the physical domain, tall people tend to be heavier than short people. The volume of a gas is related to the pressure under which it is kept. In the field of economics, there is a correlation between the price at which products are sold and the amount available for sale. So in all aspects of life, we find that there are relationships of one sort or another. It should be noted here that these relationships do not necessarily imply that one is the cause of the other. This may or may not be the case. In some situations, we find that two variables are related because they are both related to, or caused by, a third variable.

Most correlation coefficients tell us two things. First we have an indication of the magnitude of the relationship. It is worth noting at this point that a correlation of −.88 is the same size as one of +.88. The sign has nothing to do with the size of the relationship, but it does give information about the direction of the relationship. When two variables are positively related, as one increases, the other also increases. Intelligence test scores are positively related to academic grades. In general, the higher the intelligence test scores, the higher the grades received in school. Other variables are inversely related. By this we mean that as one increases, the other decreases. Think for a moment of the relationship between the speed of an automobile in high gear and the number of miles per gallon obtained from a gallon of gasoline. The faster that one drives, the worse the gasoline mileage. The absence of a relationship is denoted by a correlation coefficient of .00 or thereabouts.

## THE BIVARIATE DISTRIBUTION

In previous chapters we have been concerned with distributions involving only one variable, and in Chapter 3 we learned how to build a frequency distribution for this one variable. In this chapter on correlation we are dealing with data composed of measurements made on two variables for each individual. Our task now is to construct a bivariate frequency distribution. To do this we shall use the data in Table 7.1, the scores of 35 university students on two art judgment tests, the Meier Art Judgment Test and the Graves Design Judgment Test.

First we build a table like 7.2, which is basically composed of a series of columns and rows. Then we set up a series of intervals for the Meier Art Judgment Test. This is done on the $y$ axis using an interval size of 3 and with the first interval being 78–80. Similarly a series of intervals is laid out on the $x$ axis for the Graves test. In this case the interval size is 5 and the first interval is 25–29. The size of the class intervals used for the two distributions does not have to be the same.

Second, we tally the scores. We take the first pair, the scores of individual 1, which is an 80 on the $y$ axis and a 61 on the $x$ axis. To enter these two scores to locate individual 1 on the scatterplot, we go up the $y$ axis until we

**Table 7.1**   Scores of 35 Students in Two Art Judgment Tests

| Individual | Meier Art Judgment | Graves Design Judgment | Individual | Meier Art Judgment | Graves Design Judgment |
|---|---|---|---|---|---|
| 1 | 80 | 61 | 19 | 105 | 86 |
| 2 | 95 | 28 | 20 | 80 | 63 |
| 3 | 94 | 74 | 21 | 85 | 31 |
| 4 | 101 | 46 | 22 | 93 | 57 |
| 5 | 105 | 44 | 23 | 85 | 70 |
| 6 | 89 | 38 | 24 | 92 | 43 |
| 7 | 106 | 72 | 25 | 90 | 70 |
| 8 | 92 | 41 | 26 | 89 | 54 |
| 9 | 105 | 49 | 27 | 85 | 51 |
| 10 | 107 | 69 | 28 | 96 | 58 |
| 11 | 111 | 82 | 29 | 85 | 63 |
| 12 | 114 | 76 | 30 | 98 | 73 |
| 13 | 83 | 39 | 31 | 101 | 71 |
| 14 | 112 | 64 | 32 | 106 | 76 |
| 15 | 91 | 77 | 33 | 112 | 76 |
| 16 | 88 | 50 | 34 | 93 | 59 |
| 17 | 105 | 55 | 35 | 110 | 71 |
| 18 | 106 | 59 | | | |

**Table 7.2**   A Bivariate Frequency Distribution

x Axis—Graves Design Judgment Test

| y Axis—Meier Art Judgment Test | 25–29 | 30–34 | 35–39 | 40–44 | 45–49 | 50–54 | 55–59 | 60–64 | 65–69 | 70–74 | 75–79 | 80–84 | 85–90 | $f_y$ |
|---|---|---|---|---|---|---|---|---|---|---|---|---|---|---|
| 114–116 | | | | | | | | | | | / | | | 1 |
| 111–113 | | | | | | | | / | | | / | / | | 3 |
| 108–110 | | | | | | | | | | / | | | | 1 |
| 105–107 | | | | / | / | | // | | / | / | / | | / | 8 |
| 102–104 | | | | | | | | | | | | | | 0 |
| 99–101 | | | | | / | | | | | / | | | | 2 |
| 96–98 | | | | | | / | | | | / | | | | 2 |
| 93–95 | / | | | | | | // | | | / | | | | 4 |
| 90–92 | | | | // | | | | | | / | / | | | 4 |
| 87–89 | | | / | | | // | | | | | | | | 3 |
| 84–86 | | / | | | | | / | / | | / | | | | 4 |
| 81–83 | | | / | | | | | | | | | | | 1 |
| 78–80 | | | | | | | | // | | | | | | 2 |
| $f_x$ | 1 | 1 | 2 | 3 | 2 | 3 | 5 | 4 | 1 | 7 | 4 | 1 | 1 | 35 = N |

come to the interval containing the score of 80 and then across to the interval containing a score of 61. We enter our tally mark where these two class intervals meet. Then we take the next pair of scores, 95 and 28, and go up the y axis until we come to the interval containing 95 and then across to the interval containing 28. Again we place the tally mark in the cell where the two intervals meet. We continue this process until all pairs of scores are entered. Then we sum the tallies in each column and row and enter these sums in the column and row labeled f, frequency.

In the past, before desk calculators and computers became readily available, a correlation coefficient was computed from such a bivariate distribution as we have just constructed. Since this is a tedious task and one that is no longer of any importance, it will not be considered here. However, it is important to make a scatterplot to get a picture of the nature of the relationship between x and y. A basic condition necessary for the computation of the Pearson r is that there be a linear relationship between the two variables; to the extent that there is a departure from linearity, the calculated correlation coefficient is an underestimate of the true relationship. The running of data through a desk calculator or a computer to obtain an r does not give any indication of the nature of the relationship. One way to determine whether the condition of

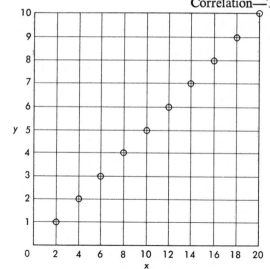

| X | Y |
|---|---|
| 20 | 10 |
| 18 | 9 |
| 16 | 8 |
| 14 | 7 |
| 12 | 6 |
| 10 | 5 |
| 8 | 4 |
| 6 | 3 |
| 4 | 2 |
| 2 | 1 |

**Figure 7.1**  A perfect positive relationship.

linearity has been met is to make a scatterplot such as that in Table 7.2. Also from such a scatterplot we can make a rough estimate of the size of the correlation coefficient. The tallies in Table 7.2 are fairly well scattered out. Hence we would infer a low correlation coefficient. Actually the computed coefficient for these data is .43.

A PERFECT POSITIVE RELATIONSHIP. Figure 7.1 has been constructed to show the relationship between two variables. Notice that for every increase of 2 units on the X variable, there is a corresponding increase of 1 unit on the Y variable. This is true for all pairs of the two variables. When pairs of values like these are plotted, they fall along a straight line, and when this straight line runs from the lower left of the scattergram to the upper right, we have an example of a perfect positive relationship. The correlation coefficient is equal to +1.00.

A PERFECT NEGATIVE RELATIONSHIP. Figure 7.2 is just the opposite of this. Notice that for these data, for every increase of 2 units on the x axis, there is a corresponding decrease of 1 unit on the y axis. Again this relationship is maintained throughout the range. This time our points again fall along a straight line which now runs from the upper left-hand part of the scatterplot to the lower right. This is the example of a perfect negative relationship, a value of −1. This means that the individual with the highest score on the X variable had the lowest score on the Y variable. The second highest scorer on the X was second lowest on Y, until at the bottom, the lowest on X was the highest on Y.

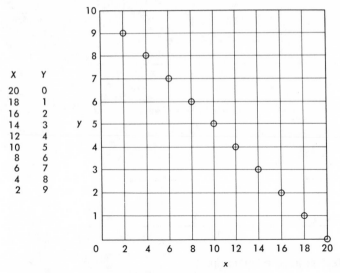

Figure 7.2  A perfect negative relationship.

OTHER RELATIONSHIPS. In actual life we usually have situations in which the relationship is not perfect. Figure 7.3 shows the scatterplot of a very high positive correlation. Notice here that although the points do not fall along a straight line, the line is still apparent. In Figure 7.4 we have the illustration of a low negative relationship. On this scatterplot the points are scattered more

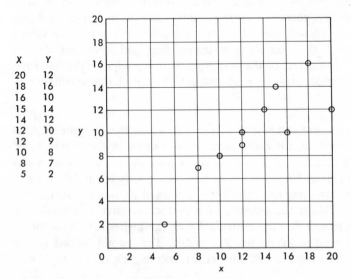

Figure 7.3  A high positive relationship (.87).

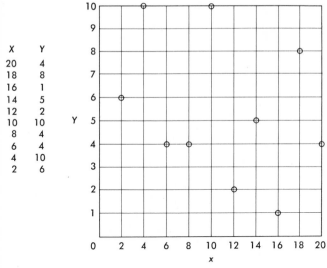

| X | Y |
|---|---|
| 20 | 4 |
| 18 | 8 |
| 16 | 1 |
| 14 | 5 |
| 12 | 2 |
| 10 | 10 |
| 8 | 4 |
| 6 | 4 |
| 4 | 10 |
| 2 | 6 |

**Figure 7.4**   A low negative relationship (−.31).

or less all over. There is an indication of the points falling in the general direction from the upper left to the lower right. When the points in the scattergram are spread evenly in all directions, we have an example of no relationship.

By making such scatterplots, the student can get an indication of the magnitude of the correlation coefficient and also an indication of the sign of the relationship. This, in part, is a rough check on the computational work.

CONDITIONS INVOLVED IN THE INTERPRETATION OF $r$. Before $r$ is computed the scatterplot should be examined and sometimes the data tested to see if two conditions exist. The first of these conditions is that we have *linear regression* (also referred to as rectilinear regression). This means that our points on the scattergram tend to fall along a straight line. This is true of the data pictured in Figures 7.1, 7.2, and 7.3. However, in Figure 7.5, notice that beginning at the lower left, as the $Y$ variable increases, so does the $X$ variable; but as $X$ keeps increasing, $Y$ begins to slow down; and further along as $X$ continues to increase, $Y$ starts and continues to decrease. This pattern of relationship is referred to as *curvilinear regression*. Such relationships are by no means rare. Consider many of our motor skills. As we get older, we are able to run faster. Then comes a period of no improvement, and then as we further increase in age, our running speed decreases.

A more exact description of linear regression is that the means of the columns and rows fall along a straight line. Suppose that you refer to a scatterplot that is similar to Table 7.2 but has more frequencies. By starting

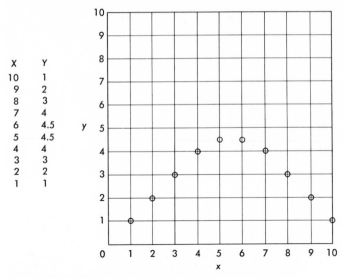

| X | Y |
|----|-----|
| 10 | 1 |
| 9 | 2 |
| 8 | 3 |
| 7 | 4 |
| 6 | 4.5 |
| 5 | 4.5 |
| 4 | 4 |
| 3 | 3 |
| 2 | 2 |
| 1 | 1 |

**Figure 7.5** An example of curvilinear regression.

at the left, it would be possible to find the mean of each of these $Y$ columns and to plot this mean on the scattergram. If there is linear regression, these means will tend to fall along a straight line. In a similar fashion, the mean of each row could be determined and then plotted. There are then two regression lines, except when $r = 1$, when they coincide. We shall discuss these lines in Chapter 9.

To the extent that data are not linear, the size of the computed $r$ is diminished. The size of the $r$ reflects the amount of variance that can be accounted for by a straight line, whether the data are essentially linear or not. It is possible that a very high, but nonlinear, relationship will appear to be very low on the basis of the Pearson $r$. However, when a bivariate relationship is primarily curvilinear, the *eta coefficient* or the correlation ratio can be used. This coefficient reflects the variance accounted for by the best-fitting line, whether it be curved or straight. This coefficient will be discussed in Chapter 8.

The second condition that we should look for is *homoscedasticity*. By this we mean that the standard deviations (or variances) of the arrays (columns and rows) tend to be equal. In Figure 7.6 there are three diagrams. In (a) the variance of the columns near the center of the $X$ distribution is smaller than that of the extreme columns. In diagram (b) all of the columns at the left have smaller variances than those at the center and right. Diagram (c) represents the situation when both conditions are met. Note that the points tend to fall in an ellipse about the regression line. When the data are not homoscedastic, the usual methods of evaluating predicted values on the $y$ axis from the corresponding values on the $x$ axis do not apply. For instance, in

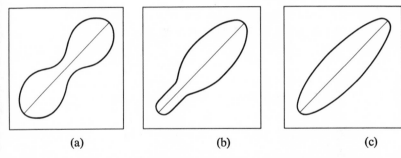

|     |     |     |
| --- | --- | --- |
| (a) | (b) | (c) |

**Figure 7.6**   Diagrams (a) and (b) lack homoscedasticity. Diagram (c) has both homoscedasticity and linear regression.

diagram (a) it would be possible to make predictions at the center of the $X$ distribution with greater accuracy than at either end.

## THE PEARSON PRODUCT-MOMENT CORRELATION COEFFICIENT

Of the various correlation coefficients in current use, the one most frequently encountered is called the Pearson product-moment correlation coefficient, the symbol of which is $r$. The size of the correlation coefficient varies from $+1$ through 0 to $-1$.

### Computing the Pearson $r$ from Standard Scores
In Table 7.3 the scores of ten individuals on two variables, $X$ and $Y$ (columns 1 and 2), are presented. Beneath each of these is the mean. In column 3 is the deviation of each of the $X$ scores from the mean of $X$, and in column 4 we find the deviation of each of the $Y$ scores from the mean of $Y$. Beneath these two columns are the standard deviations of the columns. In columns 5 and 6 are the standard scores for each of the scores in columns 1 and 2. These were obtained by dividing each score in column 3 by $s_x$ (4.34) and each value in column 4 by $s_y$ (3.71). Here we are using the usual formula for $z$, $z = x/s$. Column 7 is the product of the two $z$ scores, the product of the values in columns 5 and 6. These are summed, and the Pearson $r$ is obtained by the following formula, which describes $r$ as the mean $z$ score product:

$$r = \frac{\Sigma z_x z_y}{N} \qquad (7.1)$$

By substituting into this formula, we obtain

$$r = \frac{8.6947}{10}$$

$$r = .869 \text{ or } .87$$

**Table 7.3** Computation of the Pearson Product-Moment Correlation Coefficient by the Use of Deviations from the Means

| (1) $X$ | (2) $Y$ | (3) $x$ | (4) $y$ | (5) $z_x$ | (6) $z_y$ | (7) $z_x z_y$ | (8) $x^2$ | (9) $y^2$ | (10) $xy$ |
|---|---|---|---|---|---|---|---|---|---|
| 20 | 12 | 7 | 2 | 1.61 | .54 | .8694 | 49 | 4 | 14 |
| 18 | 16 | 5 | 6 | 1.15 | 1.62 | 1.8637 | 25 | 36 | 30 |
| 16 | 10 | 3 | 0 | .69 | .00 | .0000 | 9 | 0 | 0 |
| 15 | 14 | 2 | 4 | .46 | 1.08 | .4968 | 4 | 16 | 8 |
| 14 | 12 | 1 | 2 | .23 | .54 | .1242 | 1 | 4 | 2 |
| 12 | 10 | −1 | 0 | −.23 | .00 | .0000 | 1 | 0 | 0 |
| 12 | 9 | −1 | −1 | −.23 | −.27 | .0621 | 1 | 1 | 1 |
| 10 | 8 | −3 | −2 | −.69 | −.54 | .3726 | 9 | 4 | 6 |
| 8 | 7 | −5 | −3 | −1.15 | −.81 | .9315 | 25 | 9 | 15 |
| 5 | 2 | −8 | −8 | −1.84 | −2.16 | 3.9744 | 64 | 64 | 64 |

$\Sigma X = 130$ $\quad$ $\Sigma Y = 100$ $\qquad$ $s_x = 4.34$ $\quad$ $s_y = 3.71$ $\qquad$ $\Sigma z_x z_y = 8.6947$ $\qquad$ $\Sigma x^2 = 188$ $\quad$ $\Sigma y^2 = 138$ $\quad$ $\Sigma xy = 140$

$\bar{X} = 13$ $\qquad$ $\bar{Y} = 10$

Another way of looking at the Pearson $r$ in terms of $z$-score measures is the difference method for $r$. This formula expresses $r$ as a function of the difference between paired $z$ scores:

$$r = 1 - \frac{\Sigma(z_x - z_y)^2}{2N} \tag{7.2}$$

The expression $\Sigma(z_x - z_y)^2/2N$ increases as the variables become more dissimilar.

### Computing the Pearson $r$ from the Deviation from the Means

Since a standard score is defined as the deviation of a score from its mean, divided by its standard deviation, we can write

$$z_x = \frac{x}{s_x} \quad \text{and} \quad z_y = \frac{y}{s_y}$$

Then taking formula 7.1 and substituting each of the above for $z_x$ and $z_y$ and solving, we can reduce the formula for $r$ to

$$r = \frac{\Sigma xy}{N s_x s_y} \tag{7.3}$$

which further reduces to

$$r = \frac{\Sigma xy}{\sqrt{(\Sigma x^2)(\Sigma y^2)}} \tag{7.4}$$

since

$$s_x = \sqrt{\frac{\Sigma x^2}{N}} \quad \text{and} \quad s_y = \sqrt{\frac{\Sigma y^2}{N}}$$

If $\Sigma xy$, the numerator of equation 7.4, is divided by $N$, we have a term referred to as the *covariance*. If both factors in the denominator are divided by $N$, we have the variance of $X$ and the variance of $Y$. The Pearson $r$ can be described as the ratio of the covariance to the geometric means of the variances.

In columns 8, 9, and 10 of Table 7.3 are values for $x^2$, $y^2$, and $xy$, and the sum of each is given at the bottom of the table. After substituting these into equation 7.4, we obtain

$$r = \frac{140}{\sqrt{(188)(138)}}$$

$$= \frac{140}{\sqrt{25955}} = \frac{140}{161}$$

$$= .87$$

which is identical with the value obtained previously.

Both of the foregoing methods would be very laborious with an $N$ of large size. We shall pass on next to the two methods most frequently encountered in the computation of $r$.

### The Machine Formula for Computing $r$ from Raw Scores

We shall now go from formula 7.4 to the so-called raw score, or machine formula. We have previously written the sum of the squares as

$$\Sigma x^2 = \Sigma X^2 - \frac{(\Sigma X)^2}{N} \quad \text{and} \quad \Sigma y^2 = \Sigma Y^2 - \frac{(\Sigma Y)^2}{N}$$

and it follows by analogy that

$$\Sigma xy = \Sigma XY - \frac{(\Sigma X)(\Sigma Y)}{N}$$

When we insert the above three equations into formula 7.4 we have

$$r = \frac{\Sigma XY - [(\Sigma X)(\Sigma Y)/N]}{\sqrt{\{\Sigma X^2 - [(\Sigma X)^2/N]\}\{\Sigma Y^2 - [(\Sigma Y)^2/N]\}}}$$

After clearing this equation of fractions we have

$$r = \frac{N\Sigma XY - (\Sigma X)(\Sigma Y)}{\sqrt{[N\Sigma X^2 - (\Sigma X)^2][N\Sigma Y^2 - (\Sigma Y)^2]}} \tag{7.5}$$

The same data that appear in Table 7.3 have been reproduced in Table 7.4. The original pairs of scores appear in columns 1 and 2. In column 3 are the squares of each of the $X$ scores and in column 4 are the squares of each of the $Y$ scores. Column 5 consists of the cross-products, the product of each $X$ times each $Y$. All five columns are summed, and we enter formula 7.5 as follows:

$$r = \frac{10(1440) - 130(100)}{\sqrt{[10(1878) - 130^2][10(1138) - 100^2]}}$$

$$= \frac{14400 - 13000}{\sqrt{(18780 - 16900)(11380 - 10000)}}$$

$$= \frac{1400}{\sqrt{1880(1380)}}$$

$$= \frac{1400}{\sqrt{2594400}}$$

$$= \frac{1400}{1610}$$

$$= .87$$

which we have obtained twice previously for the same set of data.

If formula 7.5 is used in doing work by hand, a substantial amount of time is saved by using the table of squares in Appendix A. It should also

**Table 7.4**    The Pearson $r$ Computed from Raw Scores

| (1) X | (2) Y | (3) $X^2$ | (4) $Y^2$ | (5) XY |
|-------|-------|-----------|-----------|--------|
| 20 | 12 | 400 | 144 | 240 |
| 18 | 16 | 324 | 256 | 288 |
| 16 | 10 | 256 | 100 | 160 |
| 15 | 14 | 225 | 196 | 210 |
| 14 | 12 | 196 | 144 | 168 |
| 12 | 10 | 144 | 100 | 120 |
| 12 | 9 | 144 | 81 | 108 |
| 10 | 8 | 100 | 64 | 80 |
| 8 | 7 | 64 | 49 | 56 |
| 5 | 2 | 25 | 4 | 10 |
| $\Sigma X = 130$ | $\Sigma Y = 100$ | $\Sigma X^2 = 1878$ | $\Sigma Y^2 = 1138$ | $\Sigma XY = 1440$ |

be noted that much time will be saved if the data are coded. Consider the data in Table 7.1. Notice the scores in the second column. They are all large; the smallest is 80 and the highest is 114. If each of these scores is reduced by 80 and the values in the third column by 30 (this would result in one negative score, but it would be more convenient than reducing each score by 28, which is the lowest score in this distribution), the correlation coefficient computed from the coded data would be identical to that obtained from the uncoded scores, and the work would be much easier. Even if all of the work is to be done on the calculating machine, many people find that it is time saving to code the scores in the manner indicated. If automatic calculators are available, it is not necessary to go through all of the steps as shown in Table 7.4, since the machine keeps running sums of the sum of the $X$'s, the sum of the $Y$'s, the sum of the $X^2$'s, the sum of the $Y^2$'s, and twice the sum of the $XY$'s as each pair of scores is entered. When the last pair has been put into the machine, the final readings of all five values appear. The work in the solving of the formula is also considerably reduced when done on a desk calculator.

## Computing the Pearson $r$ by the Method of Differences

Another solution for the Pearson $r$ is by the method of differences, these differences being obtained by subtracting each $Y$ from its corresponding $X$, or vice versa. Letting $D$ stand for each difference, we have

$$D = X - Y$$

then

$$d = x - y$$
$$\Sigma d^2 = \Sigma(x - y)^2$$
$$\Sigma d^2 = \Sigma x^2 + \Sigma y^2 - 2\Sigma xy$$

Since by equation 7.3

$$r = \frac{\Sigma xy}{N s_x s_y}$$

or by transposing

$$\Sigma xy = N r s_x s_y$$

Substituting this in the fourth equation above gives

$$\Sigma d^2 = \Sigma x^2 + \Sigma y^2 - 2N r s_x s_y$$

Dividing by $N$

$$\frac{\Sigma d^2}{N} = \frac{\Sigma x^2}{N} + \frac{\Sigma y^2}{N} - \frac{2Nr}{N} s_x s_y$$

which may be written

$$s_d^2 = s_x^2 + s_y^2 - 2r s_x s_y \tag{7.6}$$

Stating equation 7.6 in words, we say that the variance of the difference between two measures is equal to the variance of the first plus the variance of the second minus the covariance term (this being the name applied to the third term). From equation 7.6 we transpose and obtain

$$r = \frac{s_x^2 + s_y^2 - s_d^2}{2 s_x s_y} \tag{7.7}$$

Similarly it can be shown that the variance of the sum of $X$ and $Y$ is equal to

$$s_{x+y}^2 = s_x^2 + s_y^2 + 2r s_x s_y \tag{7.8}$$

**Table 7.5**  The Pearson $r$ Computed by the Method of Differences

| (1) X | (2) Y | (3) $X^2$ | (4) $Y^2$ | (5) D | (6) $D^2$ |
|---|---|---|---|---|---|
| 20 | 12 | 400 | 144 | 8 | 64 |
| 18 | 16 | 324 | 256 | 2 | 4 |
| 16 | 10 | 256 | 100 | 6 | 36 |
| 15 | 14 | 225 | 196 | 1 | 1 |
| 14 | 12 | 196 | 144 | 2 | 4 |
| 12 | 10 | 144 | 100 | 2 | 4 |
| 12 | 9 | 144 | 81 | 3 | 9 |
| 10 | 8 | 100 | 64 | 2 | 4 |
| 8 | 7 | 64 | 49 | 1 | 1 |
| 5 | 2 | 25 | 4 | 3 | 9 |
| $\Sigma X = 130$ | $\Sigma Y = 100$ | $\Sigma X^2 = 1878$ | $\Sigma Y^2 = 1138$ | $\Sigma D = 30$ | $\Sigma D^2 = 136$ |

From this another formula for $r$ may again be obtained by transposing:

$$r = \frac{s_{x+y}^2 - s_x^2 - s_y^2}{2s_x s_y} \tag{7.9}$$

We shall illustrate the use of formula 7.7 using the data in Table 7.4, now copied into Table 7.5. Columns 1, 2, 3, and 4 are the same as in Table 7.4. Values in column 5 are obtained by subtracting each $Y$ value from its corresponding $X$ value, or vice versa. It makes no difference which way this is done as long as we are consistent. That is, if we start by subtracting $Y$ from $X$, we subtract in that order throughout the data. Sometimes the differences will be negative. In that case the sum of column 5 is the algebraic sum. Then these differences are squared (column 6) and all columns are summed.

Our task is, then, to obtain the three variances:

$$\Sigma x^2 = 1878 - \frac{130^2}{10} \qquad \Sigma y^2 = 1138 - \frac{100^2}{10} \qquad \Sigma d^2 = 136 - \frac{30^2}{10}$$

$$= 1878 - 1690 \qquad\quad = 1138 - 1000 \qquad\quad = 136 - 90$$

$$= 188 \qquad\qquad\qquad = 138 \qquad\qquad\qquad = 46$$

$$s_x^2 = \frac{188}{10} = 18.8 \qquad s_y^2 = \frac{138}{10} = 13.8 \qquad s_d^2 = \frac{46}{10} = 4.6$$

Then using equation 7.7

$$r = \frac{18.8 + 13.8 - 4.6}{2\sqrt{18.8}\ \sqrt{13.8}}$$

$$r = \frac{28}{2(16.1)}$$

$$= \frac{28}{32.2}$$

$$r = .87$$

which is the same result as was obtained previously using formula 7.5.

The formulas for the variance of the sum of or the difference between two variables are frequently encountered in statistics and measurement. The difference formulas will appear later in this book.

## CORRELATION COEFFICIENTS AND THE RANGE

In Figure 7.7 the scores of a group of individuals on a predictor test have been plotted against their ratings on a task, the criterion measure. Considering the figure as a whole, it is apparent that there is a positive relationship

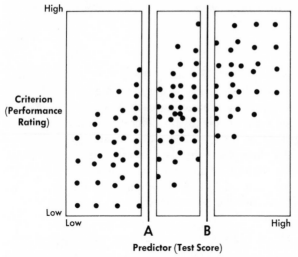

**Figure 7.7**   Diagram to illustrate the effect of restricted range on *r*. (Source: *Restriction of Range: Questions and Answers*. Test Service Bulletin No. 59, The Psychological Corporation, 1972.)

between the predictor scores and the criterion. Actually the correlation is .6. The figure has been divided into three parts, those scoring high on the predictor to the right of line B, those scoring low to the left of line A, and those scoring at an average level in the area between lines A and B. When each of the three sections is considered separately, the correlation between the two variables in each section is much lower. Notice that no ellipse is formed by the dots, the dots being scattered over the area. Such dispersion is indicative of a low correlation.

The effects of the restriction of range encountered when one is carrying on selection on the basis of test scores are both considerable and important. Suppose that a college admits only those individuals who score to the right of line B on this predictor test. Since the criterion measure, freshman grades, will be correlated only against those predictor scores of the high group, the resulting coefficient is likely to be very low. Individuals to the left of line B were not selected and hence there are no criterion measures for these individuals. When using predictor tests like this, the user must realize that the test has already done its work, since many poor students were not admitted to the college. Examples like this are numerous in educational, industrial, and military settings. When research is carried on with individuals who have gone through a selection process, correlational studies of the traits and performance of such subjects are likely to produce results that are much lower than between the same variables when individuals in unselected groups are used.

There are times in test work when correlation coefficients have to be corrected for restricted ranges. Suppose that we consider test work in the Air Force. To get into pilot training, an individual has to pass a rigorous battery of tests. The poor and low scorers are not accepted. Air Cadets are, then, a select group. After they are selected, they are given further training and testing, and during this process some drop out or are dropped. Finally, we have a more select group getting commissioned. If we were to use the group of survivors against which to study our selection tests, we would soon find that our correlation coefficients were very low. However, there are equations to take care of situations like this, and by using them we can correct the obtained coefficients for restricted range. These may be found in Guilford (1965, pp. 341–345).

The variability of the group, expressed in terms of either standard deviations or variances, should be reported when correlation coefficients are obtained. This gives some information on the nature of the group upon whom the data were collected and aids greatly in interpreting such coefficients.

The fact that it is the range and not the size of the sample that affects the size of *r* can be demonstrated by selecting an individual in the left-hand column of Figure 7.7 and then selecting a dot or two from each column as one goes across the figure. In this way, although fewer dots will be selected, those that are will be found in an ellipse, indicating a fairly high correlation coefficient.

## INTERPRETATION OF A PEARSON *r*

As we pointed out earlier in this chapter, a correlation coefficient is a measure of relationship between two variables. In everyday usage an *r* of .8 and above is considered a high coefficient, and *r* around .5 is considered moderate, and an *r* of .3 and below is considered a low coefficient. The student should bear in mind that a Pearson *r* is not a measure of causality, although in some cases a causal relationship may exist between the two variables. In other cases a relationship may exist between two variables because both are related to a third variable. This is often the case when the third variable is personal income or mental ability. Other coefficients may mean nothing. In other words, the coefficient is purely the result of chance, and if another set of data is collected and a Pearson *r* calculated, the resulting *r* would probably be meaningless. For example, suppose a statistics instructor obtained the height of each member of his class, correlated these heights with scores on the final examination in his statistics course, and obtained an *r* of .77. Obviously this coefficient would make no sense, and the best way to account for such a coefficient would be to say that it is purely the result of chance. If this instructor collected such data a hundred more times, the chances are good that such a high coefficient would not occur again. Thus the student should use care in the interpretation of any correlation coefficient.

In Chapter 17, which is related to educational and psychological measurement, some important applications of correlation coefficients are presented. One of these is described as the reliability of a measuring instrument. One definition of reliability is consistency of test scores over time. For example, if an intelligence test is administered today and again a month from today to the same individuals, we should expect a high correlation between the two sets of scores, say .9 and above, if the test is to be considered as producing reliable results. Another important use of correlation coefficients is in studying the predictive validity of a test. In this case scores on a test are correlated with a criterion measure. An example of this is the correlation of scores on the Scholastic Aptitude Test of the College Entrance Examination Board with freshman grade point average. Typically the correlation in situations like this runs between .4 and .6, although many are lower. The important point is that the size of the correlation itself is not the most important thing about it, but it is the situation in or purpose for which it is being used that determines how we evaluate it. We expect high correlation coefficients for some situations and can tolerate much lower ones in others.

Finally, the question arises as to how low a correlation coefficient may be and still be of value. This question will be answered in Chapter 16, where you will become acquainted with the technique for testing the significance of a Pearson $r$. When such a test is made, we are evaluating a coefficient to determine what the probability is that the coefficient under scrutiny came from a population in which the parameter value is not zero. Or to put it more simply, we are investigating the probability that the coefficient at hand is different from what would be expected to result by chance or if there were no relationship between the two variables. It should be emphasized that even though a correlation coefficient is statistically significant, it may be of little or no practical importance.

### EXERCISES

**1.** For each of the following relationships give an example that differs from the one in the text: high positive; medium positive; negative; curvilinear; one brought about because of an underlying third variable.

**2.** For a group of 80 the $z$-score product of two variables is 56. What is the Pearson $r$ for these data?

**3.** Find $r$ for the following scores on two tests of perceptual skills:

| Test 1 | Test 2 |
|--------|--------|
| 15 | 4 |
| 14 | 6 |
| 10 | 4 |
| 9 | 8 |
| 8 | 7 |
| 8 | 8 |
| 7 | 10 |
| 6 | 9 |
| 4 | 14 |
| 2 | 12 |

**4.** The following represent the scores on a short aptitude test with scores on a scale of a personality inventory.

| (1) | (2) | (1) | (2) |
|-----|-----|-----|-----|
| 18  | 12  | 9   | 0   |
| 16  | 10  | 8   | 2   |
| 15  | 11  | 7   | 5   |
| 14  | 8   | 6   | 7   |
| 13  | 6   | 4   | 9   |
| 12  | 4   | 2   | 11  |
| 10  | 0   | 0   | 12  |

    **a.** Make a scatterplot of these data.
    **b.** What can you infer about the size of the relationship?
    **c.** Calculate the Pearson *r*. How does this agree with your expectancy? What is wrong?

**5.** Twenty students are given a test of general ability and an English achievement test with the resulting scores:

| Mental Ability | English | Mental Ability (cont.) | English (cont.) |
|:---:|:---:|:---:|:---:|
| 54 | 203 | 44 | 181 |
| 53 | 196 | 44 | 175 |
| 51 | 202 | 44 | 168 |
| 50 | 186 | 43 | 174 |
| 48 | 204 | 40 | 162 |
| 47 | 184 | 38 | 158 |
| 47 | 196 | 37 | 170 |
| 46 | 182 | 36 | 144 |
| 45 | 170 | 34 | 141 |
| 45 | 178 | 31 | 139 |

    **a.** Compute *r* for these data.
    **b.** Also compute the two means and the two standard deviations.

**6.** Compute *r* for the data listed below and on the following page:

| Sales Attitude Score | Sales Index | Sales Attitude Score (cont.) | Sales Index (cont.) |
|:---:|:---:|:---:|:---:|
| 48 | 22 | 35 | 19 |
| 48 | 19 | 34 | 15 |
| 47 | 20 | 34 | 14 |
| 46 | 20 | 33 | 20 |
| 46 | 17 | 33 | 13 |
| 43 | 21 | 32 | 15 |
| 42 | 21 | 32 | 12 |
| 42 | 19 | 32 | 11 |
| 41 | 17 | 31 | 17 |
| 40 | 15 | 30 | 16 |
| 39 | 18 | 29 | 15 |
| 38 | 15 | 29 | 15 |
| 38 | 15 | 28 | 16 |
| 37 | 20 | 27 | 16 |
| 37 | 17 | 27 | 13 |

| Sales Attitude Score (cont.) | Sales Index (cont.) | Sales Attitude Score (cont.) | Sales Index (cont.) |
|---|---|---|---|
| 27 | 12 | 21 | 9 |
| 26 | 12 | 20 | 11 |
| 25 | 15 | 18 | 11 |
| 25 | 9 | 17 | 10 |
| 23 | 9 | 15 | 8 |
| 22 | 13 | | |

**7.** Scores of 18 deaf adolescents on the performance scale of the Wechsler adult intelligence scale and four scales of an interest inventory are as follows. Compute the correlation coefficients.

| | (1) WAIS—P | (2) Janitorial | (3) Clerical | (4) Manual Labor | (5) Paint and Handicraft |
|---|---|---|---|---|---|
| 1 | 99 | 15 | 33 | 16 | 25 |
| 2 | 103 | 24 | 20 | 20 | 40 |
| 3 | 111 | 17 | 37 | 13 | 21 |
| 4 | 116 | 5 | 42 | 8 | 20 |
| 5 | 127 | 9 | 40 | 6 | 18 |
| 6 | 117 | 5 | 48 | 3 | 17 |
| 7 | 114 | 14 | 34 | 7 | 31 |
| 8 | 113 | 13 | 33 | 13 | 35 |
| 9 | 122 | 24 | 20 | 16 | 35 |
| 10 | 113 | 15 | 32 | 12 | 27 |
| 11 | 120 | 14 | 43 | 10 | 17 |
| 12 | 108 | 12 | 36 | 11 | 28 |
| 13 | 116 | 20 | 31 | 12 | 29 |
| 14 | 106 | 20 | 19 | 19 | 37 |
| 15 | 100 | 32 | 22 | 18 | 29 |
| 16 | 96 | 25 | 21 | 16 | 39 |
| 17 | 90 | 20 | 31 | 19 | 28 |
| 18 | 97 | 23 | 16 | 21 | 38 |

# Other Correlational Techniques
## Chapter 8

Chapter 7 was devoted to the most widely used correlation coefficient, the Pearson $r$. When the data that we wish to correlate meet the assumptions basic to certain interpretations of the Pearson $r$, and when the two variables are continuous, it is the best statistic to use. However, there are times when the relationship between two variables is not linear, when one or both of the variables are not continuous, when the number of pairs of measurements is too small, or when certain assumptions cannot be made about the distribution of the trait in the population so that certain deductions from the Pearson $r$ are not applicable. In this chapter, we shall study some of the special correlational techniques that may be applied to some of the situations just described.

## LINEAR RELATIONSHIPS

### The Point-Biserial Coefficient

There are circumstances, especially in the field of test construction and test validation, where one of the variables is continuous and the other is conceived of as a dichotomy. In the usual scoring of items, the procedure is to mark the item either right or wrong. This right-wrong scoring is regarded as being a true dichotomy.

The point-biserial coefficient will be considered in some detail because it is a Pearson product-moment correlation coefficient, and it is widely used in test construction and analysis. First we shall show that the point biserial is just a special case of the product-moment correlation, and then we shall consider two methods of calculating it. Suppose that we call the test scores (the continuous variable) $Y$ and that we call the responses to the dichotomous variable $X$. In this case $X$ is either right or wrong and is scored either 1 or 0. It follows then that the $\Sigma X$ is actually the number of individuals who responded correctly to the item. This will be designated as $N_p$, the number passing the item or answering it correctly. Since each $X$ is either a 1 or a 0, the $\Sigma X^2$ will also be equal to $N_p$, and we shall call the total frequency $N_t$. Next we shall obtain the sum of the squares for $X$.

$$\Sigma x^2 = \Sigma X^2 - \frac{(\Sigma X)^2}{N}$$

$$= N_p - \frac{N_p{}^2}{N_t}$$

$$= \frac{N_p N_t - N_p{}^2}{N_t}$$

$$= \frac{N_p(N_t - N_p)}{N_t}$$

Since $N_t - N_p = N_w$, the number missing the item or responding to it incorrectly,

$$\Sigma x^2 = \frac{N_p N_w}{N_t}$$

Also in obtaining $\Sigma XY$ only those values where $X = 1$ will enter into the calculations, so actually $\Sigma XY$ may be written as $\Sigma f_p Y$, each $Y$ value multiplied by the frequency passing.

Let us now take equation 7.5 before it is cleared of fractions

$$r = \frac{\Sigma XY - (\Sigma X)(\Sigma Y)/N}{\sqrt{[\Sigma X^2 - (\Sigma X)^2/N][\Sigma Y^2 - (\Sigma Y)^2/N]}}$$

and substitute some of the information presented above into it. We now have

$$r_{pb} = \frac{\Sigma f_p Y - N_p(\Sigma fY)/N_t}{\sqrt{(N_p N_w/N_t)[\Sigma fY^2 - (\Sigma fY)^2/N_t]}}$$

When this is cleared of fractions, as in formula 7.5, it becomes

$$r_{pb} = \frac{N_t(\Sigma f_p Y) - N_p(\Sigma fY)}{\sqrt{N_p N_w[N_t(\Sigma fY^2) - (\Sigma fY)^2]}} \tag{8.1}$$

The use of this formula is illustrated by manipulating the data in Table 8.1, which presents the scores of 90 individuals on a short test of perception ($Y$) and on a test of ability to visualize the number of blocks in a geometric figure ($X$). The test was scored right or wrong, 1 or 0.

Column 1 consists of the scores on test $Y$. In column 2 are the frequencies of those who responded correctly to test $X$, tallied against the various $Y$ scores. In column 3 are the frequencies of those who failed test $X$. Column 4 presents the total frequencies for each $Y$ score. The values in column 5 ($fY$) are the

**Table 8.1** Scores on a Continuous and on a Dichotomous Variable
to Illustrate the Computation of the Point-Biserial $r$

| (1)<br>$Y$ | (2)<br>$f_p$ | (3)<br>$f_w$ | (4)<br>$f$ | (5)<br>$fY$ | (6)<br>$fY^2$ | (7)<br>$f_pY$ |
|---|---|---|---|---|---|---|
| 10 | 2 | 0 | 2 | 20 | 200 | 20 |
| 9 | 4 | 0 | 4 | 36 | 324 | 36 |
| 8 | 6 | 1 | 7 | 56 | 448 | 48 |
| 7 | 7 | 1 | 8 | 56 | 392 | 49 |
| 6 | 8 | 2 | 10 | 60 | 360 | 48 |
| 5 | 6 | 4 | 10 | 50 | 250 | 30 |
| 4 | 5 | 6 | 11 | 44 | 176 | 20 |
| 3 | 3 | 8 | 11 | 33 | 99 | 9 |
| 2 | 2 | 7 | 9 | 18 | 36 | 4 |
| 1 | 1 | 8 | 9 | 9 | 9 | 1 |
| 0 | 0 | 9 | 9 | 0 | 0 | 0 |
| | $\Sigma f_p = 44$ | $\Sigma f_w = 46$ | $\Sigma f = 90$ | $\Sigma fY = 382$ | $\Sigma fY^2 = 2294$ | $\Sigma f_pY = 265$ |

products obtained by multiplying the respective values in columns 1 and 4. Values in column 6 ($fY^2$) are obtained by multiplying the values in column 5 by their respective values in column 1. Column 7 ($f_p Y$) values are obtained by multiplying the values in column 1 by the values in column 2. Then the columns are summed and the sum entered into equation 8.1:

$$r_{pb} = \frac{90(265) - 44(382)}{\sqrt{44(46)[90(2294) - 382^2]}}$$

$$= \frac{23850 - 16808}{\sqrt{2024(206460 - 145924)}}$$

$$= \frac{7042}{\sqrt{2024(60536)}}$$

$$= \frac{7042}{11069}$$

$$= .64$$

This method is practical when the score values are low and the frequencies not too large. When values become large, a more efficient solution is to group the data and obtain the mean and the standard deviation of the total number of scores by the method using the arbitrary origin. Then these statistics are used in one of several alternate formulas for the point biserial. The derivation of these formulas is found in Guilford (1965, p. 537). The use of one of these formulas will be illustrated by an example taken from the area of the item analysis of a test. In item analysis work (see Chapter 17) the test maker is usually concerned with how well the item is separating the bright students

**Table 8.2** Worksheet for the Point-Biserial and Biserial $r$

| (1) | (2) Right $f_P$ | (3) Wrong $f_W$ | (4) $f_t$ | (5) $x'$ | (6) $f_t x'$ | (7) $f_t x'^2$ | (8) $f_p x'$ |
|---|---|---|---|---|---|---|---|
| 70–74 | 3 | 0 | 3 | 5 | 15 | 75 | 15 |
| 65–69 | 6 | 1 | 7 | 4 | 28 | 112 | 24 |
| 60–64 | 6 | 2 | 8 | 3 | 24 | 72 | 18 |
| 55–59 | 5 | 4 | 9 | 2 | 18 | 36 | 10 |
| 50–54 | 6 | 2 | 8 | 1 | 8 | 8 | 6 |
| 45–49 | 7 | 6 | 13 | 0 | 0 | 0 | 0 |
| 40–44 | 6 | 8 | 14 | −1 | −14 | 14 | −6 |
| 35–39 | 3 | 6 | 9 | −2 | −18 | 36 | −6 |
| 30–34 | 3 | 9 | 12 | −3 | −36 | 108 | −9 |
| 25–29 | 1 | 4 | 5 | −4 | −20 | 80 | −4 |
| 20–24 | 0 | 12 | 12 | −5 | −60 | 300 | 0 |

$$\Sigma f_P = 46 \quad \Sigma f_W = 54 \quad \Sigma f_t = 100$$

$$\Sigma f_t x' \qquad \Sigma f_t x'^2 \qquad \Sigma f_p x'$$
$$= -55 \qquad = 841 \qquad = 48$$

from the less bright. The ability of an item to discriminate between the two groups is frequently measured by a correlation coefficient. Dichotomously scored items may be correlated with a continuous total score or with a continuous outside criterion, such as academic grade-point indexes or other measures of achievements.

This is done by setting up a table like Table 8.2. In column 1 are the interval groupings for the total test scores. Then the papers are taken one by one and the first item on the test is examined to see whether the subject answered it correctly. Suppose that the first paper has a score of 73 and that this student answered the item correctly. Then a tally is placed in the "Right" column, column 2. Each paper is taken, and the response of the first item is tallied into either column 2 or 3. Actually, we have a scatterplot with the variable on the $y$ axis continuous, just as with the Pearson $r$, but with the variable on the $x$ axis reduced to two categories.

Column 4 is the total number of frequencies falling in each of the intervals. The sum of this column is 100, which is the sum of the number who responded correctly and the number who responded incorrectly to the item. We now proceed to compute the mean and standard deviation of the total scores. Columns 5, 6, and 7 are set up in the usual way and columns 6 and 7 are summed. We also need the mean of the individuals getting the item correct. For this we set up column 8, which contains the products of the values in columns 2 and 5.

A formula for the point biserial is as follows:

$$r_{pb} = \frac{X_p - X_t}{s_t} \sqrt{\frac{p}{q}} \tag{8.2}$$

where $X_p$ = the mean score of those answering the item correctly

$X_t$ = the mean of the total test scores

$s_t$ = the standard deviation of the test

$p$ = the proportion of the total group answering the item correctly

$q = 1 - p$

We first solve for the two means.

$$X_p = 47 + \frac{48}{46}(5) \qquad X_t = 47 + \frac{-55}{100}(5)$$

$$= 47 + 5.2 \qquad\qquad = 47 + (-2.75)$$

$$= 52.2 \qquad\qquad\quad = 44.2$$

The standard deviation must also be computed.

First, the $\Sigma x^2$:

$$\Sigma x^2 = \left(841 - \frac{-55^2}{100}\right) 5^2$$

$$= (841 - 30.25)25$$

$$= 20268.75$$

Then $s_t$:

$$s_t = \sqrt{\frac{20268.75}{100}}$$

$$= \sqrt{202.69}$$

$$= 14.2$$

$$p = \frac{46}{100} = .46$$

By substituting these values in equation 8.2, we have

$$r_{pb} = \frac{52.2 - 44.2}{14.2} \sqrt{\frac{.46}{.54}}$$

$$= \frac{8}{14.2} \sqrt{.851851}$$

$$= .563(.923)$$

$$= .52$$

If the test being analyzed contains a large number of items, the use of this computation is almost prohibitive from the point of view of time. There are data processing aids and short-cut methods that will facilitate such computations. In test validation work, the criterion may, for instance, be "pass" or "fail," "obtained his wings," "did not obtain his wings," "on probation," or "not on probation."

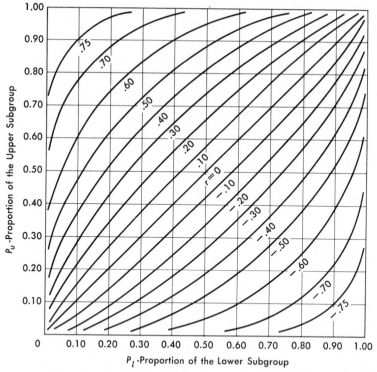

**Figure 8.1** An abac for estimates of the point-biserial $r$ when one variable is divided at the median of the distribution.

A quick and efficient method of estimating the point-biserial $r$ is to use an abac prepared by Dingman (Figure 8.1). This abac can be used when one variable is divided at the median. The proportion of subjects in the upper criterion group who pass a given item is found on the ordinate. The corresponding proportion from the "low" group is found on the abscissa. The estimated point-biserial $r$ is found at the perpendicular intersection of these values.

Let us now try out this abac for the data in Table 8.2. First the median is calculated and found to be approximately 44. In column 2 there are 33 cases above the median. This results in $p_u = 33/46 = .72$. Also in column 3 there are 15 cases above the median, and thus $p_l = 15/54 = .28$. Entering the abac with $p_u = .72$ and $p_l = .28$ results in an estimated point biserial of about .52.

## The Biserial Correlation Coefficient

Another statistic widely used in item analysis work is the biserial $r$ ($r_b$). We use this statistic when we have one continuous variable and another which is actually continuous but which has been forced into a dichotomy. Passing or failing is an example of a forced dichotomy. Achievement may be thought of as a continuum ranging from those who pass with exceedingly high honors down to those who fail miserably. Passing is made up of groups of individuals from the honor students down to the borderline cases, and failure includes all those who just barely failed down to the utter failures. We reduce this continuum to a pass-fail dichotomy, and since this is the usual procedure in test scoring, the biserial $r$ may be used as a measure of the discrimination index of an item.

To find the value of the biserial $r$, one of the easiest formulas is

$$r_b = \frac{\bar{X}_p - \bar{X}_t}{s_t}\left(\frac{p}{y}\right) \tag{8.3}$$

All symbols are the same as those defined for the point-biserial $r$ except $y$, which is the ordinate obtained from the normal probability table, cutting off above it an area equal to $p$.

We shall take the data in Table 8.2 and this time solve for the biserial $r$.

$$r_b = \frac{52.2 - 44.2}{14.2}\left(\frac{.46}{.3970}\right)$$

$$= \frac{8}{14.2}(1.159)$$

$$= .563(1.159)$$

$$= .65$$

The biserial $r$ is an estimation of the product-moment correlation coefficient. But we cannot manipulate it as we did the Pearson $r$. For instance, there is no $Z$ (Chapter 16) transformation for the biserial that could be used to avoid the skewed sampling distributions associated with high $r$'s. Also, it is not used in regression equations or in the standard error of estimate. Another peculiarity of the biserial $r$ is that sometimes it may be larger than 1 as computed. Such errors are brought about by departures from normality in the continuous or dichotomized variable, or both. When the biserial $r$ is compared with the point-biserial $r$ for the same data, the biserial $r$ is always the larger. This was demonstrated with the data in Table 8.2.

As will be shown later (Chapter 16) the biserial $r$ is a less reliable statistic than either the point biserial or the Pearson $r$. By this we mean that its size fluctuates more from sample to sample than does that of the Pearson $r$. There is little today to justify the use of this statistic. Various short-cut methods were developed to aid in obtaining the biserial. However, today with the presence of computers, these short-cut methods are no longer needed and there is very little reason left for using the biserial $r$, a statistic which, no matter how it is evaluated, is inferior to the point biserial.

**Table 8.3**  Responses of the 200 Students to Test Items

|  | Right | Wrong |  |
|---|---|---|---|
| Agree | 70<br>a | 30<br>b | 100<br>k |
| Disagree | 30<br>c | 70<br>d | 100<br>l |
|  | 100<br>m | 100<br>n | 200<br>N |

## The Fourfold Coefficient or Phi

The *phi coefficient* is used when each of the variables is a dichotomy. To illustrate this technique, let us suppose that we are making an analysis of the relation between an opinion item and an information item which have been administered to 200 students. Suppose 100 students agree with the opinion and the other 100 disagree. We set the data up as shown in Table 8.3, with the agree and disagree groups on the side and the scores of right and wrong across the top. The number in the agree and disagree groups that answered the information item correctly and the number that answered incorrectly are entered in the cells. Note that the cells are lettered and that marginal values have been computed and given letters. The formula for the phi coefficient is

$$\Phi = \frac{ad - bc}{\sqrt{klmn}} \tag{8.4}$$

where the various letters are the frequencies as shown in Table 8.3.

Solving for these data, we have

$$\Phi = \frac{70(70) - 30(30)}{\sqrt{(100)(100)(100)(100)}}$$

$$= \frac{4900 - 900}{10000}$$

$$= \frac{4000}{10000}$$

$$= .40$$

The phi coefficient, like the point-biserial coefficient, is a product-moment correlation coefficient (Edwards, 1967, p. 126). This makes it a desirable one and one that is very useful in test construction and analysis. A major limitation of this statistic is that the size of the coefficient is related to the way in which the two variables are split. When both variables are evenly divided, as are the data in Table 8.3, the maximum limits of a correlation coefficient, $\pm 1$, may be obtained. If the marginal totals are unequal, the maximum values will

**Table 8.4**  Six 2 × 2 Tables Showing How the Size of the
Phi Coefficient Varies with the Marginal Values

| (1) | | | (2) | | | (3) | | |
|---|---|---|---|---|---|---|---|---|
| 0 | 100 | 100 | 100 | 0 | 100 | 30 | 40 | 70 |
| 100 | 0 | 100 | 0 | 100 | 100 | 0 | 30 | 30 |
| 100 | 100 | 200 | 100 | 100 | 200 | 30 | 70 | 100 |
| $\Phi = -1$ | | | $\Phi = +1$ | | | $\Phi = +.43$ | | |

| (4) | | | (5) | | | (6) | | |
|---|---|---|---|---|---|---|---|---|
| 40 | 30 | 70 | 10 | 40 | 50 | 0 | 50 | 50 |
| 30 | 0 | 30 | 20 | 30 | 50 | 30 | 20 | 50 |
| 70 | 30 | 100 | 30 | 70 | 100 | 30 | 70 | 100 |
| $\Phi = -.43$ | | | $\Phi = -.22$ | | | $\Phi = -.65$ | | |

vary, but in any case they will be less than ±1. This is illustrated in Table 8.4. Parts 1 and 2 of this table show the maximum values of phi when the categories are evenly divided. In parts 3 and 4 the maximum values are shown when the two categories are divided on a .70-.30 basis. In the fifth part, again the two categories are not equally divided, and the maximum value is not obtained this time. However, this phi of −.22 might be compared with the maximum possible phi for this type of split to give some indication of the strength of the correlation between the two variables. For these data the maximum value of phi falls between −.65 and −.66 (part 6).

### The Tetrachoric Correlation Coefficient

The tetrachoric correlation coefficient also involves setting up data into a fourfold table. An assumption associated with this statistic is that both of the variables are continuous variables that have been forced into a dichotomy. In the past the tetrachoric coefficient was obtained by solving a quadratic equation of the second power for $r_t$. Since the use of these equations is laborious, various computing devices have been developed to aid in the computation of the statistic. The one that we shall demonstrate here is that proposed by Davidoff and Goheen (Appendix H).

Suppose that we have the responses of 100 individuals to a test item. A table such as Table 8.5 is set up, and the values in the various cells are tabulated. Our papers were divided into the top 50 and the bottom 50, and then the right and wrong answers for each group were counted. Notice again that the cells are lettered in the usual fashion. To use Davidoff and Goheen's table (Appendix H) we need the value $ad/bc$; then we can read from the

**Table 8.5** Responses of 100 Individuals to a Test Item, the Data Being Arranged for the Solution of the Tetrachoric Correlation Coefficient

|  | Right | Wrong |  |
|---|---|---|---|
| Upper 50 | 20 *a* | 30 *b* | 50 |
| Lower 50 | 10 *c* | 40 *d* | 50 |
|  | 30 | 70 | 100 |

table the estimation of the tetrachoric coefficient. For these data

$$\frac{(20)(40)}{(30)(10)} = \frac{800}{300} = 2.67$$

When we refer to the tables with a value of 2.67, we find an $r_t$ of .37. The equation solved for these data resulted in an $r_t$ of .36. Both of these, however, are approximations, for this quadratic equation is only solved for $r_t$ to the second power, all higher powers being ignored.

If *ad* is less than *bc,* we use the ratio *bc/ad* for the tables. The larger of the two products is always placed in the numerator. This table works best when both variables have been dichotomized on the basis of a 50-50 split. As a statistic $r_t$ is less reliable than the Pearson *r,* and currently there is little justification for its use, since the phi coefficient can be used in its place.

## NONLINEAR RELATIONSHIPS

Frequently the relationship between two sets of variables is not linear. When such a relationship occurs, the Pearson *r* and variations of this *r* as described in this chapter are inappropriate as measures of relationship. When a relationship is curvilinear, something like this happens: as one variable increases, the other may increase up to a point and then begin to decrease, as the first variable continues to increase. Examples of this were noted in the chapter on correlation. When *r* is used as a measure of relationship with data that are nonlinear, the *r* calculated is always an underestimate of the true relationship between the two variables. Sometimes this relationship is actually very high, but the resulting *r* approaches zero. Since it is difficult to tell whether or not data are linear by looking at them, a scatterplot should always be made. If a deviation from linearity is apparent or even suggested on the scattergram, then the Pearson *r* should not be used.

The correct coefficient to use when the relationship between two sets of data is curvilinear is the correlation ratio, or the eta coefficient. We shall illustrate the computation of this coefficient, using the data in Table 8.6. These data consist of measures of 200 individuals on two scales: age and scores on an information test.

### Computation of Eta

In solving for eta, we first set up a scatterplot as in Table 8.6. Here we have chronological age on the $x$ axis and scores on the information test on the $y$ axis. After the tallies have been entered, values for the $Y$ variable on the right-hand side of the scatterplot are obtained.

The most direct approach to the computation of eta is to define eta squared as the ratio of the sum of the squares of the "between" columns for variable $Y$ to the total sum-of-squares for variable $Y$:

$$\eta_{yx}^2 = \frac{\Sigma y_b^2}{\Sigma y_t^2} \tag{8.5}$$

and eta is

$$\eta_{yx} = \sqrt{\frac{\Sigma y_b^2}{\Sigma y_t^2}} \tag{8.6}$$

The total sum-of-squares for $Y$ is readily available from the data on the scatterplot:

$$\Sigma y_t^2 = \Sigma f(y')^2 - \frac{(\Sigma f y')^2}{N}$$

$$= 15298 - \frac{1648^2}{200}$$

$$= 15298 - 13579.52$$

$$= 1718.48$$

The calculation of the "between" sum-of-squares for $Y$ is shown in Table 8.7. This between sum-of-squares is determined from the means of the columns taken from the means of the entire distribution. The first column in Table 8.7 identifies the columns. It consists of $x'$ values from the bottom of the scatter-plot. Column 2 is made up of the frequencies of the various columns. In column 3, we find the sum of the $y'$ (deviation from the arbitrary reference point for $y$) for the frequencies of each of the columns. For example, find the line representing column 0 in the table and read across. Column 0 has 9 frequencies with $y'$ values of 1, 3, 3, 4, 5, 5, 6, 7, and 8, which sum to 42. In this way, all of the values in column 3 are obtained. Each of the values in column 3 is then squared, the square placed in column 4, and then each of these squares is divided by the appropriate column frequency. These are summed, and the sum of the squares for "between" columns is found as follows:

**Table 8.6** Scatterplot for Scores on a General Information Test and Ages for 200 Individuals

| y Axis: Scores | y' | x Axis: Age | | | | | | | | | | | | f | y' | fy' | f(y')² |
|---|---|---|---|---|---|---|---|---|---|---|---|---|---|---|---|---|---|
| | | 15–19 | 20–24 | 25–29 | 30–34 | 35–39 | 40–44 | 45–49 | 50–54 | 55–59 | 60–64 | 65–69 | 70–74 | | | | |
| 85–89 | 16 | | | | | 1 | | | | | | | | 1 | 16 | 16 | 256 |
| 80–84 | 15 | | | | | 1 | | | | | | | | 1 | 15 | 15 | 225 |
| 75–79 | 14 | | | 2 | | | | | | | | | | 2 | 14 | 28 | 392 |
| 70–74 | 13 | | | | 3 | 1 | 1 | 1 | | | | | | 9 | 13 | 117 | 1521 |
| 65–69 | 12 | | | 4 | 3 | 4 | | 1 | | 1 | | | | 14 | 12 | 168 | 2016 |
| 60–64 | 11 | | | 3 | 3 | 5 | 2 | 2 | 2 | | | | | 20 | 11 | 220 | 2420 |
| 55–59 | 10 | | | 3 | 5 | 3 | 4 | 3 | 3 | 1 | | | | 23 | 10 | 230 | 2300 |
| 50–54 | 9 | 1 | 1 | 1 | 2 | | 4 | 3 | 3 | 5 | 2 | | 1 | 25 | 9 | 225 | 2025 |
| 45–49 | 8 | 1 | 1 | 2 | 1 | 2 | 5 | 3 | 3 | 7 | 2 | 2 | 1 | 29 | 8 | 232 | 1856 |
| 40–44 | 7 | 1 | 1 | 2 | 2 | 3 | 3 | 3 | 2 | 4 | | | 3 | 20 | 7 | 140 | 980 |
| 35–39 | 6 | 2 | 2 | 1 | 1 | 2 | 2 | 2 | 3 | 3 | | 1 | 3 | 22 | 6 | 132 | 792 |
| 30–34 | 5 | 3 | 1 | | | | 1 | | | 2 | 3 | | 2 | 11 | 5 | 55 | 275 |
| 25–29 | 4 | 1 | | | | | | 1 | 1 | 2 | 2 | 1 | 2 | 10 | 4 | 40 | 160 |
| 20–24 | 3 | 2 | | | | | | | | 1 | 1 | 1 | | 7 | 3 | 21 | 63 |
| 15–19 | 2 | 2 | | | | | | | | 1 | 1 | | | 4 | 2 | 8 | 16 |
| 10–14 | 1 | 1 | | | | | | | | | | | | 1 | 1 | 1 | 1 |
| 5–9 | 0 | | | | | | | | | | | | 1 | 1 | 0 | 0 | 0 |
| f | | 9 | 15 | 20 | 23 | 18 | 30 | 18 | 15 | 24 | 11 | 7 | 10 | N = 200 | | Σfy' = 1648 | Σf(y')² = 15298 |
| x' | | 0 | 1 | 2 | 3 | 4 | 5 | 6 | 7 | 8 | 9 | 10 | 11 | | | | |

$$\Sigma y_b{}^2 = \sum \left[ \frac{(\Sigma y')^2}{f_x} \right] - \frac{[\Sigma(\Sigma y')]^2}{N}$$

$$= 14448.71 - \frac{1648^2}{200}$$

$$= 14448.71 - 13579.52$$

$$= 869.19$$

For these data,

$$\eta_{yx} = \sqrt{\frac{869.19}{1718.48}}$$

$$= \sqrt{.505784}$$

$$= .711$$

This can all be combined into the following formula for eta squared:

$$\eta_{yx}{}^2 = \frac{\Sigma[(\Sigma y')^2/f_x] - [\Sigma(\Sigma y')]^2/N}{\Sigma f(y')^2 - (\Sigma f y')^2/N} \tag{8.7}$$

When we were dealing with the correlation coefficient, we noted that the correlation between $X$ and $Y$ was the same as that between $Y$ and $X$ and that only one needed to be computed. With the correlation ratio there are two coefficients, one between $X$ and $Y$ and the other between $Y$ and $X$. We could obtain the other eta coefficient by inserting the corresponding values for $x$ in equation 8.7.

**Table 8.7**   Calculation of the "Between" Sum-of-Squares
for the Data in Table 8.6

| (1)<br>Column | (2)<br>$f_x$ | (3)<br>$\Sigma y'$ | (4)<br>$(\Sigma y')^2$ | (5)<br>$(\Sigma y')^2/f_x$ |
|---|---|---|---|---|
| 0 | 9 | 42 | 1764 | 196.00 |
| 1 | 15 | 73 | 5329 | 355.27 |
| 2 | 20 | 202 | 40804 | 2040.20 |
| 3 | 23 | 244 | 59536 | 2588.52 |
| 4 | 18 | 196 | 38416 | 2134.22 |
| 5 | 30 | 278 | 77284 | 2576.13 |
| 6 | 18 | 159 | 25281 | 1404.50 |
| 7 | 15 | 123 | 15129 | 1008.60 |
| 8 | 24 | 173 | 29929 | 1427.04 |
| 9 | 11 | 65 | 4225 | 384.09 |
| 10 | 7 | 43 | 1849 | 264.14 |
| 11 | 10 | 50 | 2500 | 240.00 |
| | $\Sigma f_x = 200$ | $\Sigma(\Sigma y') = 1648$ | | $\Sigma[(\Sigma y')^2/f_x] = 14448.71$ |

## Summary

If the data in a scatterplot are curvilinear, eta is larger than *r;* the discrepancy between the two is related to the size of the departure from linearity. *Eta has no sign.* An inspection of a scatterplot will show that in some parts of the range the relationship between the two variables is positive and at other parts negative. Eta then measures only the degree of the relationship. Eta is also affected by the number of columns and the frequencies within the columns. These should be large enough to give the means of the various columns stability.

## TWO SPECIAL CORRELATION COEFFICIENTS

We shall next introduce two correlation methods that involve more than two variables, the partial *r* and the multiple correlation coefficient (*R*).

### The Partial *r*

The relationship between two variables frequently is influenced by a third variable. For example, suppose that we have the relationship between intelligence test scores and arithmetic grades for a set of students and also the correlation of these same intelligence scores with grades in English. In addition, we have the relationship between the arithmetic grades and the English grades. Both of these school subjects are related to intelligence test scores, and they seem to be related to each other. Is this last relationship a true one, or is it merely the effect of these two variables being related to the common one of intelligence? With the partial correlation coefficient it is possible to control these effects of intelligence or to "partial them out." We might also ask what is the relationship between English grades and arithmetic grades with the effect of intelligence partialed out. A situation like this is referred to as a partial *r* of the first order based upon three zero-order *r*'s. The general formula for this partial-order *r* is

$$r_{12.3} = \frac{r_{12} - r_{13}r_{23}}{\sqrt{(1 - r_{13}^2)(1 - r_{23}^2)}} \qquad (8.8)$$

We would read $r_{12.3}$ as the correlation between variables one and two with the effects of variable three partialed out. Similarly, it is possible to write a comparable equation for $r_{13.2}$ and $r_{23.1}$.

Second-order partials are those in which the relationship between two variables is computed with the effects of two other variables partialed out. Since these are rarely used, they will not be discussed here.

Suppose that we have the following three variables:

1 = chronological age
2 = weight
3 = scores on an arithmetic test

For several hundred students, we compute the correlations among three variables and obtain the following:

$r_{12} = .80$
$r_{13} = .60$
$r_{23} = .50$

From this we see that we have a correlation between weight and arithmetic which, with a sample of this size, is significant. Suppose that we investigate the relationship between weight and scores on an arithmetic test with the effects of chronological age partialed out.

$$r_{23.1} = \frac{r_{23} - r_{12}r_{13}}{\sqrt{(1 - r_{12}^2)(1 - r_{13}^2)}}$$

$$= \frac{.50 - (.80)(.60)}{\sqrt{(1 - .8^2)(1 - .6^2)}}$$

$$= \frac{.50 - .48}{\sqrt{(.36)(.64)}}$$

$$= \frac{.02}{.48}$$

$$= .04$$

Now we see that with the effects of chronological age removed, there is no significant relationship between weight and scores on an arithmetic test. Since the partial $r$ is a Pearson $r$, it may be treated as such.

### Multiple Correlation

In this book we are going to discuss multiple correlation in its simplest form, the relationship between one variable and a combination of two other variables. Suppose that we have the following three coefficients based upon three variables for a large group of university freshmen:

Variable 1 = grade
Variable 2 = scores on the Ohio State Psychological Examination
Variable 3 = scores on the Cooperative Mathematics Test

Thus

$r_{12} = .50$
$r_{13} = .60$
$r_{23} = .40$

We want to compute the multiple correlation coefficient between freshmen grades and the combined effects of the two tests. The formula follows:

$$R_{1.23} = \sqrt{\frac{r_{12}^2 + r_{13}^2 - 2r_{12}r_{13}r_{23}}{1 - r_{23}^2}} \qquad (8.9)$$

$$= \sqrt{\frac{.50^2 + .60^2 - 2(.50)(.60)(.40)}{1 - .40^2}}$$

$$= \sqrt{\frac{.25 + .36 - .24}{.84}}$$

$$= \sqrt{\frac{.37}{.84}}$$

$$= \sqrt{.4405}$$

$$= .66$$

Computing charts or abacs have been developed from which the multiple $R$ can be easily read when only three variables are concerned; for an example, see Lord (1955).

## RANK CORRELATION METHODS

Many times, data are collected in the form of ranks. Sometimes one variable may be in this form and the other comprised of measurement data. There are other times when measurement data are reduced to ranks.

### Spearman Rank-Order Correlation Coefficient (Rho)

This is the most widely used of the rank correlational methods. It is particularly well suited to situations where the number of cases is 25 to 30 or less. It is also much easier and faster to compute than the Pearson $r$.

To illustrate the computation of *rho* we shall first use the data in Table 8.8 in which we have the scores on tests $X$ and $Y$ for seven individuals. This

**Table 8.8**  Calculation of the Spearman Rank-Order Correlation Coefficient—No Ties

| Individual | Test $X$ | Test $Y$ | $Rx$ | $Ry$ | $D$ | $D^2$ |
|------------|----------|----------|------|------|-----|-------|
| 1 | 18 | 24 | 1 | 4 | 3 | 9 |
| 2 | 17 | 28 | 2 | 2 | 0 | 0 |
| 3 | 14 | 30 | 3 | 1 | 2 | 4 |
| 4 | 13 | 26 | 4 | 3 | 1 | 1 |
| 5 | 12 | 22 | 5 | 5 | 0 | 0 |
| 6 | 10 | 18 | 6 | 6 | 0 | 0 |
| 7 | 8 | 15 | 7 | 7 | 0 | 0 |

$$\Sigma D^2 = 14$$

is a simple solution because there are no ties in either set of scores. To obtain rho:

**1.** Rank the scores in distribution $X$ giving the highest score a rank of 1.
**2.** Repeat the process for the scores in distribution $Y$.
**3.** Obtain the difference between the two sets of ranks. The sign of this difference is of no importance, since these differences are squared in the next operation.
**4.** Square each of these differences and sum this column of squares.
**5.** Solve for the rank-order correlation coefficient by the use of the following equation:

$$\rho = 1 - \frac{6\Sigma D^2}{N(N^2 - 1)} \qquad (8.10)$$

where $N$ = the number of pairs
$\rho$ = rho, the rank-order correlation coefficient

For this problem,

$$\rho = 1 - \frac{6(14)}{7(49 - 1)}$$

$$= 1 - \frac{84}{336}$$

$$= 1 - .25$$

$$= .75$$

THE SPEARMAN RANK-ORDER CORRELATION COEFFICIENT WITH TIED RANKS. Very often ties appear in either one or in both sets of ranks. When ties appear they are treated as illustrated in Table 8.9. Notice that in this table the ranking starts as in the previous example with the highest score in the $X$ distribution given a rank of 1. Continuing with the ranking we come to the two 49's, which are tied for the fourth and fifth positions. Instead of giving one a rank of 4 and the other a rank of 5, we average the two ranks, $(4 + 5)/2 = 4.5$, and give each a rank of 4.5. The next ties appear in the scores occupying ranks 8, 9, and 10. Taking the average of these three positions we obtain 9, which is assigned to each of the three scores. Then we proceed as before.

INTERPRETATION OF RHO. Rho is a product-moment correlation coefficient for ranked data. For all practical purposes, it may be interpreted the same as $r$.

SUMMARY. In rho we have a coefficient that makes a good substitute for $r$ when the number of cases is small. Rho is almost useless when $N$ is large, for by the time that all the data are ranked, a Pearson $r$ could have been computed.

**Table 8.9** Calculation of the Spearman Rank-Order
Correlation Coefficient—Tied Ranks

| Individual | Test $X$ | Test $Y$ | $Rx$ | $Ry$ | $D$ | $D^2$ |
|---|---|---|---|---|---|---|
| 1 | 60 | 60 | 1.0 | 2.0 | 1.0 | 1.00 |
| 2 | 54 | 68 | 2.0 | 1.0 | 1.0 | 1.00 |
| 3 | 53 | 40 | 3.0 | 7.0 | 4.0 | 16.00 |
| 4 | 49 | 52 | 4.5 | 3.0 | 1.5 | 2.25 |
| 5 | 49 | 51 | 4.5 | 4.5 | .0 | .00 |
| 6 | 47 | 38 | 6.0 | 9.0 | 3.0 | 9.00 |
| 7 | 46 | 51 | 7.0 | 4.5 | 2.5 | 6.25 |
| 8 | 45 | 32 | 9.0 | 10.0 | 1.0 | 1.00 |
| 9 | 45 | 39 | 9.0 | 8.0 | 1.0 | 1.00 |
| 10 | 45 | 41 | 9.0 | 6.0 | 3.0 | 9.00 |

$$\Sigma D^2 = 46.50$$

$$\rho = 1 - \frac{6\Sigma D^2}{N(N^2 - 1)}$$

$$= 1 - \frac{6(46.50)}{10(99)}$$

$$= 1 - \frac{279}{990}$$

$$= 1 - .28$$

$$= .72$$

## Kendall's Tau Correlation Between Ranks

Kendall's *tau,* T, can be applied wherever the Spearman rank-order coefficient is applicable. As can be seen below, it is somewhat harder to compute than rho. When there are no ties, the solution is short and simple. Data in Table 8.8 have been copied in Table 8.10. It is imperative that one of the series of

**Table 8.10** Calculation of Kendall's T Coefficient—No Ties

| Individual | Test $X$ | Test $Y$ | $Rx$ | $Ry$ | No. of Ranks Higher | No. of Ranks Lower |
|---|---|---|---|---|---|---|
| 1 | 18 | 24 | 1 | 4 | 3 | 3 |
| 2 | 17 | 28 | 2 | 2 | 4 | 1 |
| 3 | 14 | 30 | 3 | 1 | 4 | 0 |
| 4 | 13 | 26 | 4 | 3 | 3 | 0 |
| 5 | 12 | 22 | 5 | 5 | 2 | 0 |
| 6 | 10 | 18 | 6 | 6 | 1 | 0 |
| 7 | 8 | 15 | 7 | 7 | 0 | 0 |
| | | | | | $P = 17$ | $Q = 4$ |

ranks be in its natural order, that is, starting from high to low. For each individual the number of ranks below him that are higher and lower on the $Y$ variable are counted. For example, 3 persons rank higher than the first individual on test $Y$ and 3 rank lower. In the case of the second individual, 4 rank higher and 1 ranks lower. This counting continues until the columns in Table 8.10 are completed. Each column is then summed, the column with the number of ranks higher being called $P$ and the other column $Q$. Tau is then calculated:

$$T = \frac{P - Q}{N(N - 1)/2} \tag{8.11}$$

$$= \frac{17 - 4}{7(6)/2}$$

$$= \frac{13}{21}$$

$$= .62$$

When ties appear, certain adjustments have to be made. To illustrate the handling of ties, the data from Table 8.9 have been copied in Table 8.11. The number of individuals ranking higher and lower than each individual on the $Y$ variable is again determined resulting in $P = 33$ and $Q = 11$.

In handling ties we first take the $X$ distribution and for each set of ties determine $(x)(x - 1)$ where $x$ is the number tied for a particular rank. These are summed and divided by 2. For Table 8.11 we have

$$\frac{2(2 - 1) + 3(3 - 1)}{2} = \frac{2 + 6}{2}$$

$$= \frac{8}{2}$$

$$= 4$$

Repeating the process for the $Y$ distribution:

$$\frac{2(2 - 1)}{2} = \frac{2}{2} = 1$$

Next we calculate

$$\frac{N(N - 1)}{2} = \frac{10(9)}{2} = 45$$

Subtracting the correction obtained above for each distribution from this:

$$45 - 4 = 41$$
$$45 - 1 = 44$$

We next multiply the two previous terms:

$44(41) = 1804$

and take the square root of the product:

$\sqrt{1804} = 42.5$

Then

$$T = \frac{P - Q}{42.5}$$

$$= \frac{22}{42.5}$$

$$= .52$$

Kendall's tau, like rho, has many applications. For reasons beyond the scope of this text tau is preferred by many statisticians over rho. As illustrated, when both rho and tau are computed for the same data, tau is the smaller. The range of tau is the same as that of rho and both statistics are interpreted in the same way.

### Kendall's Coefficient of Concordance, W

If we wish to determine the relationship among three or more sets of ranks, one rank could be selected and a Spearman rho coefficient computed between it and all of the others, and this process could then be continued until a rho coefficient has been obtained between each set of two ranks. Then these rhos could be averaged for an overall measure of relationship.

Kendall, though, has developed a technique and a statistic which makes all of this unnecessary. Suppose that five judges ($m$) rank the projects of ten individuals ($N$) in a judging contest, and we wish to determine the overall

**Table 8.11**  Calculation of Kendall's T—Tied Ranks

| Individual | Test $X$ | Test $Y$ | $Rx$ | $Ry$ | No. of Ranks Higher | No. of Ranks Lower |
|---|---|---|---|---|---|---|
| 1 | 60 | 60 | 1.0 | 2.0 | 8 | 1 |
| 2 | 54 | 68 | 2.0 | 1.0 | 8 | 0 |
| 3 | 53 | 40 | 3.0 | 7.0 | 3 | 4 |
| 4 | 49 | 52 | 4.5 | 3.0 | 6 | 0 |
| 5 | 49 | 51 | 4.5 | 4.5 | 4 | 0 |
| 6 | 47 | 38 | 6.0 | 9.0 | 1 | 3 |
| 7 | 46 | 51 | 7.0 | 4.5 | 3 | 0 |
| 8 | 45 | 32 | 9.0 | 10.0 | 0 | 2 |
| 9 | 45 | 39 | 9.0 | 8.0 | 0 | 1 |
| 10 | 45 | 41 | 9.0 | 6.0 | 0 | 0 |
| | | | | | $P = 33$ | $Q = 11$ |

**Table 8.12**    Calculation of the Coefficient of Concordance, the
Data Consisting of the Ranking of 10 Projects by 5 Judges

| (1) Individual Project | (2) Judges' Ranks | | | | | (3) Sum of Ranks | (4) $D$ | (5) $D^2$ |
|---|---|---|---|---|---|---|---|---|
| | 1 | 2 | 3 | 4 | 5 | | | |
| 1 | 2 | 1 | 2 | 3 | 4 | 12 | 15.5 | 240.25 |
| 2 | 1 | 3 | 1 | 2 | 2 | 9 | 18.5 | 342.25 |
| 3 | 3 | 4 | 4 | 1 | 3 | 15 | 12.5 | 156.25 |
| 4 | 5 | 5 | 5 | 5 | 1 | 21 | 6.5 | 42.25 |
| 5 | 4 | 2 | 6 | 7 | 6 | 25 | 2.5 | 6.25 |
| 6 | 7 | 8 | 3 | 4 | 7 | 29 | 1.5 | 2.25 |
| 7 | 6 | 6 | 8 | 6 | 5 | 31 | 3.5 | 12.25 |
| 8 | 8 | 7 | 7 | 8 | 9 | 39 | 11.5 | 132.25 |
| 9 | 9 | 10 | 10 | 9 | 8 | 46 | 18.5 | 342.25 |
| 10 | 10 | 9 | 9 | 10 | 10 | 48 | 20.5 | 420.25 |
| | | | | | | $\Sigma = 275$ | | $\Sigma D^2 = 1696.50$ |

relationship among the ratings of the five judges. The rankings of these judges have been set up in Table 8.12. First the rankings by the five judges of each of the projects are summed. The sums appear in column 3. Column 3 is summed to give the total sum of the ranks. This can be checked for the total sum of the ranks as follows:

$$\text{Total sum of ranks} = \frac{m(N)(N + 1)}{2} \tag{8.12}$$

$$= \frac{5(10)(11) = 275}{2}$$

If there were no relationship among the ranks, we should expect the sum of the ranks for each row to be equal. For this case the sum of each would be the average sum of ranks or 275/10, which equals 27.5. We next obtain the difference of the sum of the ranks of each row from this mean and then square these differences. Then these squares are summed. This work appears in columns 4 and 5 of Table 8.12.

To compute $W$, we use the following formula:

$$W = \frac{12\Sigma D^2}{m^2(N)(N^2 - 1)} \tag{8.13}$$

$$= \frac{12(1696.5)}{25(10)(100 - 1)}$$

$$= .82$$

INTERPRETATION OF $W$. The size of this coefficient of concordance indicates that there is high agreement among these five judges in the ranking of the ten

**Table 8.13** Summary of the Major Correlation Coefficients

| | | Variables | |
|---|---|---|---|
| Coefficient | Symbol | X | Y |
| Pearson product-moment | $r$ | Continuous | Continuous |
| Point biserial | $r_{pb}$ | Continuous | True dichotomy |
| Biserial | $r_b$ | Continuous | Continuous, but forced into a dichotomy |
| Tetrachoric | $r_t$ | Continuous, but forced into a dichotomy | Continuous, but forced into a dichotomy |
| Phi or fourfold | $\Phi$ | True dichotomy | True dichotomy (see text) |
| Correlation ratio | $\eta$ (eta) | Continuous | Continuous |
| Spearman rank order | $\rho$ (rho) | Data in ranks or capable of being ranked | Data in ranks or capable of being ranked |
| Kendall's coefficient of concordance | $W$ | Used with three or more sets of ranks | |
| Kendall's tau | T | Data in ranks or capable of being ranked | |

projects. Perfect agreement is indicated by a $W = 1$ and lack of agrement by a $W = 0$.

The major facts about the various correlation coefficients have been summarized in Table 8.13.

## EXERCISES

**1.** The table below presents stanine scores on an aptitude test and successful or unsuccessful completion of a course of training. Using formula 8.1, compute the point-biserial coefficient for these data.

| Stanine | Successful | Unsuccessful |
|---|---|---|
| 9 | 9 | 1 |
| 8 | 12 | 2 |
| 7 | 13 | 4 |
| 6 | 10 | 5 |
| 5 | 12 | 4 |
| 4 | 10 | 8 |
| 3 | 6 | 10 |
| 2 | 2 | 10 |
| 1 | 0 | 2 |

**2.** A scale to measure attitudes toward Russia is being constructed. One of the items is "Do you think that the Russian people like their government?" The scale is pretested on a sample of 200. The distribution of Yes and No responses with total score on the scale is presented in the following table. Compute the point-biserial $r$ between the item and the total score.

| Total Scores | Yes | No |
|---|---|---|
| 95–99 | 1 | 0 |
| 90–94 | 6 | 0 |
| 85–89 | 18 | 1 |
| 80–84 | 22 | 1 |
| 75–79 | 31 | 3 |
| 70–74 | 20 | 5 |
| 65–69 | 18 | 9 |
| 60–64 | 12 | 13 |
| 55–59 | 6 | 10 |
| 50–54 | 4 | 8 |
| 45–49 | 1 | 5 |
| 40–44 | 0 | 3 |
| 35–39 | 1 | 0 |
| 30–34 | 0 | 1 |
| 25–29 | 0 | 1 |
|  | 140 | 60 |

**3.** With the data in problem 2 above, compute the biserial $r$ using formula 8.3.

**4.** Compute the phi coefficient for the following data:

|  | Right | Wrong |
|---|---|---|
| Upper | 65 | 35 |
| Lower | 25 | 75 |

**5.** What is the tetrachoric $r$ for the data in problem 4?

**6.** Seven instructors are rated by freshmen and sophomore students on "clarity of presentation" and the results are tabulated. What is the Spearman rho for the following:

| Instructor | Freshmen | Sophomores |
|---|---|---|
| 1 | 44 | 58 |
| 2 | 39 | 42 |
| 3 | 36 | 18 |
| 4 | 35 | 22 |
| 5 | 33 | 31 |
| 6 | 29 | 38 |
| 7 | 22 | 38 |

**7.** Compute Kendall's tau for the data of problem 6.

**8.** Four judges (parole board members) rank eight convicts on "parole readiness." By using the coefficient of concordance, indicate the degree of consistency of the judges.

| Convict | Judge 1 | 2 | 3 | 4 |
|---------|---------|---|---|---|
| 1 | 1 | 1 | 1 | 1 |
| 2 | 2 | 4 | 3 | 2 |
| 3 | 3 | 3 | 2 | 4 |
| 4 | 4 | 2 | 4 | 3 |
| 5 | 5 | 6 | 5 | 5 |
| 6 | 6 | 5 | 6 | 7 |
| 7 | 7 | 7 | 8 | 6 |
| 8 | 8 | 8 | 7 | 8 |

**9.** Compute $\eta_{xy}$ for the following data:

| Individual | Test 1 | Test 2 | Individual | Test 1 | Test 2 |
|-----------|--------|--------|-----------|--------|--------|
| 1 | 60 | 60 | 10 | 45 | 41 |
| 2 | 54 | 68 | 11 | 43 | 50 |
| 3 | 53 | 40 | 12 | 41 | 48 |
| 4 | 49 | 52 | 13 | 39 | 36 |
| 5 | 49 | 51 | 14 | 38 | 48 |
| 6 | 47 | 48 | 15 | 32 | 40 |
| 7 | 46 | 51 | 16 | 32 | 46 |
| 8 | 45 | 32 | 17 | 30 | 37 |
| 9 | 45 | 39 | | | |

**10.** In a research study on learning in arithmetic, a research worker studied the relation between spatial ability scores as well as between numerical ability scores. He produced the following correlations:

| | Spatial Ability | Numerical Ability | Concept Learning |
|---|---|---|---|
| Spatial ability | — | .522 | .496 |
| Numerical ability | | — | .254 |
| Concept learning | | | — |

  a. What is the correlation between concept learning and spatial ability with the effects of numerical ability removed?

  b. What is the correlation between concept learning and numerical ability with the effects of spatial ability partialed out?

**11.** The correlation between grades in a high school shop course and a test of mechanical ability was .5, and with a test of finger dexterity .2. The correlation between the latter two was .7. Find the multiple correlation between these shop grades and the combined effects of mechanical ability and finger dexterity scores.

**12.** Use the data in problem 7 at the end of Chapter 7 to calculate the Spearman rank-order correlations among the various sets of scores.

# Linear Regression
## Chapter 9

In Chapter 7 it was noted that one of the basic conditions to be met in using a Pearson product-moment correlation coefficient is that there be a linear relationship between the two variables being studied. In this chapter we shall examine in more detail this phenomenon of regression and the use that is made of regression equations in predicting scores on one variable from scores made on another. For example, grades in a course or a student's grade point index are predicted from a test of mental ability such as the Scholastic Aptitude Test of the College Entrance Examination Board. In this example scores on the aptitude test are the predictors, also known as the *independent variable*. That which is predicted, the grades, is referred to as the *dependent variable*. By the use of a regression equation, we can predict scores on the dependent variable from those of the independent variable. It is the usual practice to designate the independent variable by the symbol $X$ and the dependent variable by $Y$.

### The Equation for a Straight Line
We shall start by presenting the mathematical equation for a straight line:

$$Y = a + bX \tag{9.1}$$

Let us first find out what $a$ and $b$ in this equation stand for. To do this we write the equation as $Y = 4 + 2X$. Now we can substitute any value that we wish for $X$ and solve for the corresponding value of $Y$. A few such values are shown below.

| $X$ | $Y$ |
|-----|-----|
| 0   | 4   |
| 1   | 6   |
| 4   | 12  |
| 8   | 20  |

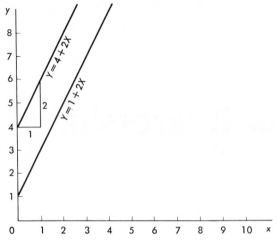

**Figure 9.1** Two lines with the same slope.

Sets of values like these may be plotted as is illustrated in Figure 9.1. Since we are dealing with a straight line, and since it takes only two points to determine a straight line, only two sets of these points need to be plotted. Let us examine this line in detail now. Notice that for an increase of 1 point on the $X$ variable there is an increase of 2 points on the $Y$ variable. The 2 in this equation is the $b$ coefficient of equation 9.1. The $b$ coefficient gives the relationship between the change in $Y$ in reference to change in $X$. This ratio of change in one variable to change in another is referred to as the *slope* of the line. In Figure 9.1 another line has been drawn, $Y = 1 + 2X$. This runs parallel to the first line, because it has the same slope. Theoretically, an infinite number of such lines with a slope of 2 can be drawn on these axes. The slope of both of these lines is positive inasmuch as the lines run from the lower left to the upper right on the graph. As was noted previously, a line located like this one is an indication of a positive relationship between the two variables.

Now examine Figure 9.2. Here we have the equation $Y = 4 \doteq .5X$. By solving for values of $Y$ we have

| $X$ | $Y$ |
|---|---|
| 0 | 4.0 |
| 1 | 3.5 |
| 2 | 3.0 |
| 3 | 2.5 |
| 4 | 2.0 |
| 8 | .0 |

This time as $X$ increases from 0 to 1, the corresponding change in $Y$ is from

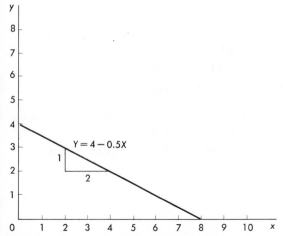

**Figure 9.2**   A line with a negative slope.

4 to 3.5. The ratio of the change in $Y$ relative to the change in X is .5 for this example. But as one variable is increasing, the other is decreasing, and for this case the $b$ coefficient is $-.5$.

As was noted earlier, there is an infinite number of lines of the same slope which can be drawn on the same set of axes. Two such lines are drawn in Figure 9.1. Notice that one of these lines crosses the y axis at 4 and the other at 1. These values of 4 and 1 are the $a$ coefficients for the two straight lines. Hence we might define the $a$ coefficient as the y *intercept*, that point where the straight line crosses the y axis. The $a$ coefficient identifies one of this infinite number of lines of the same slope.

The straight line used in regression analysis is written in a slightly altered form from that given above:

$$Y' = a + bX \qquad\qquad (9.2)$$

where $Y'$ is read as the predicted value of $Y$. $Y'$ is not usually the same as $Y$, for the score that is predicted from this equation will not be exactly the same as the one actually obtained. The values of the predicted $Y$'s will in general be closer to $\bar{Y}$ than are the values of the observed $Y$'s. Because of this relationship, this phenomenon is referred to as regression and will be discussed later.

### Obtaining the $a$ and $b$ Coefficients

The difference between the obtained score $Y$ and the predicted score $Y'$ is known as the *error of prediction*. The regression line, or the line of best fit as it is often called, is that line about which the sum of the squares of these errors of prediction is at a minimum. By starting with equation 9.2, we have

$$Y' = a + bX$$

$$Y - Y' = Y - (a + bX)$$

where the left-hand side is the error of prediction.

These errors of prediction are squared and summed

$$\Sigma(Y - Y')^2 = \Sigma[Y - (a + bX)]^2$$

To obtain values of $a$ and $b$, which minimize the sum of the squares of the errors of prediction, the expression is differentiated by means of the calculus with respect to $a$ and $b$ in turn, and each of the derivatives is equated to zero. This yields

$$b_{yx} = \frac{\Sigma XY - [(\Sigma X)(\Sigma Y)/N]}{\Sigma X^2 - [(\Sigma X)^2/N]} \qquad (9.3)$$

and

$$a_{yx} = Y - b_{yx}X \qquad (9.4)$$

the regression coefficients for predicting $Y$ from $X$ as these subscripts are read.

Previously we noted that

$$\Sigma x^2 = \Sigma X^2 - \frac{(\Sigma X)^2}{N}$$

and since

$$\Sigma xy = \Sigma XY - \frac{(\Sigma X)(\Sigma Y)}{N}$$

then the $b$ coefficient, $b_{yx}$ (equation 9.3) may be written as follows in deviation form:

$$b_{yx} = \frac{\Sigma xy}{\Sigma x^2} \qquad (9.5)$$

Since there are two regression lines except when $r = 1$, it follows that there is another set of regression coefficients, this time being used to predict $X$ from $Y$. These are written

$$b_{xy} = \frac{\Sigma XY - [(\Sigma X)(\Sigma Y)/N]}{\Sigma Y^2 - [(\Sigma Y)^2/N]} \quad \text{or} \quad b_{xy} = \frac{\Sigma xy}{\Sigma y^2} \qquad (9.6)$$

$$a_{xy} = X - b_{xy}Y \qquad (9.7)$$

AN ACTUAL PROBLEM. The calculation of the regression coefficients will now be illustrated using the data in Table 7.1. For those data, a correlation coefficient was obtained between scores on the Meier Art Judgment Test and the Graves Design Judgment Test. The data were coded by subtracting 100 from each Meier Art Judgment Test score ($Y$) and 60 from each Graves

Design Judgment Test score $(X)$. Equation 7.5 was used to compute the correlation coefficient.

The following sums and statistics resulted:

$\Sigma X = -4$          $\Sigma Y = -111$     $\Sigma XY = 2209$

$\Sigma X^2 = 7554$       $\Sigma Y^2 = 3827$

$\bar{X} = 59.9$         $\bar{Y} = 96.8$

$s_x = 14.7$          $s_y = 10.0$

$r_{xy} = .429 = .43$     $N = 35$

We shall begin by setting up the prediction equation for predicting $Y$, Meier scores, from Graves scores, $X$. First we compute the $b$ coefficient using equation 9.3:

$$b_{yx} = \frac{\Sigma XY - [(\Sigma X)(\Sigma Y)/N]}{\Sigma X^2 - [(\Sigma X)^2/N]}$$

$$= \frac{2209 - (-4)(-111)/35}{7554 - (-4)^2/35}$$

$$= \frac{2209 - 12.7}{7554 - .5}$$

$$= \frac{2196.3}{7553.5}$$

$$= .291$$

We next find the $a$ coefficient using equation 9.4:

$$a_{yx} = \bar{Y} - b_{yx}\bar{X}$$

$$= 96.8 - .291(59.9)$$

$$= 96.8 - 17.4$$

$$= 79.4$$

Then substituting in the equation for a straight line, $Y' = a + bX$, we have $Y' = 79.4 + .291X$, which is the regression equation for the $Y$ variable on $X$ for these data.

We shall next plot this line. Since two points determine a straight line, we can solve the above regression equation for a couple of points and plot these as in Figure 9.3. Suppose that we take three values of $X$ and by substituting these values in the regression equation obtain the predicted values of $Y'$.

| $X$ | $Y'$ |
|-----|------|
| 20  | 85.2 |
| 40  | 91.0 |
| 60  | 96.9 |

The plotting of these three sets of values on Figure 9.3 results in the regression line seen in that figure.

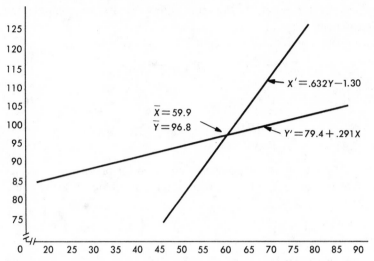

**Figure 9.3** The two regression lines for the data in Table 7.4.

As we have already noted, there are two regression lines unless $r = 1$. We shall now proceed to determine the regression line for predicting $X$ from $Y$. First the $b$ coefficient:

$$b_{xy} = \frac{\Sigma XY - [(\Sigma X)(\Sigma Y)/N]}{\Sigma Y^2 - [(\Sigma Y)^2/N]}$$

$$= \frac{2209 - (-4)(-111)/35}{3827 - (-111)^2/35}$$

$$= \frac{2209 - 12.7}{3827 - 352}$$

$$= \frac{2196.3}{3475.0}$$

$$= .632$$

Then the $a$ coefficient:

$$a_{xy} = X - b_{xy}Y$$

$$= 59.9 - .632(96.8)$$

$$= 59.9 - 61.2$$

$$= -1.3$$

$$X' = b_{xy}Y + a_{xy}$$

$$= .632Y - 1.30$$

Let us again obtain three sets of points and plot this line on the same figure as the other. By substituting in the equation for various values of $Y$, we obtain and then plot the following:

| $Y$ | $X'$ |
|-----|------|
| 120 | 74.5 |
| 100 | 61.9 |
| 80  | 49.3 |

CHECKS UPON THE WORK. There are several ways of determining if the calculations are correct. First, the product of the two $b$ coefficients should equal $r^2$:

$$b_{yx}b_{xy} = r^2$$
$$(.291)(.632) = .429^2$$
$$.183912 = .184041$$
$$.1839 = .1840$$

which checks, disregarding rounding errors.

Second, when plotted, the two regression lines should cross at a point equal to the mean of $X$ and the mean of $Y$. An inspection of Figure 9.3 shows that this is so. This is the same as saying that when the mean of $X$ is substituted in the equation for predicting $Y$, the predicted value of $Y$ will be the mean of $Y$.

By substituting for $X$, the mean of $X$, 59.9, we obtain

$$Y' = 79.4 + (.291)(59.9)$$
$$= 79.4 + 17.4$$
$$= 96.8$$

which is the mean of $Y$. When the mean of $Y$ is substituted in the other regression equation, the predicted value of $X$ will be the mean of $X$.

ALTERNATE FORMULAS FOR THE $b$ COEFFICIENT AND THE PREDICTION EQUATION. From equation 9.5, $b$ is defined

$$b_{yx} = \frac{\Sigma xy}{\Sigma x^2}$$

Multiplying both numerator and denominator of the right-hand term by $\sqrt{\Sigma x^2 \Sigma y^2}$ gives

$$b_{yx} = \frac{\Sigma xy}{\Sigma x^2} \left( \frac{\sqrt{\Sigma x^2 \Sigma y^2}}{\sqrt{\Sigma x^2 \Sigma y^2}} \right)$$

Transposing terms:

$$b_{yx} = \frac{\Sigma xy}{\sqrt{\Sigma x^2 \Sigma y^2}} \left( \frac{\sqrt{\Sigma x^2 \Sigma y^2}}{\Sigma x^2} \right)$$

Using formula 7.4:

$$b_{yx} = r_{xy} \frac{\sqrt{\Sigma x^2 \Sigma y^2}}{\Sigma x^2}$$

$$= r_{xy} \frac{1}{\Sigma x^2} \sqrt{\Sigma x^2 \Sigma y^2}$$

$$= r_{xy} \sqrt{\frac{\Sigma x^2 \Sigma y^2}{(\Sigma x^2)^2}}$$

Dividing both terms under the radical by $N$:

$$b_{yx} = r_{xy} \sqrt{\frac{\Sigma x^2 \Sigma y^2 / N}{(\Sigma x^2)^2 / N}}$$

$$= r_{xy} \frac{s_x s_y}{s_x^2}$$

Then

$$b_{yx} = r_{xy} \frac{s_y}{s_x} \tag{9.8}$$

Beginning with the regression line, $Y' = a_{yx} + b_{yx}$, and substituting for $a_{yx}$ equation 9.4 and for $b_{yx}$ equation 9.8, it follows that

$$Y' = Y - b_{yx} X + r_{xy} \frac{s_y}{s_x} X \tag{9.9}$$

$$= Y - r_{xy} \frac{s_y}{s_x} X + r_{xy} \frac{s_y}{s_x} X$$

$$= Y + r_{xy} \frac{s_y}{s_x} X - r_{xy} \frac{s_y}{s_x} X$$

$$Y' = Y + r_{xy} \frac{s_y}{s_x} (X - X)$$

The preceding equation leads to another equation based upon standard scores:

$$Y' = Y + r_{xy} \frac{s_y}{s_x} (X - X)$$

Transposing:

$$Y' - Y = r_{xy} \frac{s_y}{s_x} (X - X)$$

Dividing by $s_y$:

$$\frac{Y' - Y}{s_y} = r_{xy} \frac{X - X}{s_x}$$

Since

$$\frac{Y' - \overline{Y}}{s_y} = z_{y'} \quad \text{and} \quad \frac{X - \overline{X}}{s_x} = z_x \tag{9.9a}$$

$$z_{y'} = r_{xy} z_x$$

Then in terms of standard scores, the predicted $Y$ score is equal to the correlation coefficient times the $z$ score of the predictor.

### The Standard Error of Estimate

It has already been state that rarely will our obtained $Y$ scores be identical with the predicted $Y$ scores. There is an error in all of our predictions, and the extent of this error is measured by a statistic known as the standard error of estimate. When the size of the correlation coefficient between two variables is high, the size of the standard error of estimate is small; conversely, when the relationship between two variables is low, the size of the standard error of estimate is large. If we have a perfect relationship between $X$ and $Y$, our obtained $Y$'s will be exactly the same as the predicted ones. This is saying that all of the $Y'$ values will fall on the regression line. Since there is no variation from this line for the various values of $X$, there is no error in making estimates. When we have the opposite situation, the situation where there is no relationship between the two variables, we have a regression line parallel to the $x$ axis. The solution of the equation for a straight line when $b$ is equal to 0 results in the equation $Y' = a$. When $b$ is equal to 0, $a = \overline{Y}$. Hence when there is no relationship between two variables, the predicted value of $Y$ is equal to the mean of $Y$, that is $Y' = \overline{Y}$. When this situation prevails, the size of the standard error of estimate is at a maximum. In this case it is the equivalent of the standard deviation of the $Y$ variable. The size of the standard error of estimate ranges, then, from zero to the size of the standard deviation of the dependent variable $Y$. Most of the situations that are encountered fall between these two extremes.

Since each obtained score differs from the expected or predicted score, the discrepancy between each set of these can be obtained, squared, and then summed. If we divide this by $N - 2$ and then take the square root, the result is the standard error of estimate.

$$s_{yx} = \sqrt{\frac{\Sigma(Y - Y')^2}{N - 2}} \tag{9.10}$$

A more useful equation is

$$s_{yx} = \sqrt{\frac{\left[\Sigma Y^2 - \frac{(\Sigma Y)^2}{N}\right] - \left[\Sigma XY - \frac{(\Sigma X)(\Sigma Y)}{N}\right]^2 \bigg/ \left[\Sigma X^2 - \frac{(\Sigma X)^2}{N}\right]}{N - 2}}$$

which is the equivalent of

$$s_{yx} = \sqrt{\frac{\Sigma y^2 - [(\Sigma xy)^2/\Sigma x^2]}{N - 2}} \tag{9.11}$$

Since we have obtained all the values needed for equation 9.11 in computing the two $b$ coefficients, all that we have to do now is copy them into equation 9.11.

We shall now obtain the standard error of estimate for predicting $Y$ from $X$ for the data in Table 7.4. On substituting in formula 9.11 we obtain

$$s_{yx} = \sqrt{\frac{3475 - 2196.3^2/7553.5}{35 - 2}}$$

$$= \sqrt{\frac{3475 - 638.6}{33}}$$

$$= \sqrt{\frac{2836.4}{33}}$$

$$= \sqrt{85.9515}$$

$$= 9.27, \quad \text{or} \quad 9.3$$

Similarly, the standard error of estimate associated with predicting $X$ from $Y$ may be obtained.

An alternate formula for the standard error of estimate is often found in measurement books:

$$s_{yx} = s_y \sqrt{1 - r_{xy}^2} \tag{9.12}$$

By solving this for these data, we have

$$s_{yx} = 10.0\sqrt{1 - .429^2}$$

$$= 10.0 \sqrt{.815959}$$

$$= 10.0(.903)$$

$$= 9.03$$

Formula 9.12 is appropriately used when the data are based upon large samples. However, as the samples become smaller, less than 50, biases associated with these small samples enter into the work. This is taken into account in formulas 9.10 and 9.11. Formula 9.12 can be corrected for this bias by multiplying the obtained standard error of estimate by $\sqrt{N/(N-2)}$, which in this problem is

$$\sqrt{N/(N-2)} = \sqrt{35/33}$$

$$= \sqrt{1.0606}$$

$$= 1.029$$

Then

$$9.03(1.029) = 9.29 = 9.3$$

which is the result obtained by the use of formula 9.11.

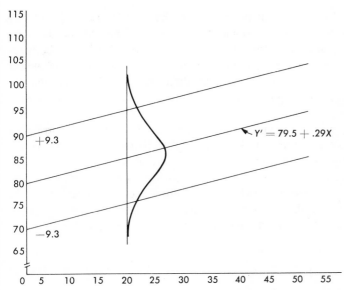

**Figure 9.4**    Regression line of $Y$ on $X$. Each of the outer parallel lines is one standard error of estimate from the regression line.

INTERPRETATION OF THE STANDARD ERROR OF ESTIMATE. Let us now take the value of 9.3 for the interpretation of this statistic. In predicting $Y$ from $X$ we can say that for any given value of $X$, the chances are two out of three that the observed cases will fall within a band which is the predicted value of $Y$ plus and minus 1 standard error of estimate. When $X$ is equal to 20, we would predict $Y$ at 85.2($Y'$), and about 68 times out of 100 in such predictions the actual values of $Y$ would be between 85.2 ± 9.3 or between 75.9 and 94.5.

This is shown in Figure 9.4. Since we assume that the standard error of estimate tends to be constant throughout the range, we could measure off 9.3 units on each side of the regression line for any other value of $X$ and then draw two lines parallel to the regression line. Two-thirds of all of the observed scores should fall between these two outer parallel lines. We could set off two lines within which 95 percent of the observed cases should fall by measuring 2 standard errors of estimate on each side of the regression line. In this kind of interpretation, we are treating the standard error of estimate as a standard deviation. To do this, we must assume that the distributions around the regression line are not only equal in variance (homoscedastic) but also normal.

## The Regression Effect
Many years ago Sir Francis Galton noted in studying the relationships between fathers and sons and between mothers and daughters that the sons of

tall fathers tended to be tall but not as tall as the fathers, and that the sons of short fathers tended to be short but not as short as the fathers; the same applied to mothers and daughters. In other words the offspring of both tall and short parents were regressing toward the mean. If the correlation between the heights of fathers and sons were perfect, there would be no regression; the points representing the height of fathers and sons would all fall along a straight line.

However, since we never have perfect correlation and since the correlation between two variables often is rather low, as in predicting academic grades from scores representing mental ability, the regression effect is very frequently encountered. Those individuals who are well above average or well below average on one variable will be less superior or less inferior on the second variable; that is, their scores on the second variable are not as extreme because they regressed toward the mean of the second variable.

## MULTIPLE PREDICTION

Up to this point we have been concerned with the prediction of a criterion $Y$ from a single predictor, $X$. A typical example of this arrangement is the prediction of freshman grades, $Y$, from a test of mental ability, $X$. Making predictions with a single predictor is not the most efficient scheme that can be arranged. In this chapter we were predicting with an $r$ of .43, which is a fairly typical validity coefficient. Notice how wide the band was when we used the standard error of estimate in making a prediction. We said that the chances were two out of three that given any $X$ score, an individual's $Y$ score would fall in a band centering on the regression line plus or minus 9.3, the error of estimate. This leaves much to be desired. However, the efficiency or accuracy of prediction may be improved by using more than one predictor to predict a single criterion. For example, in the prediction of freshman grade point index, an admission officer uses a regression equation which contains as predictors high school rank, score on the verbal part of the Scholastic Aptitude Test, and score on the mathematics part of the same test. He might add several other predictors. As was noted in Chapter 8, the extent of the relationship here could be calculated by a multiple $R$ which in this case would be written $R_{1.234}$. Here 1 is the criterion, freshman grades, and 2, 3, and 4 are the three above-named predictors. One reads this as the correlation between grades and the combined effects of variables 2, 3, and 4. Prediction efficiency usually increases up to the addition of the fourth or fifth predictor, but after that the slight gains in predictive ability are not worth the amount of time required to include them. The setting up of a regression equation with more than one predictor is beyond the scope of this book. It might be added that in the past the establishment of a multiple regression equation was a tedious chore, but that today existing computer programs make the task very simple.

A multiple $R$ has a standard error of estimate associated with it that is very similar to that associated with a single predictor:

$$s_{R_{1.23}} = s_1 \sqrt{1 - R^2} \tag{9.13}$$

## EXERCISES

**1.** Solve for $Y'$ in each of the following:

   **a.** $X = 5$    $a = -3$    $b = 2$
   **b.** $X = -10$    $a = 4$    $b = -2$
   **c.** $X = -3$    $a = 2$    $b = 3$
   **d.** $X = 7$    $a = 50$    $b = .125$

**2.** By the use of appropriate formulas, find $Y'$ value corresponding to an $X$ of 650, using the following data.

| $X$<br>College Board<br>Verbal Scores | $Y$<br>Grade-Point<br>Average |
|:---:|:---:|
| 710 | 5.50 |
| 680 | 5.70 |
| 670 | 5.20 |
| 660 | 5.10 |
| 580 | 5.00 |
| 540 | 5.00 |
| 520 | 4.80 |
| 500 | 4.90 |
| 480 | 4.40 |
| 440 | 4.60 |

**3.** What is the standard error of estimate for problem 2?

**4.** Problem 6 of Chapter 7 has already been solved.

   **a.** Set up the regression equation for predicting sales index $Y$, from attitude scale scores, $X$, from these data.

   **b.** Calculate the standard error of estimate for predicting $Y$ from $X$ for these data. Use both formulas for this statistic and compare your results.

**5.** A small college is doing research in their entrance testing. A mathematics achievement test is given as part of the entrance battery. Scores on this test and mathematics grades are given below and on the following page for 20 freshmen.

| Mathematics Test | Mathematics Grade |
|:---:|:---:|
| 120 | 90 |
| 112 | 78 |
| 87 | 61 |
| 42 | 28 |
| 56 | 48 |
| 99 | 71 |
| 22 | 18 |

| Mathematics Test (cont.) | Mathematics Grade (cont.) |
|:---:|:---:|
| 50 | 55 |
| 73 | 81 |
| 11 | 18 |
| 63 | 50 |
| 132 | 96 |
| 85 | 81 |
| 93 | 78 |
| 47 | 45 |
| 77 | 63 |
| 61 | 42 |
| 88 | 73 |
| 47 | 21 |
| 75 | 52 |

a. Compute the $a$ and $b$ coefficients for predicting grades from test scores and vice versa.

b. Construct the regression lines on a piece of graph paper.

c. Compute the standard error of estimate for each of the lines and plot them on the graph.

d. If 70 is the minimum passing grade, at which test score should candidates be denied admission to the course?

e. Do you think the math test is doing a good job?

6. Given the following data:

| Predictor $X$ | Criterion $Y$ | |
|:---|:---|:---|
| $X = 600$ | $Y = 4.8$ | $r_{xy} = .58$ |
| $s_x = 100$ | $s_y = .4$ | $N = 200$ |

a. When $X = 700$, what is the predicted value of $Y$?

b. When $X = 350$, what is the predicted value of $Y$?

c. What is the standard error of estimate for these data?

7. Suppose that a student has a $z$ score of 1.25 on predictor $X$. Predict his $z$ score on criterion $Y$ when $r$ equals:

a. .00     c. 1

b. .50     d. −.30

8. When $r_{xy} = .00$, what is the best prediction on Y that you can make from any score on $X$?

# Probability
# and the Binomial
# Distribution
## Chapter 10

In previous chapters we talked about probability and used the term in interpreting various statistics. In this chapter we shall discuss the laws of probability, the binomial distribution and its relation to probability, and some applications of probability to actual situations. Probability enters into all of our experimental work. Suppose that we have carried out a piece of experimental research to test some hypotheses. We examine the data to see if our hypotheses are supported. Procedures like this involve the use of probability, for a probability statement is associated with our findings.

Probability enters into much of our daily living. The meteorologist in forecasting weather, the agriculturist in predicting the size of next year's corn crop, the physician making a prognosis about a patient's recovery—each of these uses probability when he makes his official or professional pronouncements. The physician tells the patient that if he takes these pills, he should feel better in several days. The physician does not say, "If you do as I say, the chances are 90 in 100 that you will recover," but his statements are usually so worded that probability is inferred.

Probability values range from 1 to zero. A value of 1 stands for absolute certainty, and zero indicates there is no chance at all that the event will occur. There are very few things in life about which we can be absolutely certain. One certainty is that each of us will someday die. On the other hand the probability of the reader swimming from Seattle to Honolulu or in the course of his lifetime taking a space ship to a remote star is, in each case, zero.

Most things in life, however, have probability values of their occurrence somewhere between 1 and zero. In many predictions of outcomes, such as those of the physician or the weather forecaster, a numerical value is rarely assigned to the probability statement. In our work we shall endeavor to be more specific. Let us begin with a familiar object, a one-cent coin. If we take a new coin which in no way has been mutilated and toss it, we can state that the probability of obtaining a head is one out of two. We write this as $p = \frac{1}{2}$, or .5. It is also true that the probability of obtaining a tail is $\frac{1}{2}$, or .5. In this situation, only one of these two outcomes can occur. The toss will result in a head or in a tail. (We are not interested in the situation where the coin lands in a crack or against an object and stands on edge.) Notice that the sum of the two probabilities is equal to 1. We designate the probability of an event occurring by use of the symbol $p$ and the probability of it not occurring by the symbol $q$. The sum of $p + q$ is always equal to 1. As an additional example, let us take a die and throw it into the air. If this is an accurate and well-made die, the probabality of any single side coming up is $\frac{1}{6}$, or .167. That is, when we toss a die, the chance of throwing a six-spot is one in six. There are five chances in six chances of some other number appearing on the upturned face. In this case, $p = \frac{1}{6}$ and $q = \frac{5}{6}$, and again $p + q = 1$. In test work, the probability of a student getting an answer correct when he knows nothing about the item is, for true-false items, $\frac{1}{2}$, or .5. If we have a five-response multiple-choice item, the probability of getting the correct answer by chance is $\frac{1}{5}$, or .2.

What then is this probability that we have been talking about? It seems that it is related to the laws of chance. When chance operates, we can make statements about the expected outcomes. These statements can be expressed in terms of probability values. Probability, therefore, is the theoretical or expected frequency when the laws of chance are operating.

## LAWS OF PROBABILITY

To illustrate the laws of probability, let us use an ordinary deck of playing cards comprised of 52 cards. From this deck, after it has been well shuffled, we shall draw cards one at a time. Suppose at first we are interested in the probability of drawing an ace, deuce, or trey. The probability of drawing an ace is $\frac{4}{52}$, or $\frac{1}{13}$, that of drawing a deuce is $\frac{4}{52}$, or $\frac{1}{13}$, and the probability of drawing a trey is also $\frac{1}{13}$. The occurrence of any of these three cards is referred to as *a mutually exclusive event,* since only one of them can occur in any single trial. Thus the probability of drawing an ace, deuce, or trey from this deck of cards is $\frac{1}{13} + \frac{1}{13} + \frac{1}{13}$, or $\frac{3}{13}$, the sum of the separate probabilities. This illustrates our first law of probability. *The probability of the occurrence of any two (or more) mutually exclusive events is equal to the sum of their separate probabilities.*

Now suppose that we again draw a card from our deck. The probability of drawing an ace is $\frac{1}{13}$. If we replace the card and make another drawing, the probability of drawing an ace is again $\frac{1}{13}$. This will continue to be so, for the probability of drawing an ace each time is independent of what has gone on before. Even if we had been very lucky and had drawn five aces in a row, the probability of the next card drawn being an ace is still $\frac{1}{13}$.

We might now ask another question. What is the probability of the first, second, and third cards each being an ace when three cards are drawn from the deck with replacement? In this case $p = (\frac{1}{13})(\frac{1}{13})(\frac{1}{13})$, or $\frac{1}{2197}$. Or we might ask, what is the probability of drawing an ace, replacing the card drawn, and then drawing a deuce? In this case the probability is $(\frac{1}{13})(\frac{1}{13})$, or $\frac{1}{169}$. This illustrates a second law of probability. *The probability of the simultaneous or successive occurrence of two or more independent events is equal to the product of their separate probabilities.*

Now suppose that in the last case of drawing two cards, after the first draw the card was not replaced, and the first card drawn was not a deuce. The probability of drawing an ace followed by a deuce now becomes $(\frac{4}{52})$ $(\frac{4}{51})$, or $\frac{4}{663}$. This situation of determining the probability of one event, $X$, after another event, $Y$, has occurred is referred to as *conditional probability*. In this case it would be the conditional probability of event $X$.

## THE BINOMIAL EXPANSION

If we toss two coins at once, four possibilities may occur. We may find that two heads have come up, a head and a tail, or that both coins show tails. In terms of symbols this may be expressed like this:

| First Coin | Second Coin |
|------------|-------------|
| H          | H           |
| H          | T           |
| T          | H           |
| T          | T           |

From this we see that there are actually four possible outcomes, but two of them are the same in effect, for a head and a tail may be produced in two different ways. Since there are four possible outcomes, we can write that the probability of getting two heads is $\frac{1}{4}$, the probability of getting a head and tail is $\frac{1}{2}$, and the probability of obtaining both tails is $\frac{1}{4}$. The sum of all these probabilities is again equal to 1.

The results obtained above are what we would obtain by squaring a binomial. It may be recalled that the square of the binomial $(x + y)$ is equal to $(x^2 + 2xy + y^2)$. In the tossing of two coins, $x$ is the head, or H, and $y$ the tail, or T. If we multiply $(H + T)$ by itself, we obtain $HH + 2HT + TT$.

Suppose that we have three coins. By using the same symbols as above, we have the following possibilities:

| Row | First Coin | Second Coin | Third Coin |
|-----|-----------|-------------|------------|
| 1 | H | H | H |
| 2 | H | H | T |
| 3 | H | T | H |
| 4 | T | H | H |
| 5 | H | T | T |
| 6 | T | H | T |
| 7 | T | T | H |
| 8 | T | T | T |

Notice that now there are eight possible ways in which these three coins may fall. Inspection shows that rows 2, 3, and 4 are the same as are rows 5, 6, and 7. Actually there are four possible outcomes when three coins are tossed. So the probability of obtaining all heads is $\frac{1}{8}$, of obtaining two heads and one tail is $\frac{3}{8}$, of obtaining one head and two tails is $\frac{3}{8}$, and of all the coins being tails is $\frac{1}{8}$.

When we have only two or three coins, it is fairly easy to see what the probability will be for the various outcomes. But as the number of coins increases, we have to find more efficient ways for obtaining our probability values. We use the binomial theorem, which in its general form looks like this:

$$(p + q)^n = p^n + \frac{n}{1} p^{(n-1)}q + \frac{n(n-1)}{(1)(2)} p^{(n-2)}q^2 + \frac{n(n-1)(n-2)}{(1)(2)(3)} p^{(n-3)}q^3$$

$$+ \frac{n(n-1)(n-2)(n-3)}{(1)(2)(3)(4)} p^{(n-4)}q^4 + \cdots + q^n \tag{10.1}$$

Suppose that we have eight coins, and that we toss them simultaneously. Let us see how we can use equation 10.1. In this problem the probability of getting a head on any one coin, $p$, is $\frac{1}{2}$ and $n$, the number of coins, is 8. Then by substitution

$$\left(\frac{1}{2} + \frac{1}{2}\right)^8 = \left(\frac{1}{2}\right)^8 + \frac{8}{1}\left(\frac{1}{2}\right)^7\left(\frac{1}{2}\right) + \frac{8(7)}{(1)(2)}\left(\frac{1}{2}\right)^6\left(\frac{1}{2}\right)^2 + \frac{8(7)(6)}{(1)(2)(3)}\left(\frac{1}{2}\right)^5\left(\frac{1}{2}\right)^3$$

$$+ \frac{8(7)(6)(5)}{(1)(2)(3)(4)}\left(\frac{1}{2}\right)^4\left(\frac{1}{2}\right)^4 + \frac{8(7)(6)(5)(4)}{(1)(2)(3)(4)(5)}\left(\frac{1}{2}\right)^3\left(\frac{1}{2}\right)^5$$

$$+ \frac{8(7)(6)(5)(4)(3)}{(1)(2)(3)(4)(5)(6)}\left(\frac{1}{2}\right)^2\left(\frac{1}{2}\right)^6$$

$$+ \frac{8(7)(6)(5)(4)(3)(2)}{(1)(2)(3)(4)(5)(6)(7)}\left(\frac{1}{2}\right)\left(\frac{1}{2}\right)^7 + \left(\frac{1}{2}\right)^8$$

Simplifying, this becomes

$$\left(\frac{1}{2} + \frac{1}{2}\right)^8 = \frac{1}{256} + \frac{8}{256} + \frac{28}{256} + \frac{56}{256} + \frac{70}{256} + \frac{56}{256} + \frac{28}{256} + \frac{8}{256} + \frac{1}{256}$$

Several things should be noted from this problem. First, the sum of all of the values in the numerators must equal the value of the denominator, for

**Table 10.1**    Probabilities of
Obtaining Various
Combination of Heads
and Tails When
8 Coins are Tossed

| H | T | $p$ | $p$ |
|---|---|---|---|
| 8 | 0 | $1/256$ | .004 |
| 7 | 1 | $8/256$ | .031 |
| 6 | 2 | $28/256$ | .109 |
| 5 | 3 | $56/256$ | .219 |
| 4 | 4 | $70/256$ | .273 |
| 3 | 5 | $56/256$ | .219 |
| 2 | 6 | $28/256$ | .109 |
| 1 | 7 | $8/256$ | .031 |
| 0 | 8 | $1/256$ | .004 |
|   |   | $256/256 = 1.000$ |  |

the sum of all of the separate probabilities must equal 1. Second, after the middle term was reached, each of the other terms repeats one of the earlier ones in descending order of size. Third, the denominators are all the same. In this case they were all equal to $(\frac{1}{2})^8$. Hence after the first was computed, there was no point in multiplying the rest of the $p$ and $q$ terms. And fourth, note that there is one more term than $n$; in this case we have 9 terms in our expansion. Suppose we summarize our results in column form, showing the probabilities for the various combinations of heads and tails, as in Table 10.1. From this we see that the probability of getting all heads is .004, about 4 chances in 1000. The probability of getting either 8 heads or 8 tails is .008, the sum of the two separate probabilities. We can also answer other questions, such as the probability of getting at least 6 heads. The probability of this is the sum of the probabilities for 6 heads, 7 heads, and 8 heads; that is .109 + .031 + .004, which equals .144, or 144 in 1000.

Instead of going through all the foregoing work in expanding the binomial, time may be saved by the use of the so-called Pascal's triangle, which is seen in Table 10.2. In the problem used with the binomial we had 8 coins; $n$ is equal to 8. Going to Table 10.2, we find that for $n = 8$, the same values are obtained for the numerators of our probabilities (Table 10.1) and the column at the right gives the denominator of the probability values, which in this case is 256. Notice how the triangle is constructed. When the 6 and 15 at the top of this small triangle are added, they produce the value 21, which is one of the terms in the row for $n = 7$. By combining all of the terms in row $n = 6$ in this fashion, all of the values in row $n = 7$ are obtained. In this manner, the triangle may be expanded indefinitely. Suppose to illustrate this we compute the values for the next row, that is, when $n = 10$. The first term would be 1; then the second term, $1 + 9$, or 10; the third term, $9 + 36$, or

**Table 10.2** Pascal's Triangle

| n | Binomial Coefficients | Denominator of p |
|---|---|---|
| 1 | 1  1 | 2 |
| 2 | 1  2  1 | 4 |
| 3 | 1  3  3  1 | 8 |
| 4 | 1  4  6  4  1 | 16 |
| 5 | 1  5  10  10  5  1 | 32 |
| 6 | 1  6  15  20  15  6  1 | 64 |
| 7 | 1  7  21  35  35  21  7  1 | 128 |
| 8 | 1  8  28  56  70  56  28  8  1 | 256 |
| 9 | 1  9  36  84  126  126  84  36  9  1 | 512 |

45; the fourth term, $36 + 84$, or 120; and the fifth term, $84 + 126$, or 210; the middle term, $126 + 126$, or 252; the rest of the terms would be the same as those already computed in reverse order.

It should be noted that the binomial distribution is symmetrical only when $p = .5$. So it follows that Pascal's triangle should be used only under this condition. Suppose that $p = \frac{1}{4}$, $n = 5$, and $q = \frac{3}{4}$. The expansion of the binomial gives

$$\left(\frac{1}{4} + \frac{3}{4}\right)^5 = \frac{1}{1024} + \frac{15}{1024} + \frac{90}{1024} + \frac{270}{1024} + \frac{405}{1024} + \frac{243}{1024}$$

which is a skewed distribution. Skewness of the binomial distribution may be computed by

$$g_1 = \frac{q - p}{\sqrt{npq}} \tag{10.2}$$

From this it is seen that skewness is zero when $p = q$, that is, when $p = .5$: therefore, the greater the difference between $p$ and $q$, the greater the skewness.

### The Binomial and Normal Distribution

Figure 10.1 is a histogram representing $(p + q)$ when $p = .5$ and $n = 6$. As we might expect, there are seven columns; the height of each is in proportion to its probability. Suppose now that instead of tossing 6 coins we tossed 100 coins. This time we would have 101 columns in our graph. Now let us also suppose that for both of these graphs we have the same area and the same base line. It follows that in the second example the columns will be much narrower than those in Figure 10.1. If we have an infinite number, the columns would be very narrow and this binomial distribution would then approximate the normal distribution. Actually, the binomial distribution is always discontinuous, but as the number of cases gets larger, this binomial curve gets closer and closer to the normal curve; but unlike the latter, it never becomes continuous.

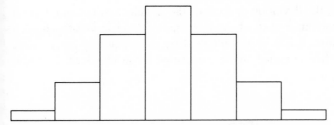

**Figure 10.1**   Curve of the binomial distribution when $n = 6$.

## Parameters of the Binomial Distribution

Before we can use this binomial distribution in any problems, we have to learn how to compute several parameters which describe the curve. Notice now that we are using the term *parameter,* because we are no longer dealing with samples but with population values. Table 10.3 shows the results of tossing 10 coins at once, or if we wish to deal in educational terms, the chance results of answering 10 true-false test items. In column 1 we have all the possible scores or outcomes. Since we had an $n$ of 10 to begin with, there are 11 values in this column. In column 2 we have the expected frequencies. These were obtained by the use of the binomial theorem or by an extension of Pascal's triangle. Column 3 is obtained by multiplying the values in each of the first two columns row by row. Column 3 is summed. If we wish to obtain the mean, we must divide this sum by the number of cases.

$$m = \frac{\Sigma f_e X}{\Sigma f_e}$$

$$= \frac{5120}{1024}$$

$$= 5$$

**Table 10.3**   Worksheet for the Computation of Binomial Parameters

| (1) X | (2) $f_e$ | (3) $f_e X$ | (4) x | (5) $x^2$ | (6) $f_e x^2$ |
|---|---|---|---|---|---|
| 10 | 1 | 10 | 5 | 25 | 25 |
| 9 | 10 | 90 | 4 | 16 | 160 |
| 8 | 45 | 360 | 3 | 9 | 405 |
| 7 | 120 | 840 | 2 | 4 | 480 |
| 6 | 210 | 1260 | 1 | 1 | 210 |
| 5 | 252 | 1260 | 0 | 0 | 0 |
| 4 | 210 | 840 | −1 | 1 | 210 |
| 3 | 120 | 360 | −2 | 4 | 480 |
| 2 | 45 | 90 | −3 | 9 | 405 |
| 1 | 10 | 10 | −4 | 16 | 160 |
| 0 | 1 | 0 | −5 | 25 | 25 |
| | $\Sigma f_e = 1024$ | $\Sigma f_e X = 5120$ | | | $\Sigma f_e x^2 = 2560$ |

In column 4 we have the deviation of each of the $X$ values in column 1 from this mean of 5. In column 5 each of these is squared, and in column 6 each of the values in column 5 has been multiplied by the frequency for its row. The last column is then summed. Next we compute the standard deviation in the following manner:

$$\sigma = \sqrt{\frac{\Sigma f_e x^2}{\Sigma f_e}}$$
$$= \sqrt{\frac{2560}{1024}}$$
$$= \sqrt{2.5}$$
$$= 1.58$$

It should be noticed that we are using different symbols for both mean and standard deviation at this time.

If we are dealing with an actual problem, it is not necessary to go through the work shown in Table 10.3. Both of these parameters can be obtained by the use of the following formulas:

$$m = np \qquad\qquad\qquad (10.3)$$

where  $m =$ the mean of the binomial distribution
$\quad\quad\ \ p =$ the probability of the event occurring
$\quad\quad\ \ n =$ the number of objects involved, or the exponent of the binomial

For these data we have

$m = np$
$\quad = 10(\frac{1}{2})$
$\quad = 5$

which was the value obtained previously. The formula for the standard deviation of the binomial distribution is as follows:

$$\sigma = \sqrt{npq} \qquad\qquad\qquad (10.4)$$

where all symbols are as previously defined.

For our problem, we find the standard deviation as follows:

$\sigma = \sqrt{10(\frac{1}{2})(\frac{1}{2})}$
$\quad = \sqrt{2.5}$
$\quad = 1.58$

which is again the same as the value previously computed. These formulas are used when the binomial distribution is taken as an approximation of the normal distribution. This should be done *only* when $np$ or $nq$ (whichever is the smaller) is equal to or greater than 5.

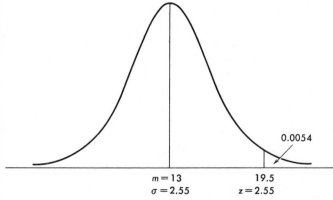

$$m = 13 \qquad\qquad 19.5$$
$$\sigma = 2.55 \qquad\qquad z = 2.55$$

**Figure 10.2**   Finding the probability of obtaining a score of 20 or more on a 26-item true-false test by chance alone.

## Use of the Formulas of the Binomial Distribution

Suppose that we construct a true–false test consisting of 26 items. We ask the question, "What is the probability of getting a score of 20 or higher by chance alone?" (that is, by guessing). We must assume that the examinee knows nothing about the material on which he is being tested. First we compute the mean and standard deviation for this distribution. Since $np$ is larger than 5, the distribution of the binomial may be taken as an approximation of the normal curve and we can use the formulas for the binomial distribution.

$$
\begin{aligned}
m &= np & \sigma &= \sqrt{npq} \\
&= 26(\tfrac{1}{2}) & &= \sqrt{26(\tfrac{1}{2})(\tfrac{1}{2})} \\
&= 13 & &= \sqrt{6.50} \\
& & &= 2.55
\end{aligned}
$$

In Figure 10.2 we have drawn a curve with this mean of 13. Note that we have placed this score of 20 off to the right. The question is exactly how much of the area of the normal curve is above this score of 20. We have worked problems like this before. The first thing that we do is to change the raw score of 20 to a standard or $z$ score as follows:

$$
\begin{aligned}
z &= \frac{X - m}{\sigma} \\
&= \frac{20 - 13}{2.55}
\end{aligned}
$$

This standard score formula is based upon a normal curve where the distribution is continuous. Since we are dealing with a discrete or noncontinuous distribution, we make a correction for the lack of continuity at this point. This is done by using, instead of the score of 20 in the formula, the lower

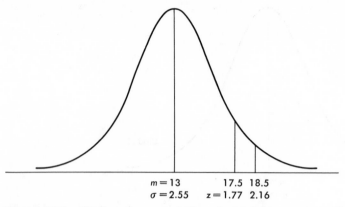

**Figure 10.3** Finding the probability of obtaining a score of 18 on a 26-item true-false test by chance alone.

limit of 20, which is 19.5. When we make this correction our work is much more accurate.

Then we have

$$z = \frac{19.5 - 13}{2.55}$$

$$= \frac{6.5}{2.55}$$

$$= 2.55$$

In the normal probability table (Appendix B), we find that .0054 of the area of the curve is above this standard score of 2.55. Then the chances are 54 in 10,000 that an individual can get 20 or more of these items correct by chance alone.

Next let us ask another question. This time we want to know what is the probability of obtaining a score of exactly 18 on this same true–false test by chance alone. This time refer to Figure 10.3. The lower limit of 18 is 17.5 and the upper limit is 18.5. We want to find the area of the curve which falls between these two points. We first change both of these values to $z$ scores and then turn to the normal probability tables to find the area cut off between the mean and each of these points. We then subtract the smaller of these areas from the larger. The remainder will be the area of the curve cut off by the two points.

$$z_1 = \frac{17.5 - 13}{2.55} \qquad z_2 = \frac{18.5 - 13}{2.55}$$

$$= \frac{4.5}{2.55} \qquad\qquad = \frac{5.5}{2.55}$$

$$= 1.76 \qquad\qquad\;\; = 2.16$$

In the normal probability table, we find the following areas between the mean and each of these points:

$z_1 = 1.76$    .4608

$z_2 = 2.16$    .4846

The difference between these two areas is .024. Then we can say that the chances are 24 in 1000 that an individual will obtain a score of exactly 18 on this 26-item true–false test.

In psychological research, the binomial distribution is used to determine the ability of individuals to distinguish between such products as different types of cola drinks, orange juices, cigarettes, margarine and butter. This is the way it is done. Suppose that we have four cola drinks and start with the hypothesis that an individual is unable to separate cola A from the three others. Suppose that over a period of days we give an individual 60 trials, and we observe that he identifies cola A correctly 20 times. The question is whether this differs significantly from chance. We solve this problem exactly as we did the previous ones. First we compute the mean and standard deviations:

$$m = np \qquad \sigma = \sqrt{60(\tfrac{1}{4})(\tfrac{3}{4})}$$
$$= 60(\tfrac{1}{4}) \qquad = \sqrt{180/16}$$
$$= 15 \qquad = \sqrt{11.25}$$
$$= 3.35$$

By changing this observed frequency to a standard score and correcting for continuity we have

$$z = \frac{19.5 - 15}{3.35}$$

$$= \frac{4.5}{3.35}$$

$$= 1.34$$

The normal probability table indicates that with a standard score of 1.34, we have .0901 of the area of the normal curve which falls to the right of an ordinate erected at this point. The chances then are 9 in 100 that this individual could identify cola A when samples of it are tasted along with three other colas. We shall discuss the real significance of this probability value in a later chapter when we discuss the general problem of testing hypotheses. It might be noted here, however, that it is the usual practice to require a probability value of 5 in 100 or less before we are willing to accept the hypothesis that an individual is doing better than we would expect by chance.

## PROBABILITY AND THE NORMAL CURVE

In Chapter 6 we observed how the normal probability table, Appendix B, is associated with the normal curve. It will be recalled that the unit normal curve has a mean of zero and a standard deviation of 1, and that the area is taken as unity. Using the normal curve with data that are continuous, questions such as the following can be answered.

Suppose that we have a test with $\bar{X} = 500$ and $s = 100$. What is the probability of selecting an individual randomly with a score of 750 or higher? First we obtain the z score.

$$z = \frac{X - \bar{X}}{s}$$
$$= \frac{750 - 500}{100}$$
$$= 2.5$$

From column 4 of Appendix B we find that the area of the normal curve above a z score of 2.50 (the smaller area of the table) is .0062. Hence the probability of randomly obtaining a person with a score of 750 or above is approximately 6 in 1000.

Using the same data we might ask what is the probability of randomly drawing two individuals with scores above 650 from the same population. This time the z is $(650 - 500)/100 = 1.5$. The corresponding value for this z score is .0668. Then applying the multiplication rule, the probability of obtaining two individuals in succession with scores above 650 is equal to .0668 × .0668, which is slightly more than 4 in 1000.

In the first example we were concerned with the probability of obtaining an individual with a score of 750 or better. Since we are dealing with only one end of the normal curve, we have in this case a one-tailed probability value. Often the question is what is the probability of obtaining a score as deviant as 750. Such a deviation of 250 points from the mean occurs on both sides of the mean and hence the probability value of .0062 has to be doubled. The probability then of obtaining a deviation of this magnitude in a random sample is .0124 or about 12 in 1000. Such a value is referred to as a two-tailed probability value. This concept will be applied later when we discuss one-tailed and two-tailed tests of significance.

## EXERCISES

**1.** When a single card is drawn from a well-shuffled deck of cards, what is the probability that it will be

**a.** An ace?
**b.** A red ace?
**c.** A diamond?
**d.** The ace of diamonds?
**e.** An ace or a diamond?

**2.** Eight dimes are well shaken and then tossed.

**a.** What is the probability of obtaining 8 heads?
**b.** What is the probability of obtaining 6 or more heads?
**c.** What is the probability of obtaining exactly 6 heads?
**d.** What is the probability that all 8 coins will show the same face?

**3.** What is the probability of the occurrence of each of the following?

**a.** Obtaining a tail on the seventh toss of a coin when each of the previous 6 tosses resulted in a tail?
**b.** Drawing the ace of diamonds from a deck of cards followed by the two of diamonds with no replacement? The first card drawn was not the two of diamonds.
**c.** Of buying a single ticket and winning the grand prize in a lottery in which 100,000 tickets were sold?
**d.** Of not drawing a club from a deck of cards? Of not drawing the ace of clubs?

**4.** Two fair dice are tossed.

**a.** List all possible outcomes.
**b.** What is the probability of getting a sum of 7?
**c.** What is the probability of getting two of a kind?
**d.** What is the probability of obtaining a 7 or a pair of 1's?
**e.** What is the probability of getting an even number?

**5.** A slot machine used in gambling has in it 3 wheels that rotate. On each of these are 6 fruits—an apple, lemon, orange, pear, plum, and strawberry—and a bell, bar, moon, and star. On the insertion of a coin, the wheels spin and each stops with one of the figures in the visual slot.

**a.** What is the probability that each will show a star?
**b.** Two stars and an apple?
**c.** A row of fruits?
**d.** A row of objects rather than fruits?
**e.** A row of lemons or a row of oranges?
**f.** A row containing no bars?

**6.** A box contains 7 red and 3 green balls.

**a.** What is the probability of drawing from this box a red ball, followed by a green ball, assuming that you replace the first ball before drawing the second?
**b.** What is the probability of drawing 2 green balls in succession, again replacing the first ball before drawing the second?

**c.** What is the probability of drawing a green ball, then a green ball, then a red ball, with no replacement? The first 2 balls drawn were red.

**7.** Six dice are tossed. What is the probability of obtaining 4 or more four-spots?

**8.** Smokers claim that they can identify the brand of cigarette which they customarily smoke. To test this an experiment is planned in which an individual is given 4 cigarettes to smoke, one of which is his favorite brand. All cigarettes are of the ordinary type. Of 110 subjects tested, 32 identify their own brand correctly. Is this a significantly larger number than would be expected by chance?

**9.** A 40-item multiple-choice test is made up of 5-response items.

**a.** If a person knows nothing about the material tested, what score could we expect by chance?
**b.** What is the probability of getting a score of 12 or higher by chance alone?
**c.** What is the probability of getting a score of 12 by chance alone?

**10.** In an experiment, 96 subjects were given a sample of each of three different cola drinks. Coca-Cola was correctly identified 39 times; Pepsi-Cola, 36 times. Is either of these identifications significantly different from what would be expected by chance?

**11.** Given a test with a mean of 84 and standard deviation of 12,

**a.** What is the probability of an individual obtaining a score of 100 or above on this test?
**b.** Of 62 and below?
**c.** What is the probability of getting a score as deviant as this score of 100?

# Sampling
## Chapter 11

One very important use of statistics is to make inferences about larger groups on the basis of information obtained from smaller groups, or, in other words, to make statements or generalizations about the population on the basis of information obtained from the study of one or more samples. The extent to which we can do this with any accuracy depends on the adequacy of our sample or samples.

## KINDS OF SAMPLE

Samples can be broken down into two basic types: nonprobability and probability. In the nonprobability type, there is no way of estimating the probability that each individual or element will be included in the sample. In probability sampling, in the most frequently encountered situations, each individual has an equal chance of becoming a part of the sample.

### Nonprobability Sampling
In much of our research, samples of this type are common. For example, the American college sophomore who happens to be taking a course in elementary psychology frequently becomes part of the sample in educational or psychological research. He is used because he is convenient. The students in a class may constitute the entire sample because they happen to be in a class whose instructor is interested in doing some research. Such samples as these are called *accidental* or *incidental* samples. Another type of nonprobability sampling is *quota* sampling. In this type of sampling, the proportions of the various subgroups in the population are determined and the sample is drawn (usually not at random) to have the same percentages in it. For example, if a population has an equal number of males and females in it, so does the sample.

Or if the population of a certain state has a certain percentage of farmers, another percentage of small town dwellers, and the rest urbanites, then a sample drawn to study the inhabitants of this state would be made up of the same percentages of the three groups. This type of sampling is very similar to the stratified random sample described below, except for the randomizing (Selltiz, et al., 1959). A third type of nonprobability sampling is known as *purposive* sampling. For example, the counties in the United States that have voted for the winner in a series of past presidential elections could be identified. We could study these counties, and from the voters' preferences in them make a prediction on the outcome of a national election.

The major advantage in the use of samples like these is that they are convenient and economical. Many students participate in experiments as a part of their work in a course. Hence no expense to the research worker is incurred. As we shall see, the drawing of probability samples is likely to be both expensive and laborious.

### Probability Samples

The basic type of probability sample is the *simple random* sample. In a simple random sample, each and every individual in the population has an equal chance of being drawn into the sample. An illustration of this type of sampling took place when the National Selective Service Act was set up. Under the regulations of this act, as each male within the prescribed age range registered with his local draft board, he was given a number which ranged from one to several thousand. Then a series of capsules was set up, each containing one of these numbers. These were then placed in a large goldfish bowl and the Secretary of War, blindfolded, reached in and drew out a capsule. The number that was drawn first placed all individuals, in the various draft boards holding this number, first in order for induction into the Armed Services. Then a second number was drawn and the process continued until all those registered received an induction number. The system of placing each number in a similar capsule, putting them all in the same bowl, mixing them, and then blindfolding the drawer made this type of sampling truly random. Only when we deal with probability samples can we know the frequency distribution of the sample statistics generated by the sampling procedure repeatedly applied to the same population. It is this knowledge that allows us to infer from a sample to its population. Randomization is essential to probability sampling and therefore to statistical inference itself.

When sampling procedures are not carried out like this, the resulting sample is said to be *biased*. Suppose that from a telephone book we select every fiftieth name as the individual to whom we are going to send a questionnaire in a study of attitudes of the citizens of a given city. This sample is biased, because it excludes all those who have no telephone or who have an unlisted number. The sample drawn may approach a random sample of those whose names are listed in the directory, and any inferences would be valid for these individuals but not necessarily for the city as a whole. The famous *Literary Digest* Poll of 1936 which predicted the election of Alfred Landon over Franklin Roosevelt sampled those individuals who had registered automobiles or listed telephones. Apparently during the Great Depression many more Republicans were on these lists than were Democrats. Another example of this on the national level was the Dewey–Truman election of 1948. Months before the election, the polls showed that Dewey was going to win by a landslide. A postmortem of this fiasco showed that there was poor sampling in the lower socioeconomic groups, and also that there was a large number of individuals interviewed who stated that they were undecided. In past polls these undecided cases were equally divided among the candidates. Apparently in 1948 a large number of these undecided individuals later voted for Truman.

The 1970 draft lottery of the Office of Selective Service was another example of sampling that went wrong (Fienberg, 1971; Rosenblatt and Filliben, 1971). In the Selective Service lottery numbers are associated with birth dates. This was done by filling each of 366 cylindrical capsules with a date. These cylinders were put into a box, mixed, and then drawn from the box one at a time. In this drawing the first capsule to be taken out contained the date September 14. All individuals born on that day were number 1 in order of induction into the Armed Forces.

A study of the results later showed that when the lottery numbers were averaged for each month, the average of these lottery numbers varied considerably from month to month. The overall average was 183.5. The monthly means from January to December were 201, 203, 226, 204, 208, 196, 182, 174, 157, 182, 149, and 122. Obviously individuals born in the last half of the year received lower draft numbers than those born in the first half. Individuals born in November and December became eligible for induction out of line with expectancies and hence were subjected to an unfavorably biased treatment, contrary to the intent of the draft lottery. Apparently what happened was that the January capsules were filled first, then the February capsules, and the sequence was followed until the capsules were prepared for all months. The January cylinders, when completed, were dumped into a wood box. Then the February capsules were poured in and mixed with the January ones. The cylinders for each ensuing month were handled in the same way. Apparently the stirring was not done too well and the capsules for the months later in the year remained on top and tended to make up a disproportionate part of the first draws.

To counteract the extensive criticism of the 1970 lottery, the 1971 lottery was set up using two different drums, each containing 366 capsules. When a capsule containing a number was drawn from one drum, another capsule containing a date was drawn from the other. The number set the sequence number of induction for the date that was drawn. The extensive work carried on in setting up random calendars and random permutations of the numbers 1 to 366 done by the U.S. National Bureau of Standards resulted in less biased results in the 1971 lottery. Moreover, great care was taken to see that the capsules in each drum were very well mixed.

A second type of probability sample is the *stratified random* sample. This is very similar to the quota sample except that after the percentages that are to be in each group are determined, individuals are drawn from each group by random sampling. Suppose that we want a picture of the attitudes of the student body of a large university on a certain campus issue. First, we would determine the number of students in each of the undergraduate classes as well as the number of graduate students. Second, we would obtain the percentage of males and females on each group. If the issue had aspects in which belonging to a fraternity or to a sorority was important, then we should know the proportion of sorority and fraternity members in each class. In another sampling, the school to which the student belonged might be important. Then we would want to know the numbers by sex and class enrolled in the various schools. Our sample should be so drawn that it is made up of the same proportions of the different classes as is the entire student body. Also the proportion of males and females and fraternity and nonfraternity members in our sample should be the same as in the entire student body. The same would be true in drawing a sample when school of enrollment was to be considered.

Next we decide how large the sample is to be. For example, on the basis of our proportions in the university population, we decide the number of freshmen, male fraternity members that are needed in our sample. Then this sample is drawn at random from all freshmen, male fraternity members. In a similar fashion our sample is drawn from the various other groups. If we wanted to study the attitudes of the citizens of any state, we could get census data from which we could determine the percentage of the population of the state that lived on farms, in small towns, and in cities. After this, we would attempt to apply random sampling within each of the three subgroups. In stratified random sampling it is *not* necessary for the size of each stratum sample to be proportional to the size of the population stratum. It is possible to weight each subsample statistic in proportion to the size of its parent subpopulation. Population estimates are then made from the subsamples combined in proper proportion.

The thoughtful student may conclude that this is a very time consuming and laborious procedure if the population is very large. In studying the attitudes of a school or university group, it is practical to draw a stratified random sample. But to attempt this on a national basis would be very difficult and expensive. A type of sample that is used on such national surveys is referred to as a *cluster* sample. In stratified random sampling, the population is categorized into groups that are distincly different from each other on relevant variables. In *cluster* sampling, the population is viewed as a collection of groups that are much the same. That is, strata are internally homogeneous, and clusters are internally heterogeneous. In stratified sampling we select at random from *within* each stratum. In cluster sampling, it is the clusters themselves that are selected at random. In practice, cluster sampling is usually conducted in conjunction with stratified sampling, or done in *stages,* or both. For example, in gathering normative data for the use of a new test on a state-wide basis, it would be possible to stratify the school districts of the state on the basis of size and then select districts within each stratum at random. In each of the selected school systems, individual schools could be chosen at random. Classes could be similarly selected within each school, and sometimes within each class individual students might be identified at random for testing. This type of sampling provides a wide coverage of districts and schools, yet is still economical and efficient in terms of the number of students tested.

We described earlier how a random sample was drawn by placing numbers in a goldfish bowl. A more usual procedure is to use a table of random numbers, such as in Appendix N. For example, suppose we have 896 answer sheets, and we wish to study a sample of these. Assume that we want to draw a random sample of 100 from these 896 papers. First, we number each answer sheet, beginning with 001 and continuing to 896. Then we select any point in the table of random numbers. This can be done by moving a pencil over the table without looking and letting the pencil down at any place. Suppose our pencil lands at the junction of column 7 and row 5 of Appendix N. This becomes our starting place. In Appendix N we find a zero at this point. We can move from this point in any direction; for example, we can proceed by taking numbers in threes downward, sideways, or obliquely. When we get to the bottom of a column, we change our direction and proceed as before. Suppose we decide to start down column 7. The digits in order are 0, 6, 4. So the paper with this number, 64, becomes the first one drawn for our sample. The next digits are 0, 0, 8; resulting in paper 8 being the second one drawn for our sample. The third paper would be 481. And so we continue until our sample of 100 is drawn. If we draw the same number twice or draw a number higher than our last paper, such as 909, we disregard the draw and continue the process.

There may be occasions when we study an entire group. You may recall that we have previously described a *population,* or *universe,* as an arbitrarily defined group. Suppose we define our population as "all the seniors majoring in home economics in a university." Since this group is likely to be small, we may use all the individuals in the group in the study. There is no sampling problem here.

## SAMPLING DISTRIBUTION OF THE MEAN

Suppose that we consider a large population, such as all 10-year-olds in the United States. With a population of this size and of this nature we can select sample after sample and study them. Suppose also that we take a group of 30 of these 10-year-olds in a certain school and give them an intelligence test for which we compute the mean and standard deviation. Then we draw another sample of the same size and compute its mean and standard deviation. We continue this until 5000 samples of size 30 are drawn, and we have 5000 means and 5000 standard deviations. A distribution of these 5000 means is referred to as a sampling *distribution*. The mean of these 5000 means is a good estimate of the parameter mean. The variability of this sampling distribution of means is measured by *the standard error of the mean*.

A distribution of the means of samples of equal size, as described above, when taken in large numbers from an infinite population will form a normal distribution. This is an example of the *central limit theorem*. If the random samples are large enough, and if there is a large number of samples, the mean of the sample means will equal the mean of the population, and the standard deviation of these sample means about the parameter mean will be equal to the standard deviation of the population divided by the square root of $N$:

$$\sigma_m = \frac{\sigma}{\sqrt{N}} \tag{11.1}$$

Thus for this application of the *central limit theorem*:

*If a population distribution (it need not be normal) has a mean m and a standard deviation $\sigma$, then the distribution of random sample means drawn from this population approaches a normal distribution with a mean of m and a standard deviation of $\sigma/\sqrt{N}$ as the sample size N increases.*

When the sample size $N$ is less than about 30 the sampling distribution of means is not accurately represented by the normal curve table (Appendix B). Instead, Appendix C (Distribution of $t$ Probability) is used when working with small samples.

Suppose that we have a population whose parameters are known. If samples are drawn from this population, probability statements may be made about the means of these samples. For example, suppose that the mean intelligence test score of high school boys is known to be 100 with a standard deviation of 16. A sample of 36 is drawn from this population. The question is then asked as to what is the probability of the mean of this sample being above or below a certain point, let us say below 96.

$$\sigma_m = \frac{\sigma}{\sqrt{N}} \qquad z = \frac{X - m}{\sigma_X}$$

$$= \frac{16}{\sqrt{36}} \qquad = \frac{96 - 100}{2.67}$$

$$= \frac{16}{6} \qquad = \frac{-4}{2.67}$$

$$= 2.67 \qquad = -1.50$$

From Appendix B we find that the smaller area for a $z$ score of 1.50 is .0668. Thus the probability of drawing a sample of size 36 with a mean of 96 or less from this population is .0668. Usually this standard deviation of the mean, which is commonly referred to as the standard error of the mean, cannot be obtained from the foregoing equation since the standard deviation of the population is usually not known. The best that we can do is use the sample standard deviation as an estimate of the parameter value.

### Correcting the Sample Standard Deviation for Bias

The sample standard deviation is considered to be a biased estimate of the parameter standard deviation. If we select a small sample, say with $N$ equal to 15, from a very large population, the chances are good that the members of our sample will tend to come from the center of the distribution, and hence the range of the sample will be less than the range of the population. It follows that the standard deviation of the sample will also be smaller than the standard deviation of the population. The bias is that sample standard deviations tend to be smaller than the parameter standard deviation. As the size of the sample increases, the chances increase of getting scores from the extremes of the distribution. The sample standard deviation becomes closer and closer to the parameter value. The sample standard deviation, as an estimate of the population value, with a correction for this bias is given by the formula

$$s = \sqrt{\frac{\Sigma x^2}{N - 1}} \qquad\qquad (11.2)$$

If we substitute the above in equation 11.1 we have

$$s_{\bar{X}} = \sqrt{\frac{\Sigma x^2}{N(N - 1)}} \qquad\qquad (11.3)$$

which gives the standard error of the mean directly from the sum of the squares. A more widely used formula for the standard error of the mean is

$$s_{\bar{X}} = \frac{s}{\sqrt{N - 1}} \qquad\qquad (11.4)$$

in which the $N - 1$ in the denominator corrects for the bias in the standard deviation in the numerator.

Suppose that we have a sample with a standard deviation of 15.6 based upon an $N$ of 145. The standard error of the mean of this sample is as follows:

$$s_{\bar{X}} = \frac{s}{\sqrt{N - 1}}$$

$$= \frac{15.6}{\sqrt{145 - 1}}$$

$$= \frac{15.6}{\sqrt{144}}$$

$$= \frac{15.6}{12}$$

$$= 1.3$$

An inspection of formula 11.4 reveals a very important characteristic of standard errors, that is, the larger the sample, the smaller the size of the standard error. This makes sense, for we would expect samples based on larger samples to be more reliable than those based on smaller ones. To be more exact, we can say that the size of the standard error of the mean is inversely proportional to the square root of the number of cases in the sample and directly proportional to the standard deviation. This is what we see in formula 11.4. We can generalize from this and say that the size of the standard error of any statistic is inversely proportional to the number of cases in the sample upon which the statistic was computed.

The sample mean is usually described as being an unbiased estimate of the population mean. That is to say, any given sample mean may be higher or lower than the population mean. If we take enough of these samples and average them out, the result will be an unbiased estimate of the population; in other words, this result will tend to be systematically neither too large nor too small when compared with the parameter mean. This is in contrast with the sample standard deviation, which we have shown to be a biased estimate of the parameter standard deviation.

All statistics have sampling distributions, and hence all have standard errors. Each of these gives an indication of the reliability of the statistic. When the size of the standard error is small in relation to the units of measurement, it follows that our statistics will tend to vary less from sample to sample, and thus we can have more confidence in our results.

THE STANDARD ERROR OF THE MEDIAN. The standard error of the median is obtained by the following formula:

$$s_{mdn} = \frac{1.253s}{\sqrt{N}} \tag{11.5}$$

where all terms are as previously defined. When the formula is compared with that for the standard error of the mean, we see that the standard error of the median is about 25 percent larger than that of the mean. It may be recalled that it was stated in an early chapter that the mean is the most reliable average. By this we meant that it has the smallest standard error of the various measures of central tendency.

STANDARD ERROR OF A PROPORTION. This is estimated by the following formula:

$$s_p = \sqrt{\frac{pq}{N}} \tag{11.6}$$

where $q = 1 - p$. The use and limitations of this formula will be discussed in Chapter 13.

STANDARD ERROR OF A PERCENTAGE. Since a percentage is a 100 times a proportion we can write the formula for estimating the standard error of a percentage as follows in terms of proportions:

$$s_p = 100 \sqrt{\frac{pq}{N}} \tag{11.7}$$

or, in terms of percentage,

$$s_P = \sqrt{\frac{PQ}{N}} \tag{11.8}$$

where $Q = 100 - P$.

STANDARD ERROR OF A FREQUENCY. Since a proportion is obtained by dividing a frequency by the number of cases ($p = f/N$) or by rearranging the terms ($f = Np$), it follows that the standard error of a frequency ($s_f$) is equal to $\sqrt{pq/N}$ multiplied by $N$, which may be expressed

$$s_f = \sqrt{Npq} \tag{11.9}$$

As we noted previously, all statistics have standard errors. However, since many statistics are very infrequently used and others have very special applications, no more will be introduced at this point.

## ESTIMATING PARAMETER VALUES

There are two aspects to statistical inference: the first, estimating parameter values, and the second, hypothesis testing. In the remainder of this chapter, we will be concerned with estimation of parameter values, and the next three chapters will be devoted mainly to the testing of hypotheses.

Suppose that we have drawn a sample of 145 vocational high school students and have administered a test of mechanical ability to them. We have computed both the mean and the standard deviation for our sample, these being 82 and 14, respectively. From these statistics we now wish to make statements or inferences about the parameter value of the mean.

Neither the population mean nor standard deviation is known. However, we have pointed out that the sample mean was an unbiased estimate of the population mean and that the sample standard deviation was a biased estimate of the parameter standard deviation (this bias, however, being taken into account in equation 11.4). By using equation 11.4, we obtain the standard error of the mean as estimated from this sample:

$$s_{\bar{X}} = \frac{s}{\sqrt{N-1}}$$

$$= \frac{14}{\sqrt{145-1}}$$

$$= \frac{14}{\sqrt{144}}$$

$$= \frac{14}{12}$$

$$= 1.17$$

Using the information that we have, it is now possible to set up confidence intervals for this mean. A *confidence interval* is a band about which a probability statement can be made. For example, the 95 percent confidence interval for a mean gives the limits within which we can be 95 percent confident that the population mean falls. Since 1.96 standard deviations (standard errors in this case) taken on each side of the mean include 95 percent of the cases, it follows that the 95 percent confidence interval is equal to 1.96 × 1.17 or 2.29 units taken on each side of the sample mean of 82. In short this becomes:

$$95\% \text{ confidence interval} = \bar{X} \pm 1.96 \, s_{\bar{X}} \tag{11.10}$$

$$= 82 \pm 1.96(1.17)$$

$$= 82 \pm 2.29$$

$$= 79.71 - 84.29$$

In a similar fashion, the 99 percent confidence interval is established. If we consult the normal probability table, Appendix B, we find that a $z$ score of 2.58 cuts off .005 of the area in the smaller portion. When this is doubled to include the area cut off by a deviation of the same size on both sides of the mean, the result includes 1 percent of the area of the curve. This procedure is similar to that above, except that 2.58 is used instead of 1.96.

$$99\% \text{ confidence interval} = \bar{X} \pm 2.58 \, s_{\bar{X}} \tag{11.11}$$

$$= 82 \pm 2.58(1.17)$$

$$= 82 \pm 3.02$$

$$= 78.98 - 85.02$$

In the foregoing fashion, confidence intervals can be set up for proportions, percentages, Pearson $r$'s, and so forth. The technique above is for large samples. As we shall see in the next chapter, when we have small samples we substitute for the $z$ values a statistic called $t$ and proceed as above.

In actual practice the 95 percent interval is frequently encountered. If the size of the 99 percent interval above is compared with that of the 95 percent interval, it is seen that the former is approximately 1.32 times as large. Thus as the level of confidence increases from 95 to 99 percent, the size of the confidence interval also increases. It follows that we can then be more certain about any statements made. When we use the 99 percent interval, we have the chance of only 1 in 100 of being wrong. On the other hand, when we use the 95 percent interval, we have 5 chances in 100 of being incorrect.

From all of this, we can see that the smaller the standard error, the more confidence we can have in our statistic. We say that our statistics are more reliable, and by this we mean that our statistics tend to vary less from sample to sample. It should be kept in mind that the size of any standard error is a function of the size of the sample. As sample sizes increase, the sizes of the standard errors decrease. Thus large samples produce more reliable statistics than small samples.

## EXERCISES

**1.** Which of the following selection techniques will result in random samples?

**a.** Population: all the residents of a given city.
Sampling technique: for one week stopping every third person who passes by a busy downtown street corner.

**b.** Population: all the students in a large primary school.
Sampling technique: selecting the first 75 students reporting to school on a Monday morning.

**c.** Population: all tomato plants in a field.
Sampling technique: blindfolded, selecting one plant from each square yard.

**d.** Describe a situation other than those listed above that would result in (1) a random sample; (2) a biased sample.

**2.** Suppose there are 1300 fifth-grade students in a school system. You are given the task of estimating their arithmetic achievement.

**a.** Define precisely the population.
**b.** How would you sample this population?
**c.** Would your sampling cost less than the population measure?
**d.** It has been said that an investigator can learn more from a sample than from a universe. In a practical situation, does this make sense? Why?

**3.** What is a sampling distribution?

**a.** What is plotted on the vertical axis of a sampling distribution?
**b.** What is plotted on the horizontal axis?

**4.** Assume a normal distribution with a mean of 94 and a standard deviation of 4.2. What is the probability that a sample of 81 drawn from this population will have a mean greater than 95.1?

**5.** If we assume that intelligence test scores are normally distributed with a mean of 100 and a standard deviation of 15, what is the probability that a sample of 49 drawn from this normal population will have a mean less than 98?

**6.** Suppose that in problem 5 an infinitely large number of samples of size 36 is drawn from the population. Within what two points would you expect the middle 90 percent of the sample means to fall?

**7.** Given a distribution with a mean of 80, a standard deviation of 12, and based upon an $N$ of 626, set up the 95 percent and 99 percent confidence intervals for the mean.

**8.** Given that 42 out of 146 respond "Agree" to an item on a questionnaire, set up the 99 percent confidence interval for this proportion.

# Testing Hypotheses: Tests Related to Means
## Chapter 12

In this and the following chapter we shall consider the testing of two kinds of hypothesis. The first is related to whether or not a certain sample was drawn from a given population with known parameters. The second is whether or not the difference between two sample means is merely a chance variation or a true difference. We would expect real differences to appear again between future samples.

### MAKING A STATISTICAL TEST

Suppose that we illustrate the making of a statistical test by using a sample drawn from a population with known parameters, the mean being 100 and the standard deviation 15. The sample being studied, based upon 49 cases, has a mean of 106. The question to be answered is whether or not this sample mean was drawn from a population with the given parameters.

We begin by setting up a null hypothesis ($H_0$) that the mean of the sample does not differ significantly from the population mean. It should be noted that we always start a statistical test with a null hypothesis. We cannot apply statistical tests to other types of hypothesis. Our null hypothesis is written

$$H_0 : m = 100$$

This null hypothesis implies that the sample mean of 106 is a random sample from the given population with a mean of 100. Each null hypothesis has an accompanying alternate hypothesis, $H_1$. This is written

$$H_1 : m \neq 100$$

Such an alternate hypothesis implies that the population mean from which this sample with a mean of 106 is drawn is not 100. In other words, this sample with a mean of 106 is drawn from a different population than the sample with the mean equal to 100. When making a statistical test, if we reject $H_0$, then we accept the alternate hypothesis $H_1$ at a given probability level.

It is customary for the research worker to state ahead of time the level at which he is going to test his hypothesis. For example, he might decide to use the .05, or 5 percent, level of significance. This is called the *alpha level* at which he is working. Another worker might decide to use a more stringent test, such as the 1 percent level. As we shall see soon, $z$ scores are used in making tests of significance. We know that a $z$ score of 1.96 taken at each end of the normal curve cuts off 5 percent of the total area and a $z$ score of 2.58 similarly taken cuts off 1 percent of the area. Any time that we obtain a $z$ score between these two points, we reject our null hypothesis at the 5 percent level. If the computed $z$ is greater than 2.58, we reject our null hypothesis at the 1 percent level of significance.

Suppose that we set our alpha level at the .05 significance level, and that we obtain a $z$ of 2.27. On the basis of our information about the normal curve, we would reject the null hypothesis at the 5 percent level, since the obtained $z$ falls between 1.96 and 2.58. If we had set our alpha at the .01 level, we would have been unable to reject the null hypothesis, since the obtained $z$ is less than 2.58.

When we reject a null hypothesis at the 5 percent level, there are 5 chances in 100 that we are wrong, that we are rejecting the null hypothesis when it is actually true. This is known as a *Type I error*. Errors of this type may be reduced by making a more rigorous test, such as putting the alpha value at the 1 percent level. Then there would be only 1 chance in 100 of being wrong. If we so desire, we can further reduce our chances by going to the .1 percent level, and so on. But as we reduce our chances of making a Type I error, we are increasing our chances of making a *Type II error*. This error consists of *not* rejecting the null hypothesis when it should be rejected. So as we decrease the possibility of making one type of error, we increase the possibility of making the other type. Most workers prefer to be cautious, and they try to limit the probability of making a Type I error.

The fact that a research worker always rejects a hypothesis with a certain probability of being wrong should lead him to be extremely cautious about any claims that he makes concerning results that are significant. The least that he should do is to run the experiment again. The literature is filled with experimental results that cannot be replicated. Perhaps it would be a good idea if journal editors accepted only experimental reports that had been replicated. Along these lines, McNemar (1969) notes that psychologists who argue for the .05 level quote Fisher as their authority, but they fail to mention that all of Fisher's work was done in agriculture or biology, where sampling is better controlled than in the social sciences. We have seen students who, in striving for significance in their research, begin talking about the 15, 20, or 25 percent level of significance. For many years a z score of 3 (critical ratio), as it was then called, was arbitrarily used as a test of significance. This has a p of approximately .003 associated with it.

Very frequently, the student finds that he must make several hundred or more tests of significance, such as when he is evaluating the significance of indices of discrimination (Chapter 17). Now suppose that a student has made 200 such evaluations with his alpha level at .05. He finds that 11 of his items are significant at this level or beyond. By chance alone, he would expect 10 of these indices to be significant, that is, 5 percent of 200. All that he has demonstrated is the working of chance. If he selected the 11 items that appeared to be significant and built a short test with them, he would probably find the test worthless.

This discussion might lead one to believe that we are always interested in the rejection of the null hypothesis. Although this is the usual case in experimental work, there are occasions when we hope to demonstrate that two groups are the same so that we can combine them. In such cases we would be interested in having the null hypothesis stand.

It cannot be overemphasized that we never make definite or absolute statements when testing hypotheses. When we reject a hypothesis at the 1 percent level, we are saying that the chances are 99 in 100 that it is false. There is still that 1 chance in 100 that the hypothesis is true.

Returning to our example, we first use formula 11.1 to get an estimate of the standard error of the mean:

$$s_{\bar{X}} = \frac{\sigma}{\sqrt{N}}$$

$$= \frac{15}{\sqrt{49}}$$

$$= \frac{15}{7}$$

$$= 2.14$$

Next we compute z:

$$z = \frac{X - m}{s_{\bar{X}}}$$

$$= \frac{106 - 100}{2.14}$$

$$= \frac{6}{2.14}$$

$$= 2.80$$

Since this z of 2.80 is greater than 2.58, the 1 percent value, we reject $H_0$ at the 1 percent level. Our conclusion is that the chances are 99 in 100 that this sample with a mean of 106 was not drawn from the population with a mean of 100. Suppose that we have a second sample with a mean of 102.5. Again we set up a null hypothesis that this sample mean does not differ from the population mean—$H_0: m = 100$. Once more we compute $z$:

$$z = \frac{X - m}{s_{\bar{X}}}$$

$$= \frac{102.4 - 100}{2.14}$$

$$= \frac{2.5}{2.14}$$

$$= 1.17$$

Since this value is less than 1.96, the critical score, we have no evidence that this sample was not drawn from the given population and thus we retain $H_0$.

### Two-Tailed Tests

The preceding two examples are actually two-tailed tests. In the first case we were testing the probability of a sample with a mean as extreme as 106, that is, 6 units above the parameter mean, being part of the distribution of sample means with $m = 100$. There is also another mean, 94, which deviates 6 units from $m = 100$. Thus in a two-tailed test we are testing the probability of a mean that deviates 6 units from $m$ of being a part of the sampling distribution of $m$. In a two-tailed test we are concerned only with whether or not the difference is a significant one. We do not care whether it is larger or smaller than the parameter mean. Notice in Figure 12.1 that in a two-tailed test at the 5 percent level we have 2.5 percent of the area in each tail of the normal curve. In such a test a negative $z$ score is interpreted in the same manner as a positive one.

### One-Tailed or Directional Tests

In the above illustration we would be making a one-tailed or directional test if we set up our hypothesis concerning the probability of a mean of 106 or higher being drawn from the given population. There are times in experi-

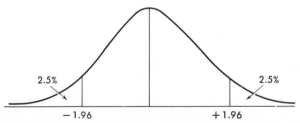

**Figure 12.1**   Two-tailed test at the 5 percent level.

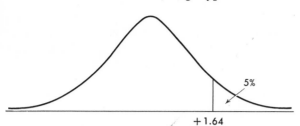

**Figure 12.2**    One-tailed test at the 5 percent level.

mental work when such directional hypotheses are justified. Figure 12.2 illustrates a one-tailed test. Note now that 5 percent of the area is cut off in one tail of the normal curve and that a $z$ score of 1.645 is the one that cuts off 5 percent of the area. It might be mentioned here that the corresponding 1 percent value is a $z$ score of 2.33. When we have a one-tailed test we have the usual null hypothesis

1.    $H_0 : m = 100$

and three alternate hypotheses

2.    $H_1 : m \neq 100$

3.    $H_1 : m > 100$

4.    $H_1 : m < 100$

If it is shown that 1 and 4 of the above are not tenable, this results in the support of 2 and 3. This means that the probability is high that the means differ significantly and that the sample mean is higher than the parameter mean. The alternate hypotheses are competitive. One cannot have all three.

## DIFFERENCE BETWEEN THE MEANS—UNCORRELATED DATA

Data that are uncorrelated are also known as being independent. Suppose that we have two groups, an experimental and a control group. Then after a treatment has been applied to the experimental group, we administer a measure of some sort such as a test. Next means and standard deviations are computed for both groups. With data such as these, population values are unknown and we have to obtain an unbiased estimate of the parameter variance as in the section on pooling the variance. Further, with data such as these we use a $t$ test instead of the $z$ test.

### The $t$ Ratio or Student's $t$

When the number of cases is large, the distribution of $z$ is normal. However, when $N$ is small, especially below 30, the distribution of $z$ departs from normal and we use then the $t$ ratio or Student's $t$ in evaluating statistical results. These ratios were published by W. S. Gosset in papers that were signed only "Student." An inspection of the $t$ table in Appendix C shows that when degrees of freedom $(df)$ is large,[1] the size of the 1 percent and 5 percent values of the $t$ are the same as that for $z$. By the time $df$ has decreased to 60, the 1 percent value is 2.660 and the 5 percent is 2. From this point on down there is considerable difference between the values of $t$ and those of $z$. In the discussion of the central limit theorem (Chapter 11), it was explained that the nature of sampling distributions changes when small $N$'s are used.

The $t$ ratio is defined in the same fashion as $z$. In other words, it is a deviation divided by a standard deviation; the difference between the means is the deviation, and the standard error of the difference between the means is the standard deviation. It follows that it is not the computation of our work that must change when we use small samples but the interpretation of our results. This is so because this $t$ statistic is not normally distributed when $N$ is small. As the number of cases decreases, the sampling distribution of $t$ has the extremes of the tails lifted from the base line, allowing for more cases in the tails. Thus we need different sampling distributions for the different sample sizes.

### Degrees of Freedom

Notice that Appendix C does not list the size of the sample but the number of degrees of freedom $(df)$. Degrees of freedom means freedom to vary. Suppose we have six scores, and the mean of these six scores is to be 10. This sixth score makes adjustments in the variation brought about by the first five scores and assures that the mean of the scores will be 10. For example, suppose we have five scores 10, 12, 18, 16, and 4. For the mean to be equal to 10, a sixth score must be zero. In another series, 2, 8, 4, 6, and 10, the sixth score must this time be 30 if the mean of the six scores is to be 10. In each of these cases we have 5 degrees of freedom. Five of the scores in the series may have any value, but the size of the sixth is determined, because the mean in each case is 10.

Suppose we take a case of 44 scores. First we compute the mean, and then we take the deviation of each score from the mean and compute the standard deviation of the sample. In computing this mean, we have used up 1 degree of freedom. We had 44 degrees of freedom to begin with, but now after computing the mean we have $(N - 1)$, or 43, degrees of freedom.

When data are paired, as they are in Table 12.2, the number of degrees of freedom is equal to one less than the number of pairs. In Table 12.2 we then have 19 degrees of freedom.

---

[1] This is explained in the following section.

**Pooling the Variances**

The pooling of the variances and the formula for the corresponding standard error of the difference is shown below.

We start with formula 12.1, the standard error of the difference between two means.

$$s_{D\bar{X}} = \sqrt{s_{\bar{X}_1}{}^2 + s_{\bar{X}_2}{}^2} \qquad (12.1)$$

We know that

$$s_{\bar{X}_1} = \frac{s_1}{\sqrt{N_1}} \quad \text{and} \quad s_{\bar{X}_2} = \frac{s_2}{\sqrt{N_2}}$$

By substituting the above in formula 12.1 we have

$$s_{D\bar{X}} = \sqrt{\frac{s_1{}^2}{N_1} + \frac{s_2{}^2}{N_2}} \qquad (12.1a)$$

The unbiased estimate of the variance is given by

$$s^2 = \frac{\Sigma x^2}{N - 1}$$

If we substitute in 12.1a,

$$s_{D\bar{X}} = \sqrt{\frac{\Sigma x_1{}^2/(N_1 - 1)}{N_1} + \frac{\Sigma x_2{}^2/(N_2 - 1)}{N_2}} \qquad (12.1b)$$

The pooling of the two variances gives

$$\frac{\Sigma x_1{}^2 + \Sigma x_2{}^2}{N_1 + N_2 - 2}$$

The two sample variances are pooled to get a better estimate of the population variance than would be obtained by using each sample variance separately.

When this combination is substituted for each sample variance in 12.1b,

$$s_{D\bar{X}} = \sqrt{\frac{(\Sigma x_1{}^2 + \Sigma x_2{}^2)/(N_1 + N_2 - 2)}{N_1} + \frac{(\Sigma x_1{}^2 + \Sigma x_2{}^2)/(N_1 + N_2 - 2)}{N_2}}$$

and factored, the result is the usual formula for the difference between the means for small samples:

$$s_{D\bar{X}} = \sqrt{\frac{\Sigma x_1{}^2 + \Sigma x_2{}^2}{N_1 + N_2 - 2}\left(\frac{1}{N_1} + \frac{1}{N_2}\right)} \qquad (12.2)$$

When $N_1 = N_2$, this further reduces to

$$s_{D\bar{X}} = \sqrt{\frac{\Sigma x_1{}^2 + \Sigma x_2{}^2}{N(N - 1)}} \qquad (12.3)$$

**Table 12.1** Scores of Two Groups of
Individual on the Same Test

| $X_1$ | $X_2$ | $X_1{}^2$ | $X_2{}^2$ |
|---|---|---|---|
| 26 | 38 | 676 | 1444 |
| 24 | 26 | 576 | 676 |
| 18 | 24 | 324 | 576 |
| 17 | 24 | 289 | 576 |
| 18 | 30 | 324 | 900 |
| 20 | 22 | 400 | 484 |
| 18 | | 324 | |

$\Sigma X_1 = 141$  $\Sigma X_2 = 164$  $\Sigma X_1{}^2 = 2913$  $\Sigma X_2{}^2 = 4656$

$\overline{X}_1 = 20.14$  $\overline{X}_2 = 27.33$

$\Sigma x_1{}^2 = 73$  $\Sigma x_2{}^2 = 173$ ?

$N_1 = 7$  $N_2 = 6$

Appendix C

As an example, Table 12.1 presents the scores of two groups of individuals on a short test. The third and fourth columns are the squares of the first two columns. The means and the sums of the squares are shown at the bottom of the table. The sum of the squares for each distribution was obtained by the use of this formula:

$$\Sigma x_2 = \Sigma X^2 - \frac{(\Sigma X)^2}{N}$$

It is necessary that these sums be computed, because they are used later in the computation.

We start with the usual null hypothesis of no difference between the population means. This time we use the $t$ test, and this is defined as the ratio of the difference between the means divided by the standard error of the difference:

$$t = \frac{\overline{X}_1 - \overline{X}_2}{s_{D\overline{X}}}$$

By substituting the data from Table 12.1 in equation 12.2 we have

$$s_{D\overline{X}} = \sqrt{\frac{73 + 173}{7 + 6 - 2}\left(\frac{1}{7} + \frac{1}{6}\right)}$$

$$= \sqrt{\frac{246}{11}\left(\frac{1}{7} + \frac{1}{6}\right)}$$

$$= \sqrt{22.3636(.309524)}$$

$$= \sqrt{6.92207}$$

$$= 2.63$$

We proceed with the $t$ test:

$$t = \frac{\overline{X}_1 - \overline{X}_2}{s_{D\overline{X}}}$$

$$= \frac{20.14 - 27.33}{2.63}$$

$$= \frac{-7.19}{2.63}$$

$$= -2.73$$

For this problem the number of degrees of freedom is $(N_1 + N_2 - 2)$, or 11. The $t$ table tells us that with a $df$ of 11, the 5 percent value is 2.201. Hence we can reject our null hypothesis at the 5 percent level.

In making tests such as this we have to make certain assumptions. First, we assume that the two samples are random samples independently drawn from distributions that are normal. Second, we assume that the variances of the populations from which the samples are drawn are the same. This second condition is referred to as *homogeneity of variance*. The first problem is easily solved. Through the use of a table of random numbers the research worker can be assured that his sample or samples are random ones and are correctly drawn. The problem of homogeneity of variance is not as easily taken care of. There has been considerable research related to this problem over the years and today definite statements cannot be made about it to the satisfaction of all statisticians.

A major study was conducted by Boneau (1960). He concluded that the use of the $t$ test will often result in probability statements that are extremely accurate, despite the fact that the assumption of homogeneity of variance and normality of the underlying distributions are untenable. The many cases he cites are characterized by (1) being samples of equal or nearly equal size, and (2) having the same or nearly the same shape of the underlying population distributions. In giving an example of how the violation of the homogeneity of variance assumption affects results, Boneau states that if the preceding two conditions are met, samples as small as 5 will produce results for which the true probability of rejecting $H_0$ at the 5 percent level will tend to be within .03 of this level; and if the sample sizes are 15, the true probabilities are likely to be within .01 of this level. If sample sizes are unequal and variances are basically the same, no problem is encountered. However, unequal sizes and unequal variances create a problem which requires another solution. (This problem is discussed below.) Boneau adds that if the two distributions are not of the same shape, there again is little or no problem provided that both distributions are symmetrical. But if they are skewed, the resulting $t$ ratios tend also to be skewed, and this will lead to biased results. Using samples of larger size tends to remove this skew. Edwards (1967) notes that if the $t$ test is applied to two independent random samples of size 25 or more, the $t$ test is relatively unaffected by rather severe violations of the assumptions of homogeneity of variance and normality of the distributions in the population. For this reason the $t$ test is said to be robust; that is, it is a test relatively unaffected by violations of its underlying assumptions.

Homogeneity of variance is tested for using the $F$ test, which is defined as follows:

$$F = \frac{s_1^2}{s_2^2} \qquad (12.4)$$

where $s_1^2$ = the larger of the two sample variances
$\quad s_2^2$ = the smaller of the two sample variances

With this $F$ test we are evaluating the null hypothesis of no difference between the two population variances. If the $F$ is not significant, the null hypothesis stands, and we pool the variances. If the $F$ is found to be significant, a procedure that will be discussed later should be used.

For this problem we compute the two variances as follows:

$$s_1^2 = \frac{73}{7} \qquad\qquad s_2^2 = \frac{173}{6}$$

$$s_1^2 = 10.429 \qquad\qquad s_2^2 = 28.833$$

$$F = \frac{28.833}{10.429} = 2.76$$

Consulting the $F$ table, Appendix E, with a $df$ of 5 for the larger variance (the greater mean square) and a $df$ of 6 for the smaller variance (the lesser mean square), we find that our $F$ of 2.76 is below the 5 and 1 percent values. Hence the null hypothesis of no difference between the two population variances stands, and we pool the variances.

### $t$ Test When the Variances Differ

When the variances differ as shown by an $F$ test, $t$ is computed by the usual formula. However, this $t$ cannot be interpreted by entering the $t$ table in the usual fashion. Cochran and Cox (1950) developed a formula for testing the significance of the computed $t$. This formula (assuming that we are working at the 5 percent level of significance) is

$$t_{.05} = \frac{s_{\bar{X}_1}^2(t_1) + s_{\bar{X}_2}^2(t_2)}{s_{\bar{X}_1}^2 + s_{\bar{X}_2}^2} \qquad (12.5)$$

where $t_1$ = the 5 percent value for $t$ at $N_1 - 1$ degrees of freedom
$\quad t_2$ = the 5 percent value for $t$ at $N_2 - 1$ degrees of freedom

Suppose that we have the following data:

| Group I | | Group II | |
|---|---|---|---|
| $X_1$ | 33 | $\bar{X}_2$ | 40.4 |
| $s_1^2$ | 144 | $s_2^2$ | 289 |
| $N_1$ | 31 | $N_2$ | 41 |

For these data we compute $F$:

$$F = \frac{289}{144} = 2.01$$

which for these degrees of freedom is significant at the 5 percent level. Next we compute the squared standard errors of the means, or the *variance error* as this is called,

$$s_{\bar{X}_1}^2 = \frac{144}{30} = 4.80 \qquad s_{\bar{X}_2}^2 = \frac{289}{40} = 7.225$$

To substitute into equation 12.5 we first obtain $t_1$, which is the .05 value of $t$ for $N_1 - 1 = 30$ degrees of freedom. This we find to be 2.042. Similarly, $t_2$, the .05 value for $N_2 - 1 = 40$ degrees of freedom, is found to be 2.021.

$$t = \frac{4.80(2.042) + 7.225(2.021)}{4.80 + 7.225}$$

$$= \frac{24.4033}{12.025}$$

$$= 2.03$$

The 1 percent value of $t$ $(t_{.01})$ is similarly obtained by taking equation 12.5 and putting the respective .01 values of $t$ in the places occupied by $t_1$ and $t_2$. Computed for these data, $t$ is

$$t = \frac{40.4 - 33}{\sqrt{4.80 + 7.225}}$$

$$= \frac{7.4}{3.47}$$

$$= 2.13$$

Since our computed value of $t$ (2.13) is greater than 2.03, we reject the null hypothesis at the 5 percent level.

If $N_1 = N_2$, this procedure can be shortened by computing $t$ in the usual way and using the $t$ table with one-half the usual number of degrees of freedom.

### Setting up Confidence Intervals for Small Samples
We previously discussed the method for setting up confidence intervals for a large sample. To do this for the 95 percent level, the confidence interval of the mean was

$$\bar{X} \pm 1.96 \, s_{\bar{X}}$$

Suppose now that we have the following statistics for a set of data: $\bar{X} = 40$, $s = 6$, $N = 26$. We wish to set up the 95 percent confidence interval for this mean. We would compute the standard error of this mean, which is $6/\sqrt{25}$,

**Table 12.2** Testing the Significance of the
Difference Between Means
When Data Are Correlated

| (1)<br>X | (2)<br>Y | (3)<br>D | (4)<br>$D^2$ |
|---|---|---|---|
| 18 | 20 | 2 | 4 |
| 16 | 22 | 6 | 36 |
| 18 | 24 | 6 | 36 |
| 12 | 10 | −2 | 4 |
| 20 | 25 | 5 | 25 |
| 17 | 19 | 2 | 4 |
| 18 | 20 | 2 | 4 |
| 20 | 21 | 1 | 1 |
| 22 | 23 | 1 | 1 |
| 20 | 20 | 0 | 0 |
| 10 | 10 | 0 | 0 |
| 8 | 12 | 4 | 16 |
| 20 | 22 | 2 | 4 |
| 12 | 14 | 2 | 4 |
| 16 | 12 | −4 | 16 |
| 16 | 20 | 4 | 16 |
| 18 | 22 | 4 | 16 |
| 20 | 24 | 4 | 16 |
| 18 | 23 | 5 | 25 |
| 21 | 17 | −4 | 16 |
| | | $\Sigma D = 40$ | $\Sigma D^2 = 244$ |

which equals 1.20. Instead of using 1.96 for the 5 percent value of $z$, we enter the $t$ table with the appropriate number of degrees of freedom (25 in this case) and obtain the $t$ value. We find this to be 2.06. So the 95 percent confidence interval for these data becomes

$40 \pm 2.06(1.20)$

$40 \pm 2.47$

$37.53 - 42.47$

The 99 percent confidence interval is set up in the same manner, using a $t$ ratio of 2.79, which is the 1 percent value for a $df$ of 25. These confidence intervals are interpreted in a way similar to those for large samples.

## DIFFERENCE BETWEEN MEANS—CORRELATED DATA

We previously presented a formula for the standard error of the difference between two means for uncorrelated data (12.1):

$$s_{D\bar{X}} = \sqrt{s_{\bar{X}_1}^2 + s_{\bar{X}_2}^2}$$

When data are correlated another term is added under the radical as follows:

$$s_{D\bar{X}} = \sqrt{s_{\bar{X}_1}^2 + s_{\bar{X}_2}^2 - 2r(s_{\bar{X}_1})(s_{\bar{X}_2})} \qquad (12.6)$$

Inspection shows that equation 12.1 is a special case of 12.6 existing only when the correlation between the two variables being studied is zero. It is possible, although time consuming, to find the Pearson $r$ and the other statistics needed to solve equation 12.6. Fortunately, we have an easier method that gives identical results.

This solution will be demonstrated using the data in Table 12.2. The following steps are required in this test:

**1.** Set up column 3, which is the difference between column 1 and column 2. It makes no difference which way columns 1 and 2 are subtracted, except that whatever the direction started, it must be continued throughout the entire process.

**2.** Sum column 3. Add the negative values and then subtract this sum from the sum of the positive values. Divide this sum by the number of pairs to compute the mean difference. For these data the mean difference is equal to 40 divided by 20, which is 2. It should be noted that this mean difference is identical to the difference between the means.

We are now going to take this mean difference, compute its standard error, and make the usual $t$ test. We first need, however, the sum of the squares for $D$ and then the standard deviation for $D$, as described in steps 3 through 5.

**3.** Square all of the values in column 3 and enter these squares in column 4. Then sum column 4.

**4.** Compute the sum of the squares for $D$

$$\Sigma d^2 = \Sigma D^2 - \frac{(\Sigma D)^2}{N}$$

$$= 244 - \frac{(40)^2}{20}$$

$$= 244 - 80$$

$$= 164$$

**5.** Find the standard deviation of these differences.

$$s_D = \sqrt{\frac{\Sigma d^2}{N}}$$

$$= \sqrt{\frac{164}{20}}$$

$$= \sqrt{8.20}$$

$$= 2.863$$

**6.** Find the standard error of the mean difference.

$$s_{\bar{D}} = \frac{s_D}{\sqrt{N-1}}$$

$$= \frac{2.863}{\sqrt{19}}$$

$$= \frac{2.863}{4.3588}$$

$$= .657$$

**7.** Compute the $t$ ratio.

$$t = \frac{\text{mean difference}}{\text{standard error of the mean difference}}$$

$$= \frac{2}{.657}$$

$$= 3.04*$$

When data are correlated we interpret $t$ with $N - 1$ ($N$ equals the number of pairs) degrees of freedom. In this case there are 20 pairs and hence 19 degrees of freedom. Consulting Appendix C, the $t$ distribution shows the .01 value of $t$ to be 2.861. Hence we conclude that this $t$ of 3.04 is significant at the 1 percent level.

It is to the advantage of the research worker to use the correct formula when testing differences between the means. Examination of the formula for the standard error of the difference between two means (12.6) shows that when the data are correlated, the size of the standard error is reduced, depending upon the size of the correlation coefficient. When the standard error of the difference decreases, the size of the $z$ score computed is larger, with the same value remaining in the numerator. Thus the student who uses the formula for uncorrelated data when he actually has correlated data is applying an unnecessarily stringent test to his data.

**EXERCISES**

**1.** Suppose that a standardized test for elementary statistics exists with a mean of 124 and a standard deviation of 12. A class of 30 students takes this test, and the resulting mean is 120. Is this class below the norm group? Use alpha $= .05$.

**2.** On the Stanford-Binet test the mean IQ is 100 with a standard deviation of 16. A class of 25 kindergarten children is tested with a resulting mean of 108. Does this group have a mean significantly higher than the norm group? Use alpha $= .01$.

**3.** Below are the scores of the members of two groups on a short spelling test:

---

* If we had subtracted scores in column 2 from those in column 1, the mean difference would have been negative and hence our $t$ ratio would have been negative.

| X | Y |
|----|----|
| 16 | 14 |
| 14 | 8 |
| 12 | 7 |
| 12 | 6 |
| 10 | 4 |
| 8 | 4 |
| 6 | 12 |
| 4 | |

**a.** Test $H_0$: $m_1 = m_2$.
**b.** Test $H_0$: $\sigma_1^2 = \sigma_2^2$.

**4.** A group of 24 seniors in a technical high school achieved the following scores on the DAT Numerical Ability Test:

| | | | | | |
|----|----|----|----|----|----|
| 25 | 17 | 29 | 29 | 26 | 24 |
| 27 | 33 | 23 | 14 | 21 | 26 |
| 20 | 27 | 26 | 32 | 20 | 32 |
| 17 | 23 | 20 | 30 | 26 | 12 |

A group of 24 seniors taking the college-entrance curriculum in another high school in the same city obtained these scores on the same test:

| | | | | | |
|----|----|----|----|----|----|
| 21 | 26 | 28 | 31 | 14 | 27 |
| 29 | 23 | 18 | 25 | 32 | 23 |
| 16 | 21 | 17 | 20 | 26 | 23 |
| 7 | 18 | 29 | 32 | 24 | 17 |

Suppose that you are interested in knowing if the mean of the technical high-school group is higher than the college-entrance group. State the appropriate hypothesis and test it statistically.

**5.** Given $X_1 = 12.8$, $X_2 = 16.9$, $N_1 = 11$, $N_2 = 9$, $\Sigma x_1^2 = 61$, $\Sigma x_2^2 = 51$, assume that these data are uncorrelated and compute $t$.

**6.** Given a mean of 17.5, with a standard deviation of 5.2, $N$ being 10, determine the 99 percent confidence interval for this mean.

**7.** A group of subjects was given an attitude test on a controversial subject. Then they were shown a film favorable to the subject and the attitude test was then readministered. State and test an appropriate hypothesis for the following scores:

| $X_1$ | $X_2$ |
|----|----|
| 16 | 24 |
| 18 | 20 |
| 20 | 24 |
| 24 | 28 |
| 24 | 30 |
| 22 | 20 |
| 20 | 24 |
| 18 | 22 |
| 10 | 18 |
| 8 | 18 |
| 20 | 24 |

**8.** A group of students was given a test in addition. Then a state of "anxiety" was induced and the arithmetic test was readministered. The results are:

| Pretest | End Test |
|---------|----------|
| $X = 70$ | $X = 67$ |
| $s = 6$ | $s = 5.8$ |
| $N = 30$ | $N = 30$ |

The $r$ between pretest and end-test scores is .82.

    **a.** Is there a significant difference between the two sets of scores?

    **b.** Test the hypothesis that the population mean on the end test is significantly lower than the population mean on the pretest.

**9.** The following sets of scores were made by 16 individuals in a laboratory experiment on perception: Is there a significant difference between the means of the two distributions?

| Test 1 | Test 2 |
|--------|--------|
| 18 | 16 |
| 12 | 14 |
| 8 | 8 |
| 6 | 8 |
| 3 | 8 |
| 12 | 10 |
| 16 | 8 |
| 7 | 14 |
| 8 | 2 |
| 12 | 8 |
| 15 | 14 |
| 7 | 4 |
| 5 | 6 |
| 12 | 6 |
| 3 | 0 |
| 11 | 7 |

# Testing Differences Between Proportions
## Chapter 13

We previously noted that the standard error of a proportion is estimated by the formula

$$s_p = \sqrt{\frac{pq}{N}} \tag{13.1}$$

where $q = 1 - p$.

This formula is limited in its use. The sampling distribution of $p$ is not the same for all values of $p$. It may be recalled that $p$ can never be larger than 1. Then it follows that if the population values of $p$ are either large or small, that is, in either extreme of the distribution, it is impossible for the sampling distribution to be a normal one. This is seen in Figure 13.1, where the sampling distributions of values of .10, .50, and .95 are illustrated. From Figure 13.1 it can be inferred that the closer $p$ is to .50, the more normal is the curve of the sampling distribution. As the size of samples increases, the fact that the curves are cut off at 0 and 1 is of less importance, because the sampling distributions become very narrow and these truncations are of less importance. However, when $N$ is small the departure of the sampling distribution may be very important. It is not recommended that this formula for the standard error of a proportion be used whenever the product of $N$ times $p$ (or $Nq$, if it is smaller) is less than 10.

## TESTING THE DIFFERENCE BETWEEN TWO PROPORTIONS OR TWO PERCENTAGES FOR UNCORRELATED DATA

We start with this basic formula as the estimate of $\sigma_p$, the standard error of a proportion:

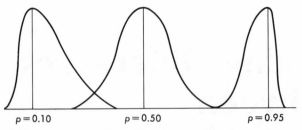

**Figure 13.1** Sampling distributions of various values of $p$.

$$s_p = \sqrt{\frac{pq}{N}}$$

Then our test of significance would be

$$z = \frac{p_1 - p_2}{\sqrt{s_{p_1}^2 + s_{p_2}^2}} \qquad (13.2)$$

The denominator may be simplified as follows:

$$s_{D_p} = \sqrt{s_{p_1}^2 + s_{p_2}^2}$$

$$s_{D_p} = \sqrt{(\sqrt{p_1 q_1/N_1})^2 + (\sqrt{p_2 q_2/N_2})^2}$$

This reduces to

$$s_{D_p} = \sqrt{\frac{p_1 q_1}{N_1} + \frac{p_2 q_2}{N_2}} \qquad (13.3)$$

Equation 13.3 is acceptable when $N$'s are large and the proportions are not extreme, but it should not be used when $N$'s are small and the proportions high or low. When this formula is used, two independent proportions, $p_1$ and $p_2$, are used separately as an estimate of the population proportion. It is recommended that instead of using each of these two sample proportions as an estimate of the parameter proportion, another statistic, $p$, be used. This estimated proportion, $p$, is defined as the proportion in the two groups combined:

$$p = \frac{N_1 p_1 + N_2 p_2}{N_1 + N_2} \qquad (13.4)$$

or

$$p = \frac{f_1 + f_2}{N_1 + N_2} \qquad (13.5)$$

This $p$ is sometimes referred to as the weighted mean of the proportions, or as the pooled estimate of the parameter proportion.

For example, suppose we take the responses of two groups to an item on an attitude scale. Suppose that 90 Ph.D. candidates in psychology and 80 Ph.D. candidates in biology respond to an item concerning the usefulness of the foreign language requirement for the Ph.D. The results are that 30 psychology candidates and 55 biology candidates agree with the statement. We now wish to test the significance of the difference between the proportion in each group that responded "Agree." We do this by making the usual two-tailed test of the null hypothesis.

From the formula for the best estimate of the parameter proportion (13.5), we have

$$p = \frac{f_1 + f_2}{N_1 + N_2}$$

$$= \frac{30 + 55}{80 + 90}$$

$$= \frac{85}{170}$$

$$= .5$$

Before we go on with the problem we substitute $p$ for $p_1$ and $p_2$ in equation 13.3,

$$s_{D_p} = \sqrt{\frac{p_1 q_1}{N_1} + \frac{p_2 q_2}{N_2}}$$

Then

$$s_{D_p} = \sqrt{\frac{pq}{N_1} + \frac{pq}{N_2}} \tag{13.6}$$

or

$$s_{D_p} = \sqrt{pq\left(\frac{1}{N_1} + \frac{1}{N_2}\right)} \tag{13.7}$$

By substituting with the given data we have

$$s_{D_p} = \sqrt{(.5)(.5)\left(\frac{1}{90} + \frac{1}{80}\right)}$$

$$= \sqrt{(.25)(.02361)}$$

$$= \sqrt{.0059025}$$

$$= .0768$$

Next we compute $p_1$, the proportion in psychology responding "Agree": $p_1 = 30/90 = .333$. We compute $p_2$, the proportion in biology responding "Agree": $p_2 = 55/85 = .687$.

Then making the test of significance, we have

$$z = \frac{p_1 - p_2}{s_{D_D}}$$

$$= \frac{.333 - .687}{.0768}$$

$$= \frac{-.354}{.0768}$$

$$= -4.609$$

Since this value far exceeds the 1 percent value (2.58), we can reject $H_0$ (no significant difference between these two proportions) with great confidence.

A $z$ computed this way may be interpreted using the unit normal curve (Appendix B) provided that both $p$ or $q$ and $N$ are reasonably large. A practical rule is that $p$ or $q$, whichever is the smaller, multiplied by $N_1$ or $N_2$, whichever is the smaller, must result in a product greater than 5. In the example here, $.50 \times 80 = 40$, which is much larger than 5. Second, when this product falls between 5 and 10, a correction for continuity must be applied by reducing the absolute value of the numerator, $p_1 - p_2$, by the quantity

$$\frac{1}{2}\left(\frac{N_1 + N_2}{N_1 N_2}\right)$$

before $z$ is calculated.

If we are dealing with smaller samples and are making a $t$ test rather than a $z$ test, we obtain the number of degrees of freedom with the following:

$$df = (N_1 - 1) + (N_2 - 1)$$

In the next chapter it will be seen that an alternate method exists for testing the significance of the difference between two proportions. This is through the use of the statistic *chi-square*. Many students find this method easier than the one just described.

In certain types of work, we find that we have to make a large number of tests of significance between percentages or proportions. This is especially true in item analysis work (see Chapter 17). Figure 13.2, which is referred to as the Lawshe-Baker nomograph, was constructed to facilitate such work. This is very simple to use. Notice that the left-hand column is $p_1$ and the right-hand $p_2$. We place a straightedge between $p_1$ and $p_2$—the percentages or proportions whose difference is being tested. The center line is the *omega value* ($\omega$). Notice that there are two $t$'s presented, one to be used when $N_1 = N_2$ and the other when $N_1 \neq N_2$. Suppose we had a series of tests, all of which were based upon 100 in each group. We would first solve the equation for $t = \omega\sqrt{N}$ for the 5 percent level of significance; that is, $1.96 = \omega\sqrt{100}$ or $\omega = .196$. Similarly, the 1 percent value could be obtained by $2.58 = \omega\sqrt{100}$ or $\omega = .258$. By rounding these we would have the 5 percent value of omega at .20 and the 1 percent at .26. As we read our omega values from the nomograph, we could immediately determine whether or not each was significant and if so, at what level.

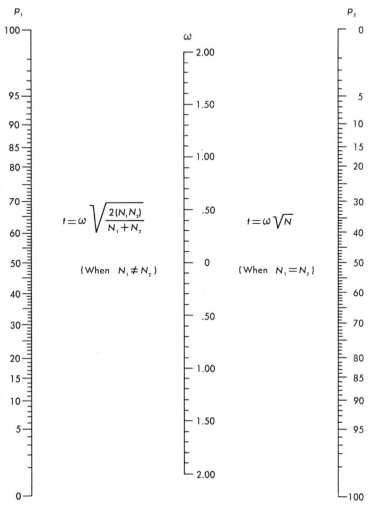

**Figure 13.2**   Lawshe-Baker nomograph for testing the significance of the difference between two percentages.

## TESTING THE DIFFERENCE BETWEEN PROPORTIONS
## FOR CORRELATED DATA

The student should note that the preceding solutions were for uncorrelated data, $r = .00$. When the data are correlated, the formula for the standard error of the difference between two proportions becomes

$$s_{Dp} = \sqrt{(s_{p_1}^2 + s_{p_2}^2) - 2(r_{12})(s_{p_1})(s_{p_2})} \qquad (13.8)$$

McNemar (1969) developed a technique for testing the difference between two percentages that does not necessitate the computation of the correlation

**Table 13.1** Responses of 120
Individuals to Two
Attitude Scale Items

|  |  | Item 2 | | |
|---|---|---|---|---|
|  |  | No | Yes | |
|  | Yes | a | b | |
| Item 1 |  | 33 | 47 | 80 |
|  | No | c | d | |
|  |  | 25 | 15 | 40 |
|  |  | 58 | 62 | 120 |

coefficient between the two variables. The data have to be tabulated into a two-by-two table, such as Table 13.1. Suppose that we are interested in the differences of the responses that the same group of individuals make to two different attitude scale items. The responses of each individual have to be taken separately and each one entered into the table. Suppose that the first individual responded Yes to the first and second items. A tally would then be placed in cell *b*. The next individual responded No to the first item and Yes to the second. This tally would be placed in the *d* cell. In this fashion the responses to all of the items by all of the individuals are tallied. Notice that the tallies placed in the *b* and *c* responses are those individuals who responded in the same direction to the two items. Cells *a* and *d* represent the individuals who answered the two items differently. Table 13.1 presents the responses of 120 individuals to two such items tallied in the manner just described.

The test of significance is made in the following manner:

$$z = \sqrt{\frac{(a-d)^2}{a+d}} \qquad\qquad\qquad (13.9)$$

$$= \sqrt{\frac{(33-15)^2}{33+15}}$$

$$= \sqrt{\frac{18^2}{48}}$$

$$= \sqrt{\frac{324}{48}}$$

$$= 2.60$$

which is significant at the 1 percent level.

As a concluding note, since proportions and percentages are similar, all techniques applied here to proportions can be used equally well with percentages.

**EXERCISES**

**1.** For each of the following compute $p$, the weighted proportion:

| | $N_1$ | $f_1$ | $N_2$ | $f_2$ |
|---|---|---|---|---|
| a. | 40 | 20 | 30 | 17 |
| b. | 60 | 32 | 80 | 53 |
| c. | 115 | 25 | 145 | 65 |

**2.** Thirty-two in a group of 40 and 18 in a group of 50 respond Yes to a certain attitude test item. Do the responses of the two groups differ significantly?

**3.** In the United States Senate there were once 11 members of the scholastic honorary society, Phi Beta Kappa. Nine of these were Democrats and 2 were Republicans. There were 49 Democrats and 47 Republicans in the Senate. Is the difference between the proportions of Phi Beta Kappas statistically significant?

**4.** In a survey it was found that 60 out of 170 females and 32 out of 128 males preferred a certain TV program over two other programs. Is there a sex difference in the preference for this program?

**5.** Given the following responses of 100 individuals to two items on a personality scale, determine whether there is a significant difference in the responses to the two items.

|  |  | Item 1 | |
|---|---|---|---|
|  |  | Disagree | Agree |
| Item 2 | Agree | 40 | 20 |
|  | Disagree | 25 | 15 |

**6.** The following are the responses of 100 students to two items on a test:

| | Item 1 | Item 2 |
|---|---|---|
| Right | 78 | 68 |
| Wrong | 22 | 32 |

    **a.** Assume that the two items are uncorrelated, and test to see if there is a significant difference in the responses to the two items.

    **b.** Suppose that the correlation between item 1 and 2 is .32, again test the difference using this information.

**7.** Given that a proportion based upon an $N$ of 144 is .62, set up the 95 and 99 percent confidence intervals for this proportion.

# $x^2$—Chi-Square
## Chapter 14

The previous tests of significance discussed, $z$ and $t$, and the $F$ test, discussed in the next chapter, are all based upon an assumption of a normal distribution in the population studied. Such statistical tests are referred to as *parametric tests*. In this chapter we introduce a statistic, *chi-square*, that is free from such an assumption. Because of this freedom, we shall see that chi-square is a very useful statistic. Such statistics as chi-square are known as *nonparametric, or distribution-free* statstics. As we shall find later (Chapter 18), chi-square is but one of a large group of such statistics.

Chi-square is used as a test of significance when we have data that are expressed in frequencies or data that are in terms of percentages or proportions and that can be reduced to frequencies. Many of the applications of chi-square are with discrete data; however, any continuous data may be reduced to categories and the data so tabulated that chi-square may be applied. For example, scores on a test of mental ability and a dexterity test could be tabulated as shown below, called a *contingency table*.

| | Dexterity Test Scores | | |
| Mental Ability Scores | 12–20 | 21–29 | 30–38 |
|---|---|---|---|
| 140 and up | | / | /// |
| 120–139 | // | 𝖳𝖧𝖫 | // |
| 100–119 | 𝖳𝖧𝖫 | /// | / |
| 80–99 | /// | | |

To use the chi-square statistic, the data must be independent; that is, no response is related to any other response. Also the categories into which data are placed must be mutually exclusive; that is, a frequency may be placed in one and only one category. And, finally, all data must be used. Note in the problem below that we are concerned not with just the 40 heads obtained by tossing the 100 coins but also with the 60 tails. All of the observed data *must* be used in a chi-square problem.

In this chapter are presented the most frequent uses of chi-square. The student who takes additional work in statistics will find that chi-square has many additional applications.

## TESTING GOODNESS OF FIT

Chi-square is used to determine if a certain distribution differs from some predetermined theoretical distribution. For example, suppose we toss a die or a coin a number of times, and each time we record the results of the toss. We study these observed results against the frequencies we would expect to occur by chance. Or in the field of genetics we study the outcomes of a certain cross to see if they conform to the expectations based on Mendel's laws.

As a simple illustration, suppose that we toss a fifty-cent coin in the air 100 times and record our results. We observe that 40 heads and 60 tails appear. We refer to these frequencies as the *observed frequencies.* In a previous chapter we gave these the symbol $f_o$. However in this chapter we shall, for the sake of convenience, refer to these observed frequencies with the symbol $O$. Next we make the usual null hypothesis that this distribution of 40 heads and 60 tails does not differ from what we would expect by chance, that is, 50 heads and 50 tails. These frequencies are called the *expected, theoretical,* or *hypothetical frequencies.* In this chapter we shall refer to these using the symbol $E$.

We can set all these data up in a contingency table like this:

|       | $O$ | $E$ |
|-------|-----|-----|
| Heads | 40  | 50  |
| Tails | 60  | 50  |
|       | 100 | 100 |

Notice that the sum of the observed frequencies equals the sum of the expected frequencies. Also, it is important to note that we use all of the data. We do not consider merely the number of observed heads but also the "nonheads" or tails.

We test our hypothesis by using the general formula for chi-square:

$$\chi^2 = \sum \frac{(O - E)^2}{E} \tag{14.1}$$

$$= \frac{(40 - 50)^2}{50} + \frac{(60 - 50)^2}{50}$$

$$= \frac{10^2}{50} + \frac{10^2}{50}$$

$$= \frac{100}{50} + \frac{100}{50}$$

$$= 2 + 2$$

$$= 4$$

Formula 14.1 demands that we take each observed frequency, subtract from it the corresponding expected frequency, square the difference, and divide the result by the expected frequency. The sum of these is chi-square.

Appendix D is used for the interpretation of chi-square. We enter this table with a $df$ of 1 (this will be explained later) and find that the value of chi-square for 1 degree of freedom is 3.841. Our computed value of 4 is larger than this but less than the 2 and 1 percent values; hence we reject our hypothesis of no difference at the 5 percent level of significance. We can be fairly confident that these results are different from those produced by chance alone.

This problem, or any chi-square problem, may also be solved in a tabular form as below.

|       | $O$ | $E$ | $O - E$ | $(O - E)^2$ | $(O - E)^2/E$ |
|-------|-----|-----|---------|-------------|---------------|
| Heads | 40  | 50  | 10      | 100         | 2             |
| Tails | 60  | 50  | 10      | 100         | 2             |
|       | 100 | 100 |         |             | $\Sigma = 4 = \chi^2$ |

Suppose that we take another example. This time we toss a die 144 times, and we observe that a five-spot appears 36 times. We organize our observations as follows:

|                | $O$ | $E$ | $(O - E)$ | $(O - E)^2$ | $(O - E)^2/E$ |
|----------------|-----|-----|-----------|-------------|---------------|
| Five-spots     | 36  | 24  | 12        | 144         | 6.0           |
| Not five-spots | 108 | 120 | 12        | 144         | 1.2           |
|                | 144 | 144 |           |             | $\chi^2 = 7.2$ |

Since there were 36 five-spots, it follows that there are 108 observations that were not five-spots. Our expected frequencies are based upon chance. Since the probability of tossing a five-spot is $\frac{1}{6}$, we put down our expected frequencies in that proportion, which in this case gives us 24 and 120. It is important to note that in a chi-square problem, the sum of the observed frequencies has to equal the sum of the expected frequencies. For these data in which *df* again equals 1 the obtained chi-square of 7.2 is significant at the 1 percent level.

## An Alternate Formula

Chi-square can also be obtained by the formula

$$\chi^2 = \sum \frac{O^2}{E} - N \tag{14.2}$$

If we use the above data in our example about the tossing of a die, we have

$$\chi^2 = \left( \frac{36^2}{24} + \frac{108^2}{120} \right) - 144$$

$$= \left( \frac{1296}{24} + \frac{11664}{120} \right) - 144$$

$$= (54 + 97.2) - 144$$

$$= 151.2 - 144$$

$$= 7.2$$

To illustrate the application of chi-square to a problem in genetics suppose that we make a typical Mendelian cross. In peas a round, smooth skin is dominant over wrinkled skin, and yellow color is dominant over green. In a crossing resulting in 1120 peas the following were obtained: 620 smooth, yellow peas; 220 smooth, green peas; 200 wrinkled, yellow; and 80 wrinkled, green. According to Mendel's laws we would expect these combinations to appear in a $9:3:3:1$ ratio. The data appear below:

| $O$ | $E$ | $(O-E)$ | $(O-E)^2$ | $(O-E)^2/E$ |
|---|---|---|---|---|
| 620 | 630 | 10 | 100 | .167 |
| 220 | 210 | 10 | 100 | .476 |
| 200 | 210 | 10 | 100 | .476 |
| 80 | 70 | 10 | 100 | 1.428 |
| 1120 | 1120 | | | $\chi^2 = 2.547$ |

This time there are 3 degrees of freedom since with the numbers in four categories having to sum to 1120, three can vary, but the fourth is fixed by the size of the other three. The obtained chi-square of 2.547 is not significant and it follows that these observed frequencies do not depart from what was expected according to Mendel's laws.

Another use of chi-square is testing a set of data to see if the data are normally distributed, that is, if they fit the normal distribution. In Chapter 6 we learned how to normalize a set of data. The student who wishes to review at this point is referred to Table 6.2 and the discussion associated with it. In Table 6.2 we set up the expected frequencies for the distribution in the table. Both the observed and expected frequencies from Table 6.2 are reproduced in the first two columns of Table 14.1. It should be apparent by now that

**Table 14.1**   Testing Goodness of Fit of Data to the Normal Distribution Model

| | (1) O | (2) E | (3) O | (4) E | (5) O − E | (6) (O − E)² | (7) (O − E)²/E |
|---|---|---|---|---|---|---|---|
| 90–94 | 1 | 1.8 | | | | | |
| 85–89 | 3 | 4.1 | | | | | |
| 80–84 | 8 | 8.2 | 12 | 14.1 | 2.1 | 4.41 | .313 |
| 75–79 | 12 | 13.8 | 12 | 13.8 | 1.8 | 3.24 | .235 |
| 70–74 | 28 | 19.6 | 28 | 19.6 | 8.4 | 70.56 | 3.600 |
| 65–69 | 36 | 23.6 | 36 | 23.6 | 12.4 | 153.76 | 6.515 |
| 60–64 | 12 | 24.1 | 12 | 24.1 | 12.1 | 146.41 | 6.075 |
| 55–59 | 18 | 20.8 | 18 | 20.8 | 2.8 | 7.84 | .377 |
| 50–54 | 10 | 15.2 | 10 | 15.2 | 5.2 | 27.04 | 1.779 |
| 45–49 | 8 | 9.5 | 8 | 9.5 | 1.5 | 2.25 | .237 |
| 40–44 | 8 | 5.0 | 14 | 8.4 | 5.6 | 31.36 | 3.733 |
| 35–39 | 5 | 2.2 | | | | | |
| 30–34 | 1 | 1.2 | | | | | |

$\Sigma = 150$   $\Sigma = 149.1$     $\chi^2 = 22.864$

any time that we have a set of observed and a set of expected frequencies, we can apply the chi-square test.

In columns 1 and 2 of Table 14.1 we have the observed and expected frequencies. In columns 3 and 4 are the same frequencies, this time with the frequencies in the extreme class intervals combined so that none of the expected frequencies is less than 5. In column 5 are the differences between each $O$ and $E$, in column 6 the square of these differences, and in column 7 the square of the differences divided by the expected frequencies. These are then summed, and a chi-square of 22.864 results.

The null hypothesis that we have here is that the distribution of observed scores is a chance variation from a normal population. Basically, the number of degrees of freedom for this situation is the number of intervals minus three. We placed three restrictions upon our data when we normalized them. We noted that the best-fitting normal curve for a set of data has the same mean, standard deviation, and number of cases as the original data. One degree of freedom was lost for each of these restrictions. In our problem we reduced the number of class intervals to 9 when we combined those in the tails having small frequencies. For these data it follows that there are 6 degrees of freedom, that is, 9 categories minus 3. In the chi-square table we find for a $df$ of 6 that the 1 percent value of chi-square is 16.812. Since our chi-square value is larger than this, we reject the null hypothesis at the 1 percent level. In other words, differences this large would be expected 1 time in 100 with repeated sampling from a normal curve. We do not believe this *is* that once in a hundred occurrence and therefore choose to believe that the population from which this sample was drawn is not normal. It might be noted from Appendix D that this chi-square of 22.864 is actually significant slightly beyond the .001 level.

## TESTS OF INDEPENDENCE

### In a 2 x 2 Table
A more frequently encountered use of chi-square is found in so-called tests of independence. For example, a random sample of 68 males and 68 females were asked whether or not they had voted in a recent election. Their responses are shown in Table 14.2. On the assumption that chance alone is operating, the expected frequencies are set up by assigning half of the Yes responses and half of the No responses to each of the two groups since the groups are equal in size. It may be noted that it is not necessary to have the groups of equal size for the use of this technique. Chi-square is computed for these data in the same manner as it was for the previous problem, except that now we have four sets of cells to compare.

| $O$ | $E$ | $(O-E)$ | $(O-E)^2$ | $(O-E)^2/E$ |
|---|---|---|---|---|
| 40 | 45 | 5 | 25 | .555 |
| 28 | 23 | 5 | 25 | 1.087 |
| 50 | 45 | 5 | 25 | .555 |
| 18 | 23 | 5 | 25 | 1.087 |
| 136 | 136 | | | $\chi^2 = 3.285$ |

With $df$ equals 1 (see below) the obtained chi-square of 3.285 is not significant. Here we are testing $H_0$ that sex and whether or not a person voted are not related—that is, that they are independent. Hence we conclude that the evidence is insufficient to make us believe that sex and voting performance are related.

Another method for finding the value of chi-square in a 2 × 2 table is by the use of the following formula, which avoids the computation of the ex-

**Table 14.2**  Chi-Square in a 2 × 2 Table

| | $O$ | | | $E$ | | |
|---|---|---|---|---|---|---|
| | Yes | No | | Yes | No | |
| Group 1 | 40 | 28 | 68 | 45 | 23 | 68 |
| | a | b | k | | | |
| Group 2 | 50 | 18 | 68 | 45 | 23 | 68 |
| | c | d | l | | | |
| | 90 | 46 | 136 | 90 | 46 | 136 |
| | m | n | | | | |

pected frequencies:

$$\chi^2 = \frac{N(ad - bc)^2}{klmn} \tag{14.3}$$

where the various letters are shown in the left-hand contingency table in Table 14.2 and $N$ is the total number of frequencies. If we combine the data in Table 14.2 and formula 14.3,

$$\chi^2 = \frac{136[40(18) - 28(50)]^2}{(68)(68)(90)(46)}$$

$$= \frac{136(720 - 1400)^2}{19,143,360}$$

$$= \frac{62,886,400}{19,143,360}$$

$$= 3.285$$

which is identical to the answer computed by the other method.

## Degrees of Freedom

In a chi-square problem degrees of freedom are determined generally by the use of the following formula:

$$df = (r - 1)(c - 1) \tag{14.4}$$

where $r$ = the numbers of rows in the contingency table
$c$ = the number of columns in the contingency table

For the data in Table 14.2 this reduces to

$$df = (2 - 1)(2 - 1) = 1 \times 1 = 1$$

The student will note that as soon as one of the cell values is determined, the marginal values remaining the same, the frequencies of the other cells are then fixed. For this table, then, there is only 1 degree of freedom.

## Obtaining the Expected Frequencies

To illustrate how the expected frequencies are obtained, we shall use the material in Table 14.3. The cell frequencies of this 3 × 3 contingency table are designated by the letters within the cells. The rows and columns are summed and the total number of frequencies is represented by $T$. We are going to obtain these expected frequencies by manipulating the margins. To obtain the expected frequency for the extreme upper-left cell, we multiply $M$ by $R$ and divide this by the total number of cases, $T$. The $E$ for the cell below this is obtained by multiplying $M$ by $S$ and dividing by $T$. After we finish the left-hand column of the contingency table, we move to the right, and to obtain the $E$ for the top middle cell we multiply $N$ by $R$ and divide this by $T$. In

**Table 14.3**   Obtaining the Expected Frequencies

| $O$ | | | | | $E$ | | | |
|---|---|---|---|---|---|---|---|---|
| $a$ | $b$ | $c$ | $R$ | | $\dfrac{MR}{T}$ | $\dfrac{NR}{T}$ | $\dfrac{PR}{T}$ | $R$ |
| $d$ | $e$ | $f$ | $S$ | | $\dfrac{MS}{T}$ | $\dfrac{NS}{T}$ | $\dfrac{PS}{T}$ | $S$ |
| $g$ | $h$ | $i$ | $W$ | | $\dfrac{MW}{T}$ | $\dfrac{NW}{T}$ | $\dfrac{PW}{T}$ | $W$ |
| $M$ | $N$ | $P$ | $T$ | | $M$ | $N$ | $P$ | $T$ |

this manner all the expected cell frequencies in the contingency table may be determined. If the student wishes he may go back to Table 14.2, and, if the method just described is applied, he will compute the same expected frequencies as are noted there. This method results in theoretical frequencies that are proportional to both margins.

## Chi-Square in a Table Larger than 2 x 2

Suppose that a research worker asks a single question of a random sample of university students, "Do you favor having a student representative on the University board of trustees?" The responses were recorded as "Strongly Agree," "Agree," "No opinion," "Disagree," and "Strongly Disagree." The frequencies of the responses for the three groups are shown in Table 14.4. There is nothing different in the solution of the chi-square for these data than that for the 2 × 2 table, other than the amount of work involved. First the marginal values for the observed frequencies are obtained. Then, using these margins, the expected frequencies are determined. For this problem these were carried to the nearest tenth. The chi-square component for each cell is then determined, these components are summed, the chi-square table in Appendix D is entered with $(3 - 1)(5 - 1)$, or 8 degrees of freedom, and the appropriate conclusion is drawn. In the above, ellipses are used to represent the missing twelve components.

The resulting chi-square is large, and we reject the null hypothesis of no significant difference among the three groups at the 1 percent level. We now have strong reason to believe that there is a significant difference among the three groups. But where? Further tests have to be made to see if each group differs from the other or if two of them are similar, both differing from a third. A chi-square test can be made, taking one group at a time versus another group. If the overall chi-square had not been significant, we could have stopped at that point. But if we have evidence that differences are present, we should take additional steps to find out where they are.

**Table 14.4**   Chi-Square in a Large Table

|  | *O* | | | | | |
|---|---|---|---|---|---|---|
|  | SA | A | NO | D | SD |  |
| Group 1 | 12 | 18 | 4 | 8 | 12 | 54 |
| Group 2 | 48 | 22 | 10 | 8 | 10 | 98 |
| Group 3 | 10 | 4 | 12 | 10 | 12 | 48 |
|  | 70 | 44 | 26 | 26 | 34 | 200 |

|  | *E* | | | | | |
|---|---|---|---|---|---|---|
| Group 1 | 18.9 | 11.9 | 7.0 | 7.0 | 9.2 | 54 |
| Group 2 | 34.3 | 21.6 | 12.7 | 12.7 | 16.6 | 97.9 |
| Group 3 | 16.8 | 10.5 | 6.3 | 6.3 | 8.2 | 48 |
|  | 70 | 44 | 26 | 26 | 34 | 200 |

$$\chi^2 = \frac{(12 - 18.9)^2}{18.9} + \frac{(48 - 34.3)^2}{34.3} + \cdots + \frac{(12 - 8.2)^2}{8.2}$$

$$= 2.519 + 5.472 + \cdots + 1.761$$

$= 34.209$ which with $df = 8$ is significant for
$\chi^2$ at the 1% level $= 20.09$

## Small Frequencies in Contingency Tables

When $df$ equals 1 and any expected frequency is small, less than 10, a chi-square computed for such data is likely to be an overestimate and may lead to erroneous conclusions. In the past a correction called Yates' correction for lack of continuity was applied to the computations because the distribution of chi-square is discrete, whereas the values obtained by the use of the formulas result in a continuous probability model. When Yates' correction was used formula 14.1 became

$$\chi^2 = \sum \frac{(|O - E| - .5)^2}{E} \tag{14.5}$$

and formula 14.3 became

$$\chi^2 = \frac{N(|ad - bc| - N/2)^2}{klmn} \tag{14.6}$$

**Table 14.5**   The Use of Yates' Correction in Chi-Square

|         | O Yes | O No |     | E Yes | E No |     |
|---------|-------|------|-----|-------|------|-----|
| Group 1 | 23 *a* | 2 *b* | 25 *k* | 19 | 6 | 25 |
| Group 2 | 15 *c* | 10 *d* | 25 *l* | 19 | 6 | 25 |
|         | 38 *m* | 12 *n* | 50 *N* | 38 | 12 | 50 |

The use of Yates' correction is illustrated by manipulating the data in Table 14.5.

$$\chi^2 = \frac{(23 - 19 - .5)^2}{19} + \frac{(2 - 6 - .5)^2}{6} + \frac{(15 - 19 - .5)^2}{19} + \frac{(10 - 6 - .5)^2}{6}$$

$$= \frac{3.5^2}{19} + \frac{3.5^2}{6} + \frac{3.5^2}{19} + \frac{3.5^2}{6}$$

$$= .6447 + 2.0417 + .6447 + 2.0417$$

$$= 5.3728$$

When formula 14.6 is used, this problem is worked as follows:

$$\chi^2 = \frac{50[|23(10) - 15(2)| - 50/2]^2}{(25)(25)(38)(12)}$$

$$= \frac{50(200 - 25)^2}{285000}$$

$$= \frac{1,531,250}{285,000}$$

$$= 5.37$$

which, with *df* equal to 1, is significant at the 5 percent level. It might be noted here that the chi-square when computed for these data, not using Yates' correction, is 7.017, which is significant at the 1 percent level.

It has been noted that Yates' correction overcorrects and some statisticians have recommended that Yates' correction not be used (Plackett, 1964). More recently Pirie and Hamden (1972) criticized those who would do away with a correction for lack of continuity. They proposed a correction to be used in place of Yates' correction, one that does not change the value of chi-square as greatly; that is, it makes a much smaller correction. Their formula is

$$\chi^2 = \frac{N[(ad - bc) - 1/2]^2}{klmn} \qquad\qquad (14.7)$$

Applying this formula to the data in Table 14.5 we have

$$\chi^2 = \frac{50[23(10) - 15(2) - .5]^2}{(38)(12)(25)(25)}$$

$$= \frac{50[(230 - 30) - .5]^2}{285000}$$

$$= \frac{50(200 - .5)^2}{285000}$$

$$= \frac{50(199.5^2)}{285000}$$

$$= \frac{50(39800.25)}{285000}$$

$$= \frac{1990012.5}{285000}$$

$$= 6.98$$

As noted above the uncorrected value of chi-square is 7.02. It can be seen that the use of the method of Pirie and Hamden results in only a slight change in the size of chi-square. Using this method the null hypothesis is rejected at the 1 percent level in contrast with the rejection at the 5 percent level when Yates' correction is used with the same data. It is recommended that the Pirie-Hamden formula be used when correcting for lack of continuity in 2 × 2 tables.

The question arises as to how small frequencies can be and still permit chi-square to be used. There are limits to the use of correction formulas such as those described above. A good rule to follow is not to use chi-square when expected frequencies are very small, 2 or less. When such frequencies occur and a test of significance has to be made, a test called Fisher's exact method should be used (McNemar, 1969).

When there is more than 1 degree of freedom, such correction factors are *not* used. However, in these cases when any expected frequency is 2 or less, certain changes are in order, as will be discussed below.

Suppose, using the same terminology as used in Table 14.4, we have the following responses of two groups to a question:

|         | SA | A  | NO | D  | SD |
|---------|----|----|----|----|----|
| Group 1 | 2  | 15 | 12 | 18 | 1  |
| Group 2 | 1  | 14 | 8  | 13 | 2  |

Before we start we should combine the frequencies in the Strongly Agree and Agree categories and those also in the two Disagree categories and apply chi-square to a contingency table that looks like this:

|         | A  | NO | D  |
|---------|----|----|----|
| Group 1 | 17 | 12 | 19 |
| Group 2 | 15 | 8  | 15 |

Another situation like this may arise:

|         | SA | A  | NO | D  | SD |
|---------|----|----|----|----|----|
| Group 1 | 24 | 46 | 2  | 14 | 8  |
| Group 2 | 16 | 13 | 3  | 34 | 16 |

For these data it would be best to discard the No Opinion category and perform the chi-square test using the other four columns.

The next situation contains the responses of students in five different schools of a university to a five-response attitude test item.

|                | SA | A  | NO | D  | SD |
|----------------|----|----|----|----|----|
| Liberal arts   | 16 | 32 | 12 | 37 | 18 |
| Engineering    | 23 | 15 | 7  | 23 | 5  |
| Home economics | 6  | 14 | 6  | 13 | 6  |
| Fine arts      | 9  | 23 | 6  | 7  | 8  |
| Education      | 2  | 3  | 4  | 10 | 1  |

In the above contingency table it will be noted that the number of respondents in education is small. There are three possibilities in handling data like this. First, the education data can be dropped from the analysis. Second, the frequencies of the education group can be combined with those of some other group if there is any logic for such a combination. In this situation education might be combined with liberal arts, for on some campuses these groups are found together. Probably the third and best solution to the problem is to increase the number in education in the sample.


## THE NATURE OF THE CHI-SQUARE DISTRIBUTION

Although the origin and development of chi-square is beyond the scope of this text, it can be noted that the distribution of chi-square is a function of the number of degrees of freedom. These distributions vary considerably in shape when the number of degrees of freedom is small. As the degrees of freedom approach 30, the shape of the chi-square distribution approaches that of the normal curve.

### Chi-Square, a Nondirectional Test

From the above we can get some idea of the nature of the chi-square test. It has been noted that only the right-hand side of Appendix D is used, and the reader might infer from this that we are to make a one-tailed test. This is not so; chi-square is a nondirectional test. Since the statistic is arrived at by squaring the difference between observed and expected frequencies, it has no sign. Consider the .05 value for 1 degree of freedom in Appendix D. This value is 3.841. The square root of this is $\pm 1.96$, which is the .05 value of $z$ for a two-tailed test. Similarly, from this same table for the same number of degrees of freedom we find the .01 value of $z$ to be 6.635. The square root of this is $\pm 2.58$, the .01 value of $z$ for a two-tailed test.

As can be seen from the examples we have worked in this chapter, the usual type of test made with chi-square is a two-tailed test. However, there may be cases where chi-square is to be used in making a one-tailed test. In this case we double the $p$ value. The 5 percent level becomes 2.706 and the 1 percent value becomes 5.412.

## USING CHI-SQUARE IN TESTING THE SIGNIFICANCE OF THE DIFFERENCE BETWEEN PROPORTIONS

### Uncorrelated Data
In Chapter 13 we made a $z$ test to test the significance of the difference between two proportions using the following data:

$p_1 = 30$ psychology students out of $90 = .333$

$p_2 = 55$ biology students out of $80 \quad = .687$

and we found $z$ to be 4.609, which was significant beyond the 1 percent level.

It is possible to take the same data and set them up in a $2 \times 2$ table as follows:

|  | Agree | Disagree |  |
|---|---|---|---|
| Psychology | 30 | 60 | 90 |
| Biology | 55 | 25 | 80 |
|  | 85 | 85 | 170 |

Using formula 14.4 we have

$$\chi^2 = \frac{N(ad - bc)^2}{klmn}$$

$$= \frac{170[30(25) - 60(55)]^2}{(85)(85)(90)(80)}$$

$$= \frac{(170)(-2550)^2}{52020000}$$

$$= \frac{1105425000}{52020000}$$

$$= 21.25$$

This, with a $df$ of 1, is significant beyond the .1 percent level. When the obtained $z$ is squared, the result is 21.24. This is as it should be, within rounding error, since with a $df$ of 1, $\chi^2 = z^2$.

## Correlated Proportions
In Chapter 13, we made a $z$ test (or a $t$ test) for the difference between two proportions when the data are correlated. The data used in Chapter 13 are reproduced below:

|          |     | Item 2 No | Item 2 Yes |
|----------|-----|-----------|------------|
| Item 1   | Yes | $a$ 33    | $b$ 47     |
|          | No  | $c$ 25    | $d$ 15     |

In this table we have the responses of 120 individuals to two test items. The data are arranged in the table by taking into account the agreement and disagreement of the responses of the individuals to the two items.

The formula for chi-square for this type of problem is

$$\chi^2 = \frac{(a - d)^2}{a + d}$$

$$= \frac{(33 - 15)^2}{33 + 15}$$

$$= \frac{18^2}{48}$$

$$= \frac{324}{48}$$

$$= 6.76$$

In Chapter 13, we found a $z$ of 2.60 for this problem. This value squared is equal to 6.76, which it should be, since with 1 degree of freedom, $z^2$ is equal to chi-square.

## The Coefficient of Contingency
This statistic is a measure of relationship similar to the phi coefficient (Chapter 8). This phi coefficient will also be shown to be related to chi-square (Chapter 16). When the phi coefficient is used, data are in two classes, or categories. When we have data in more than two categories, the *coefficient of contingency, C,* may be used as a measure of relationship. $C$ is then a measure of association for nominal data or for data that may be placed in categories.

**Table 14.6**  The Use of the Contingency Coefficient

|  | Succeed | Condition | Fail |  |
|---|---|---|---|---|
| Some college | 20 ⑱ | 10 ⑧ | 10 ⑭ | 40 |
| High school graduate | 60 ㊺ | 10 ⑳ | 30 ㉟ | 100 |
| Some or no high school | 10 ㉗ | 20 ⑫ | 30 ㉑ | 60 |
|  | 90 | 40 | 70 | 200 |

Thus $C$ may be used without making any assumption about the nature of the distribution of the variables used. The contingency coefficient, $C$, does not have 1 as an upper limit, the upper limit being related to the number of categories. For a table made up of an equal number of columns and rows, $k$ by $k$, the upper limits is $\sqrt{(k-1)/k}$. Thus for a 3 × 3 table the upper limit is $\sqrt{2/3} = .82$, for a 4 × 4 table, $\sqrt{3/4} = .87$, and so on. When the number of columns and rows differ, as in 3 × 4, the upper limit follows that of the smaller number.

The contingency coefficient is obtained by the following formula:

$$C = \sqrt{\frac{\chi^2}{N + \chi^2}} \tag{14.8}$$

The data in Table 14.6 illustrate the use of this coefficient. Here a group of 200 young men are divided into three groups on the basis of the amount of education attained. Then their attainment in a course of study is tabled against educational achievement. The expected frequencies are obtained using the cell margins, and then chi-square is computed. The expected frequencies are shown in circles in the lower left corner of each square.

| $O$ | $E$ | $O - E$ | $(O - E)^2$ | $(O - E)^2/E$ |
|---|---|---|---|---|
| 20 | 18 | 2 | 4 | .222 |
| 60 | 45 | 15 | 225 | 5.000 |
| 10 | 27 | 17 | 289 | 10.704 |
| 10 | 8 | 2 | 4 | .500 |
| 10 | 20 | 10 | 100 | 5.000 |
| 20 | 12 | 8 | 64 | 5.333 |
| 10 | 14 | 4 | 16 | 1.143 |
| 30 | 35 | 5 | 25 | .714 |
| 30 | 21 | 9 | 81 | 3.857 |
| 200 | 200 |  |  | $\chi^2 = 32.473$ |

This $\chi^2$ with $df$ equal to 4 is significant beyond the .001 level.

Then

$$C = \sqrt{\frac{32.473}{200 + 32.473}}$$

$$= \sqrt{\frac{32.473}{232.473}}$$

$$= .37$$

This $C$ value is not directly comparable to $r$, rho, tau, or any other correlation coefficient. Nor should $C$'s computed from unlike tables be directly compared. Although $C$ has no sign, if direction is important in any relationship, its sign can be determined by an inspection of the data. However, as a measure of the relationship between two sets of attributes, $C$ is easy to compute, requires that no assumption be made about the population distribution, and can be applied to data that are normal or skewed, continuous or discrete, and nominal or ordinal.

The quickest way to test the significance of $C$ is to test the significance of $\chi^2$. If the latter is significant, so is $C$.

Another measure of relationship that does not have the limitations of the contingency coefficient in respect to its upper limits is Cramér's statistic, $\phi'$.

$$\phi' = \sqrt{\frac{\chi^2}{N(L - 1)}} \tag{14.9}$$

where $L$ is the smaller of *either* the number of columns or the number of rows. Applying this formula to the data in Table 14.6, we have

$$\phi' = \sqrt{\frac{32.473}{200(2)}}$$

$$= \sqrt{\frac{32.472}{400}}$$

$$= \sqrt{.081182}$$

$$= .285$$

Cramér's statistic varies from zero to 1 and its statistical significance may be tested by testing the significance of the chi-square statistic as is done with the contingency coefficient.


**EXERCISES**

1. A coin is tossed 80 times, resulting in 50 heads and 30 tails. Does this differ from what is expected by chance?

**2.** A die was tossed 96 times and a one-spot was observed 20 times. Does this differ from what is expected by chance?

**3.** Using a well-shuffled deck of playing cards, 100 hundred cards were drawn, each being replaced after being drawn, and then the deck was reshuffled. The results are below. Test each one to see if it differs from chance.

Red card        40      Ace, King, Queen, or Jack    25
Black card      60      Numbered card                75

**4.** In an experiment involving 100 males and 100 females, subjects were asked to state a preference between frozen orange juice and a newly developed type of preserved juice. Do the preferences of the two groups differ?

|         | Frozen | Preserved |
|---------|--------|-----------|
| Males   | 62     | 38        |
| Females | 42     | 58        |

**5.** In a survey conducted with university students on a controversial issue the following results were obtained. Do the responses of the two groups differ?

|          | Agree | Disagree |
|----------|-------|----------|
| Seniors  | 68    | 122      |
| Freshmen | 170   | 240      |

**6.** Two lots of rats were used in testing the effectiveness of a new serum in combating a certain disease. Both were inoculated with the causative organism, but only one lot was previously given the preventive serum. The results are below. Can anything be said about the effectiveness of the serum?

|           | Serum | No Serum |
|-----------|-------|----------|
| Recovered | 18    | 4        |
| Died      | 4     | 8        |

**7.** Thirty-two in a group of 40 and 18 in a group of 50 respond Yes to a certain question. Do the responses of the two groups differ significantly?

**8.** In a survey it was found that 60 out of 170 females and 32 out of 128 males preferred a certain TV program over another one. Is there a sex difference in the preference for this program?

**9.** Six coins were tossed with the following results. Are the results different from what would be expected by chance?

| 6 heads, 0 tails | 0 times  |
|------------------|----------|
| 5 heads, 1 tail  | 8 times  |
| 4 heads, 2 tails | 16 times |
| 3 heads, 3 tails | 16 times |
| 2 heads, 4 tails | 13 times |
| 1 head, 5 tails  | 8 times  |
| 0 heads, 6 tails | 3 times  |

**10.** A group of students was classified into "Above Average" and "Average and Below." They were asked how easy they found it to talk with their teachers. Do the following responses of the two groups differ significantly?

| | Sometimes to Never | Usually or Always |
|---|---|---|
| Above Average | 41 | 83 |
| Average and Below | 151 | 113 |

**11.** Is there a significant difference in the responses of the two groups to a certain question?

| | Strongly Disagree | Disagree | No Opinion | Agree | Strongly Agree |
|---|---|---|---|---|---|
| Northern students | 30 | 60 | 10 | 68 | 32 |
| Southern students | 60 | 80 | 12 | 38 | 40 |

**12.** A smaller group responded to the same question with the following results. Is there a significant difference between these responses?

| | Strongly Disagree | Disagree | No Opinion | Agree | Strongly Agree |
|---|---|---|---|---|---|
| N | 1 | 14 | 1 | 12 | 2 |
| S | 2 | 8 | 0 | 15 | 1 |

**13.** Below are the opinions of three groups on an attitude scale.

| | Agree | No Opinion | Disagree |
|---|---|---|---|
| Single | 10 | 15 | 25 |
| Married | 20 | 5 | 20 |
| Divorced | 30 | 10 | 15 |

**a.** Compute a contingency coefficient for these data.
**b.** Compute Cramér's statistic for the same data.

# An Introduction to the Analysis of Variance
## Chapter 15

In the previous chapter we were concerned with the $z$ and $t$ tests for testing differences between two groups. In actual practice, however, it often happens that more than two groups are involved in a study. For instance, one of the authors carried on a study involving a series of tests given to students in five different colleges of a large university. He was interested in differences in performance of the students in the different colleges on these different tests. By means of the $z$ test, he could have taken the colleges two at a time and tested for differences between each two, but for the five different colleges this would have amounted to ten different $z$ tests. Frequently there are more than five groups in a research study, and the number of comparisons that would have to be made to cover all possible tests would be too cumbersome to carry out. The general formula for determining the number of combinations to be made, taking the groups two at a time, is $N(N - 1)/2$, where $N$ is the number of groups. For example, if there are 15 groups, we would have to run 105 separate $z$ tests.

In addition to the nearly prohibitive number of calculations involved in comparing subsamples one by one, there is another, and more important, limitation. When we analyze our data this way, we ignore the fact that these subsamples exist in a set. The elements of such sets are known to *interact*. This interaction should be taken into account in our analysis. Analysis of variance does not ignore interaction. In more complex statistical designs estimates of variance due to interaction are made. These estimates often prove to be extremely important in the interpretation of the statistical analysis.

To avoid these limitations, statisticians have designed the so-called *analysis of variance* techniques in which all of the data are treated at once and a general null hypothesis of no difference among the means of the various groups is tested. In this simple type of analysis of variance, we are concerned with two types of variation.

Suppose that we have IQ scores on five samples of adults. The mean and variance ($s^2$) of each group is

| | | | Groups | | |
|---|---|---|---|---|---|
| | 1 | 2 | 3 | 4 | 5 |
| $\bar{X}$ | 102 | 123 | 100 | 108 | 121 |
| $s^2$ | 15 | 12 | 12 | 14 | 10 |

We can see that the means of the groups vary. This variation of group means from the total or grand mean of all groups is referred to as *between groups variance*. The average variability of the scores within each group is called *within groups variance*.

Now let us suppose that we throw all the IQ scores into one big pot and mix thoroughly. We can forget for the moment which scores belong in which groups. These scores will vary. The variation of these individual scores is called *total variance*.

The heart of analysis of variance lies in the following fact: If the groups are random samples from the same population, the two variances, *within* and *between,* are unbiased estimates of the same population variance. We can test for the significance of the difference of the two types by use of the *F* test.

## ASSUMPTIONS UNDERLYING THE USE OF THE ANALYSIS OF VARIANCE

When the analysis of variance technique is used, the following assumptions should be met:

**1.** The individuals in the various subgroups should be selected on the basis of random sampling from normally distributed populations.

**2.** The variance of the subgroups should be homogeneous ($H_0$: $\sigma_1^2 = \sigma_1^2 = \cdots = \sigma_5^2$).

**3.** The samples comprising the groups should be *independent*. Unless the samples are independent, and thereby yield independent variance estimates, the ratio of *between* to *within* variances will not have the *F* distribution.

It has been shown that the analysis of variance technique is robust in respect to these assumptions. This means that when the analysis of variance is used, the results will be accurate even if the homogeneity assumption is violated. However, sample sizes should be the same or very similar in number. Likewise, the assumption of normality of distribution may be violated provided the departure from normal is not too large. It may be recalled that similar remarks were made about the *t* test in Chapter 12. Tests for the homogeneity of variance do exist, but they are beyond the scope of this book. The interested student is referred to Bartlett (1937), Cochran (1947), and Edwards (1968, 1960).

## SINGLE CLASSIFICATION ANALYSIS OF VARIANCE

We shall illustrate the basic solution of an analysis of variance problem by using the three sets of data in Table 15.1. Here we have the scores of 7 individuals in three groups, A, B, and C, to a short test. In the three columns at the right, we have the squares of each of these scores.

### The Total Sum-of-Squares

The total sum-of-squares could be obtained by finding the mean of the 21 scores, taking the deviation of each score from this mean, and squaring and summing these squared deviations. It may be recalled that we can obtain the sum of the squares by the use of the following equation:

$$\Sigma x^2 = \Sigma X^2 - \frac{(\Sigma X)^2}{N} \qquad (15.1)$$

This would mean

$$\Sigma x^2 = (12^2 + 18^2 + 16^2 + \cdots + 12^2 + 14^2) - \frac{250^2}{21}$$

Or from the data in Table 15.1

$$\Sigma x_i^2 = 1068 + 1726 + 640 - \frac{(82 + 108 + 60)^2}{21}$$

$$= 3434 - \frac{250^2}{21}$$

$$= 3434 - \frac{62500}{21}$$

$$= 3434 - 2976.2$$

$$= 457.8$$

**Table 15.1** Example of Single Classification Analysis of Variance

| Group A X | Group B X | Group C X | | | Group A $X^2$ | Group B $X^2$ | Group C $X^2$ | |
|---|---|---|---|---|---|---|---|---|
| 12 | 18 | 6 | | | 144 | 324 | 36 | |
| 18 | 17 | 4 | | | 324 | 289 | 16 | |
| 16 | 16 | 14 | | | 256 | 256 | 196 | |
| 8 | 18 | 4 | | | 64 | 324 | 16 | |
| 6 | 12 | 6 | | | 36 | 144 | 36 | |
| 12 | 17 | 12 | | | 144 | 289 | 144 | |
| 10 | 10 | 14 | | | 100 | 100 | 196 | |
| 82 | 108 | 60 | | | 1068 | 1726 | 640 | $\Sigma = 3434$ |
| 11.71 | 15.43 | 8.57 | $\Sigma X = 250$ | | | | | |
| | | | $X_t = 11.90$ | | | | | |

## The Between Sum-of-Squares

The sum of the squares between the various groups can be found by taking the mean of each group, getting its deviation from the total mean, squaring this deviation, and then multiplying each of these by the number of individuals in each group ($n$), as follows:

$$\Sigma x_b^2 = \Sigma(\bar{X} - \bar{X}_t)^2 n \tag{15.2}$$

$$= (11.71 - 11.90)^2(7) + (15.43 - 11.90)^2(7) + (8.57 - 11.90)^2(7)$$

$$= (-.19)^2(7) + (3.53)^2(7) + (-3.33)^2(7)$$

$$= (.0361)(7) + (12.4609)(7) + (11.0889)(7)$$

$$= .2527 + 87.2263 + 77.6223$$

$$= 165.1013$$

A more direct method of obtaining the so-called between sum-of-squares is as follows:

$$\Sigma x_b^2 = \sum \frac{(\Sigma X)^2}{n} - \frac{(\Sigma X_T)^2}{N} \tag{15.3}$$

$$= \left(\frac{82^2}{7} + \frac{108^2}{7} + \frac{60^2}{7}\right) - \frac{250^2}{21}$$

$$= \frac{6724}{7} + \frac{11664}{7} + \frac{3600}{7} - \frac{62500}{21}$$

$$= 960.6 + 1666.3 + 514.3 - 2976.2$$

$$= 3141.2 - 2976.2 = 165.0$$

This agrees within rounding error with the value obtained above. This value will be used in the following calculations.

## The Within Sum-of-Squares

To obtain the within sum-of-squares we could find the sum-of-squares of each group as follows:

For group A

$$\Sigma x^2 = \Sigma X^2 - \frac{(\Sigma X)^2}{n}$$

$$= 1068 - \frac{82^2}{7}$$

$$= 1068 - 960.6$$

$$= 107.4$$

For group B

$$\Sigma x^2 = 1726 - \frac{108^2}{7}$$
$$= 1726 - 1666.3$$
$$= 59.7$$

For group C

$$\Sigma x^2 = 640 - \frac{60^2}{7}$$
$$= 640 - 514.3$$
$$= 125.7$$

Summing for all three groups

$$\Sigma x_w^2 = 107.4 + 59.7 + 125.7$$
$$= 292.8$$

The within sum-of-squares added to the between sum-of-squares should total the total sum-of-squares.

$$165.0 + 292.8 = 457.8$$

It follows then that the within sum-of-squares can be obtained directly by subtracting the between sum-of-squares from the total sum-of-squares.

$$\Sigma x_w^2 = \Sigma x_t^2 - \Sigma x_b^2 \tag{15.4}$$

### Degrees of Freedom

Since there are 21 cases in the problem that we are working, we have $N - 1$, or 20 degrees of freedom. In group A there are 7 cases; hence there are 6 degrees of freedom for this group, and since in this problem the number of cases is the same in each problem, there are 6 degrees of freedom in each of the other groups. So far we have accounted for 18 of the total number of degrees of freedom. We have three groups. Then it follows that there are 2 degrees of freedom for the groups. To generalize:

$df$ for total groups = number of cases in total ($N$) minus 1
$df$ for groups between = number of groups ($k$) minus 1
$df$ for groups within = sum of the number of cases within each subgroup ($n$) minus 1. $(n_1 - 1) + (n_2 - 1) + \cdots + (n_k - 1)$

### The Analysis of Variance

The usual technique at this point is to set up a table similar to Table 15.2. In the appropriate column in this table we place the number of degrees of freedom, the sum of the squares for each of the three categories, and in the last column the so-called *mean-square* values. These mean-squares are ob-

**Table 15.2**   Analysis of Variance for the Data in Table 15.1

| Source of Variation | df | Sum-of-Squares | Mean-Square |
|---|---|---|---|
| Between groups | 2 | 165.0 | 82.5 |
| Within groups | 18 | 292.8 | 16.3 |
| Total | 20 | 457.8 | |

tained by dividing each of the sum-of-squares by its respective number of degrees of freedom. Such a ratio or division results in a variance. The between and the within mean-squares are then two estimates of the population variance.

## The $F$ Test

We made an $F$ test previously when testing the difference between two variances to see whether or not we should pool them. The analysis of variance table is evaluated by making the following $F$ test:

$$F = \frac{\text{mean-square for between groups}}{\text{mean-square for within groups}} \tag{15.5}$$

$$= \frac{82.5}{16.3}$$

$$= 5.06$$

$F$ ratios are interpreted by use of the $F$ table (Appendix E). This table is entered with the number of degrees of freedom for the greater mean-square across the top and with the number of degrees of freedom for the lesser mean-square on the left-hand side. For this problem, we go over to 2 and down to 18. In that location we observe that the value of $F$ needed for significance at the 5 percent point is 3.55. Since our obtained $F$ is greater than this we reject the null hypothesis of no difference among these means at the 5 percent point. It is important to remember that though our test is a ratio of variances, the null hypothesis is that the means of the populations from which the samples are selected are equal ($H_0$: $m_A = m_B = m_C$).

There are times when the value of the $F$ ratio will be less than 1. There is no point in solving for the value of such a ratio, since such ratios are not significant.

## Tests after the $F$ Test

Since in the previous problem we found that there were significant differences between the means, the next task is to see where the difference or differences are. Winer (1971) summarized a half dozen different methods for doing this. Some of these methods are more rigorous than others and reduce the probability of making a Type I error. One of these tests (Scheffé, 1957) will be introduced here. Although this is one of the more rigorous methods, it is also one of the easier ones to compute.

In our problem there are three means; thus three comparisons may be made:

$\bar{A}$ vs. $\bar{B}$

$\bar{A}$ vs. $\bar{C}$

$\bar{B}$ vs. $\bar{C}$

First for each group an $F$ ratio like the following is computed:

$$F = \frac{(\bar{X}_1 - \bar{X}_2)^2}{s_w{}^2(N_1 + N_2)/N_1 N_2}$$

For distributions A and B:

$$F = \frac{(11.71 - 15.43)^2}{16.3(14)/49}$$

$$= \frac{3.72^2}{4.66}$$

$$= \frac{13.8384}{4.66}$$

$$= 2.97$$

For distributions A and C:

$$F = \frac{(11.71 - 8.57)^2}{16.3(14)/49}$$

$$= \frac{3.14^2}{4.66}$$

$$= \frac{9.8596}{4.66}$$

$$= 2.12$$

For distributions B and C:

$$F = \frac{(15.43 - 8.57)^2}{16.3(14)/49}$$

$$= \frac{6.86^2}{4.66}$$

$$= \frac{47.0596}{4.66}$$

$$= 10.1$$

As pointed out earlier, the 5 percent level of $F$ for 2, 18 degrees of freedom, is 3.55. This value is multiplied by $(k - 1)$, where $k$ is the number of groups or treatments. In this case we have $(3 - 1)(3.55)$, which is $(2)(3.55)$, which equals 7.10. Then the 1 percent value of $F$ for 2, 18 degrees of freedom, which is 6.01, is multiplied by $(k - 1)$, giving $(2)(6.01)$, or 12.02. Each of the three $F$'s computed above is then compared with these values of 7.10 and 12.02. One of them is larger than the 5 percent value and this is the 10.1, the $F$ computed between groups B and C. Then it follows that the mean of B differs significantly from that of C at the 5 percent level and there is no significant difference between each of the other comparisons. In practice one would set alpha at either .01 or .05 and would compute only the one value that is necessary.

## THE ANALYSIS OF VARIANCE WITH ONLY TWO GROUPS

The analysis of variance may be applied to only two groups. When this is done, a relationship of the $F$ ratio and the $t$ ratio becomes apparent. In Table 15.3 are two sets of scores representing the responses of two groups to a short test. In the table we are shown the various sums and means. The sum of the squares is computed as follows:

Sum-of-squares for total

$$\Sigma x_t^2 = 5219 - \frac{305^2}{20}$$

$$= 5219 - 4651.25$$

$$= 567.75$$

The between sum-of-squares

**Table 15.3**   Analysis of Variance for Two Groups

| $X$ | $X_1$ | $X^2$ | $X_1^2$ |
|---|---|---|---|
| 22 | 12 | 484 | 144 |
| 18 | 16 | 324 | 256 |
| 24 | 10 | 576 | 100 |
| 22 | 10 | 484 | 100 |
| 16 | 4 | 256 | 16 |
| 18 | 6 | 324 | 36 |
| 13 | 17 | 169 | 289 |
| 18 | 14 | 324 | 196 |
| 19 | 14 | 361 | 196 |
| 22 | 10 | 484 | 100 |
| $\Sigma X = 192$ | $\Sigma X_1 = 113$ | $\Sigma X^2 = 3786$ | $\Sigma X_1^2 = 1433$ |
| $X = 19.2$ | 11.3 | | |

$$\Sigma x_b^2 = \frac{192^2}{10} + \frac{113^2}{10} - \frac{305^2}{20}$$

$$= 3686.4 + 1276.9 - 4651.25$$

$$= 4963.3 - 4651.25$$

$$= 312.05$$

The within sum-of-squares

$$\Sigma x_w^2 = 567.75 - 312.05 = 255.7$$

The analysis of variance table is organized as shown in Table 15.4 and the $F$ test is made. This time we find that the $F$ ratio is equal to 21.96.

Suppose that we next compute a $t$ ratio for the preceding data. We shall use the following formula for $t$, the formula for small samples:

$$t = \frac{X - X_1}{\sqrt{\dfrac{\Sigma x^2 + \Sigma x_1^2}{N_1 + N_2 - 2}\left(\dfrac{1}{N_1} + \dfrac{1}{N_2}\right)}}$$

$$= \frac{19.2 - 11.3}{\sqrt{\dfrac{99.6 + 156.1}{10 + 10 - 2}\left(\dfrac{1}{10} + \dfrac{1}{10}\right)}}$$

$$= \frac{7.9}{\sqrt{\dfrac{255.7}{18}(.1 + .1)}}$$

$$= \frac{7.9}{\sqrt{14.20(.2)}}$$

$$= \frac{7.9}{\sqrt{2.84}}$$

$$= \frac{7.9}{1.685}$$

$$= 4.69$$

Both this $F$ of 21.96 and the $t$ of 4.69 are significant beyond the 1 percent level of significance. Both techniques lead to the same conclusion. It may be

**Table 15.4**  Analysis of Variance for the Data in Table 15.3

| Source of Variation | df | Sum-of-Squares | Square Mean- |
|---|---|---|---|
| Between groups | 1 | 312.05 | 312.05 |
| Within groups | 18 | 255.7 | 14.21 |
| Total | 19 | 567.75 | |

noted that when $df$ for between groups $= 1$, $\sqrt{F} = t$ or, putting it the other way around, $t^2 = F$.

$$\sqrt{21.96} = 4.69 = t$$

## TWO-WAY CLASSIFICATION ANALYSIS OF VARIANCE

So far in this chapter the analysis of variance technique that was carried out was done in respect to only one variable. In each case we examined two or more group means by applying the analysis of variance method to see if the means of the groups differed significantly as a result of some treatment applied to the groups. In the pages that follow we shall analyze data that have been classified according to two variables. For example, suppose we draw a sample of students from each of the four classes of a university and study these samples in respect to their attitudes toward certain controversial issues. A single classification analysis of variance could be applied to scores made on a scale constructed to measure their attitudes to see if the means of the four groups differed significantly. Now suppose that we also classify our subjects on the basis of whether or not they were fraternity members. In this case we now have two sets of means for each class instead of the original four. The two-way analysis of variance makes it possible to answer questions as to whether members of the four classes differ on this attitude scale, whether fraternity men differ from non-fraternity men, and finally if there is inter-action between class and fraternity–non-fraternity status in respect to the attitudes measured. This latter means that there are differences over and beyond those brought about by belonging to a certain class of fraternity membership. These differences are the result of the interaction of class and fraternity membership.

To illustrate a two-way analysis of variance, we shall use a methodology worked out by Ebel (1951), dealing with the analysis of ratings made by a group of judges. In Table 15.5 we have the results of the rating of 6 trainees

**Table 15.5**   The Ratings of 6 Trainees by 4 Supervisors

| | | Rater | | | | | Squares of Ratings | | |
|---|---|---|---|---|---|---|---|---|---|
| Trainee | A | B | C | D | Σ of rows | A | B | C | D |
| 1 | 10 | 6 | 8 | 7 | 31 | 100 | 36 | 64 | 49 |
| 2 | 4 | 5 | 3 | 4 | 16 | 16 | 25 | 9 | 16 |
| 3 | 8 | 4 | 7 | 4 | 23 | 64 | 16 | 49 | 16 |
| 4 | 3 | 4 | 2 | 2 | 11 | 9 | 16 | 4 | 4 |
| 5 | 6 | 8 | 6 | 7 | 27 | 36 | 64 | 36 | 49 |
| 6 | 9 | 7 | 8 | 7 | 31 | 81 | 49 | 64 | 49 |
| | 40 | 34 | 34 | 31 | 139 | 306 | 206 | 226 | 183 Σ = 921 |

by 4 supervisors (judges) using a ten-point scale. Note that in the table various squaring and summing has been carried out. This illustration is one of the simpler two-way analysis of variance applications in that although there are two variables, rater and ratee, there is only one observation recorded in each cell or for each combination of the two variables. Such a model is referred to as a *random model,* since both trainees and supervisors are assumed to be random samples from normal populations.

The analysis of variance proceeds as follows:

**1.** The total sum-of-squares ($SS_t$) is found in the usual manner.

$$SS_t = \Sigma X^2 - \frac{(\Sigma X)^2}{N}$$

$$= 921 - \frac{139^2}{24}$$

$$= 921 - 805$$

$$= 116$$

**2.** The sum of the squares for rows, Trainees ($SS_r$), is next obtained.

$$SS_r = \frac{31^2 + 16^2 + 23^2 + 11^2 + 27^2 + 31^2}{4} - \frac{139^2}{24}$$

$$= \frac{3557}{4} - 805$$

$$= 889.25 - 805$$

$$= 84.25$$

**3.** The sum of the squares for columns, Raters ($SS_c$), is then calculated.

$$SS_c = \frac{40^2 + 34^2 + 34^2 + 31^2}{6} - \frac{139^2}{24}$$

$$= \frac{4873}{6} - 805$$

$$= 812.16 - 805$$

$$= 7.16$$

**4.** After the $SS$ for both rows and columns have been obtained, this leaves the residual or error sum of squares, $SS_e$. This error term is a combination of interaction and error effects.

$$SS_e = SS_t - (SS_r + SS_c)$$

$$= 116 - (84.25 + 7.16)$$

$$= 116 - 91.41$$

$$= 24.59$$

Finally a table is set up:

| Source of Variation | df | SS | MS | F | p |
|---|---|---|---|---|---|
| Rows (trainees) | 5 | 84.25 | 16.85 | 10.27 | $<.01$ |
| Columns (raters) | 3 | 7.16 | 2.39 | 1.46 | $>.05$ |
| Error | 15 | 24.59 | 1.64 | | |
| Total | 23 | 116.00 | | | |

The first $F$ in this table is obtained by dividing the variance for rows by the error variance $(MS_r/MS_e)$. This obtained $F$ of 10.27 is highly significant for 5, 15 degrees of freedom, and is far beyond the 1 percent point. This indicates that the supervisors did discriminate among the trainees. The second $F$ is obtained by dividing the variance for columns, Raters, by the error variance $(MS_c/MS_e)$. With 3, 15 degrees of freedom the obtained $F$ of 1.46 is not significant, indicating that the supervisors did not differ significantly among themselves in their ratings of the trainees. In other words, their ratings were reliable.

Data from this technique are often used in determining the reliability of raters. Reliability is discussed in some detail in Chapter 17. By the use of the following formula, the reliability of the ratings of a single rater is obtained:

$$r_{xx} = \frac{MS_r - MS_e}{MS_r + (k - 1)MS_e}$$

where $k$ is the number of columns and the other terms are as used above.

$$r_{xx} = \frac{16.85 - 1.64}{16.85 + (4 - 1)1.64}$$

$$= \frac{15.21}{16.85 + 4.92}$$

$$= \frac{15.21}{21.77}$$

$$= .70$$

The reliability of all four raters taken together is obtained by

$$r_{tt} = \frac{MS_r - MS_e}{MS_r}$$

$$= \frac{16.85 - 1.64}{16.85}$$

$$= \frac{15.21}{16.85}$$

$$= .90$$

**Table 15.6**  Scores of 36 Individuals of Both Sexes on a Learning
Task under Three Different Dosages of a Drug

| Factor 1. Sex | Factor 2. Drug Dosage | | | | Squares | | | |
|---|---|---|---|---|---|---|---|---|
| | Group 1, 10 mg. | Group 2, 50 mg. | Group 3, 100 mg. | | | | | |
| | $X_1$ | $X_2$ | $X_3$ | Σ of rows | $X_1$ | $X_2$ | $X_3$ | |
| | 18 | 14 | 8 | 40 | 324 | 196 | 64 | |
| | 16 | 13 | 12 | 41 | 256 | 169 | 144 | |
| Males | 18 | 12 | 10 | 40 | 324 | 144 | 100 | |
| | 19 | 10 | 8 | 37 | 361 | 100 | 64 | |
| | 20 | 16 | 6 | 42 | 400 | 256 | 36 | |
| | 14 | 18 | 12 | 44 | 196 | 324 | 144 | |
| | 105 | 83 | 56 | 244 | 1861 | 1189 | 552 | Σ = 3602 |
| | $X_4$ | $X_5$ | $X_6$ | | $X_4$ | $X_5$ | $X_6$ | |
| | 24 | 16 | 14 | 54 | 576 | 256 | 196 | |
| | 22 | 17 | 10 | 49 | 484 | 289 | 100 | |
| Females | 18 | 14 | 12 | 44 | 324 | 196 | 144 | |
| | 20 | 20 | 13 | 53 | 400 | 400 | 169 | |
| | 16 | 19 | 15 | 50 | 256 | 361 | 225 | |
| | 22 | 18 | 16 | 56 | 484 | 324 | 256 | |
| | 122 | 104 | 80 | 306 | 2524 | 1826 | 1090 | Σ = 5440 |

$$\Sigma X_t = 550$$
$$\Sigma X_t{}^2 = 9042$$

## A FACTORIAL DESIGN

To illustrate what can be done with the analysis of variance technique in a
more complex design we shall give an example of a factorial design. In Table
15.6 are some fictitious data showing results obtained with a group of males
and females in a learning situation under the effects of varying amounts of
the same drug. In this experiment there are two factors—sex, male or female,
and dosage, the amount of the drug measured in milligrams. In our analysis
we are concerned with differences in learning related to (1) sex, (2) the
amount of the drug taken, and (3) the interaction effect of sex and drug.

We proceed as we did when undertaking the single classification analysis
of variance by obtaining the various sums of squares.

**1.** The total sum of the squares for all scores.

$$SS_t = \Sigma X_T{}^2 - \frac{(\Sigma X_T)^2}{N_T}$$

$$= 9042 - \frac{550^2}{36}$$

$$= 9042 - 8402.8$$

$$= 639.2$$

**2.** The between sum-of-squares.

$$SS_b = \left[ \frac{(\Sigma X_1)^2}{N_1} + \frac{(\Sigma X_2)^2}{N_2} + \frac{(\Sigma X_3)^2}{N_3} + \frac{(\Sigma X_4)^2}{N_4} + \frac{(\Sigma X_5)^2}{N_5} + \frac{(\Sigma X_6)^2}{N_6} \right] - \frac{(\Sigma X_T)^2}{N_T}$$

$$= \left( \frac{105^2}{6} + \frac{83^2}{6} + \frac{56^2}{6} + \frac{122^2}{6} + \frac{104^2}{6} + \frac{80^2}{6} \right) - \frac{550^2}{36}$$

$$= (1837.5 + 1148.2 + 522.7 + 2480.7 + 1802.7 + 1066.7) - 8402.8$$

$$= 8858.5 - 8402.8$$

$$= 455.7$$

**3.** The within sum-of-squares (error).

$$SS_w = SS_T - SS_b$$

$$= 639.2 - 455.7$$

$$= 183.5$$

**4.** $SS$ for factor 1, sex.

$$SS_{f1} = \frac{(\Sigma \text{ of rows})^2}{\text{number in rows}} - \frac{(\Sigma X_T)^2}{N_T}$$

$$= \frac{244^2}{18} + \frac{306^2}{18} - \frac{550^2}{36}$$

$$= 3307.6 + 5202.0 - 8402.8$$

$$= 8509.6 - 8402.8$$

$$= 106.8$$

**5.** $SS$ for factor 2, dosage.

$$SS_{f2} = \frac{(\Sigma \text{ of columns})^2}{\text{number in columns}} - \frac{(\Sigma X_T{}^2)^2}{N_T}$$

$$= \frac{227^2}{12} + \frac{187^2}{12} + \frac{136^2}{12} - \frac{550^2}{36}$$

$$= 4294.1 + 2914.1 + 1541.3 - 8402.8$$

$$= 8749.5 - 8402.8$$

$$= 346.7$$

**6.** *SS* for interaction between factor 1 and 2.

$$SS_{(1 \times 2)} = SS_b - SS_{f1} - SS_{f2}$$
$$= 455.7 - 346.7 - 106.8$$
$$= 2.2$$

The above sums of squares are summarized in the analysis of variance table, Table 15.7.

As far as degrees of freedom are concerned note that we have a total of 35 degrees of freedom, the total number of cases minus 1. Each subgroup of 6 individuals has 5 degrees of freedom associated with it. Since there are 6 such subgroups, this results in 30 degrees of freedom associated with the within variance (6 × 5). Subtracting this from the total number of degrees of freedom leaves 5 degrees of freedom for the between variance. This between variance is broken down into 1 degree of freedom for factor 1, sex, $(2 - 1)$; 2 degrees of freedom for factor 2, dosage, $(3 - 1)$; and the remaining 2 degrees of freedom are associated with interaction. The mean-squares in Table 15.7 are obtained by dividing each sum-of-squares by the appropriate number of degrees of freedom.

The next step is to make the appropriate *F* tests for the various null hypotheses. The first null hypothesis is that the factor of sex had no results on the learning.

$$F = \frac{MS_{f1}}{MS_w}$$
$$= \frac{106.8}{6.12}$$
$$= 17.45$$

This *F* with 1, 30 degrees of freedom is found from Appendix E to be far larger than the 1 percent value, 7.56. Hence we reject $H_0$ at the 1 percent level and we can be very confident that the two sexes do differ in their learning performance in the experiment.

**Table 15.7**   Analysis of Variance Table for the Data in Table 15.6

| Source of Variation | df | SS | MS | F | p |
|---|---|---|---|---|---|
| Within | 30 | 183.5 | 6.12 | | |
| Between | (5) | | | | |
| Factor 1 | 1 | 106.8 | 106.80 | 17.45 | <.01 |
| Factor 2 | 2 | 346.7 | 173.35 | 28.32 | <.01 |
| Interaction 1 × 2 | 2 | 2.2 | 1.1 | .18 | >.05 |
| Total | 35 | 639.2 | | | |

The second null hypothesis is that the three groups did not differ in their performance in relation to the amount of drug administered. Again

$$F = \frac{MS_{f2}}{MS_w}$$

$$= \frac{173.35}{6.12}$$

$$= 28.32$$

With 2, 30 degrees of freedom we again find that this $F$ is significant beyond the .01 level. From Appendix E, $F_{.01}$, 2, 30 degrees of freedom is 5.39. Again we reject $H_0$ and conclude that the size of the dosage of the drug did have a significant effect on performance.

The third $H_0$ states that there is no interaction effect. By interaction we mean that the combined effects of sex and dosage are related to the learning task.

$$F = \frac{MS_{1 \times 2}}{MS_w}$$

$$= \frac{1.1}{6.12}$$

$$= .18$$

With a $df$ of 2, 30 we see that this $F$ is far below the table value for the .05 level, 3.32. Hence we conclude that we have no reason for rejecting $H_0$ and that the interaction effect is not significant.

In this chapter we have described only a few of the simple uses of the analysis of variance. To go further is beyond the scope of this book. The interested student will find in more advanced texts the large number of experimental designs that can be analyzed by the application of the analysis of variance. See Hicks (1964) or Winer (1971).

## EXERCISES

**1.** Below are the scores of 10 individuals on test $X$ and 10 others on test $Y$.

| X | Y | X | Y |
|---|---|---|---|
| 40 | 20 | 22 | 6 |
| 46 | 10 | 20 | 12 |
| 35 | 20 | 31 | 12 |
| 17 | 15 | 18 | 6 |
| 11 | 18 | 22 | 11 |

  **a.** Use an $F$ test and test the difference between the means of groups $X$ and $Y$.
  **b.** Test the significance of the difference between these two means using a $t$ test.

**2.** The following represent the scores of three groups of students on a short test.

| A | B | C |
|----|----|----|
| 10 | 4 | 5 |
| 9 | 6 | 2 |
| 8 | 4 | 8 |
| 7 | 8 | 7 |
| 12 | 6 | 10 |
| 10 | 4 | 2 |
| 11 | 5 | 1 |
| 12 | 7 | 4 |
| 9 | 10 | 3 |
| 8 | 11 | 4 |

**a.** Do the means of the three groups differ significantly?

**b.** If so, apply the Scheffé test to find out where the differences are.

**3.** Given the following four random samples, test the hypothesis that they are from the same population.

| Sample A | Sample B | Sample C | Sample D |
|----------|----------|----------|----------|
| 15 | 24 | 20 | 25 |
| 20 | 22 | 22 | 18 |
| 26 | 20 | 30 | 16 |
| 26 | 21 | 27 | 32 |
| 24 | 34 | | 24 |
| | 18 | | |

**4.** In a learning experiment 3 groups of subjects made the following number of correct responses in a series of 8 trials.

| | Groups | | |
|-------|----|----|----|
| Trial | A | B | C |
| 1 | 8 | 4 | 2 |
| 2 | 6 | 3 | 6 |
| 3 | 8 | 4 | 8 |
| 4 | 10 | 8 | 8 |
| 5 | 12 | 6 | 8 |
| 6 | 14 | 10 | 9 |
| 7 | 16 | 8 | 10 |
| 8 | 20 | 11 | 8 |

**a.** Apply an analysis of variance to see if the groups differed significantly in their performance.

**b.** If the results showed significant differences apply Scheffé's technique to find out where the differences are.

**5.** Six judges rated 7 subjects as follows on a ten-point scale:

| | Judge | | | | | |
|---------|----|----|----|----|----|----|
| Subject | A | B | C | D | E | F |
| 1 | 3 | 2 | 4 | 3 | 2 | 3 |
| 2 | 5 | 6 | 8 | 4 | 6 | 5 |
| 3 | 7 | 6 | 8 | 6 | 7 | 8 |
| 4 | 4 | 1 | 5 | 2 | 2 | 3 |
| 5 | 9 | 6 | 8 | 7 | 8 | 7 |
| 6 | 4 | 6 | 4 | 7 | 4 | 6 |
| 7 | 8 | 6 | 9 | 8 | 7 | 8 |

**a.** Did the judges discriminate among the subjects in their ratings?
**b.** Did the judges differ among themselves in their ratings?
**c.** What is the reliability of the ratings of all the judges combined?

**6.** Below are the scores of a group of fraternity and non-fraternity men broken down by class on an attitude scale. Analyze these data following the factorial design and make the appropriate $F$ tests.

### Fraternity Men

| Freshmen | Sophomores | Juniors | Seniors |
|----------|-----------|---------|---------|
| 12 | 22 | 26 | 30 |
| 8 | 18 | 20 | 26 |
| 10 | 12 | 18 | 20 |
| 6 | 20 | 16 | 22 |
| 8 | 8 | 10 | 18 |

### Non-Fraternity Men

| | | | |
|----------|-----------|---------|---------|
| 18 | 20 | 22 | 24 |
| 16 | 18 | 21 | 22 |
| 8 | 12 | 19 | 16 |
| 12 | 18 | 17 | 18 |
| 14 | 16 | 18 | 24 |

# Testing the Significance of Correlation Coefficients
## Chapter 16

After a correlation coefficient is computed, the next question is whether or not the *r* is significant. That is, does it represent a real correlation, or is the computed *r* merely brought about by chance? We ask whether the *r* in question is a chance deviation from a population *R* of zero. Also there are times when we have two or more *r*'s, and we wish to know if significant differences exist among them. Both of these questions are answered in this chapter.

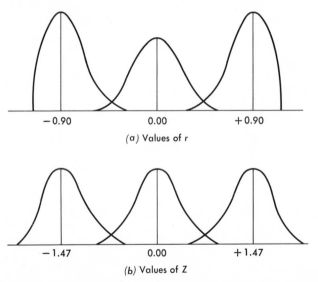

(a) Values of r

(b) Values of Z

**Figure 16.1** Sampling distribution of *r* when the population values are .00 and ± .90, and the sampling distribution of the *Z* statistics are equivalent to the *r*'s.

THE SAMPLING DISTRIBUTION OF $r$. As we have already learned, the Pearson product-moment correlation varies between $+1$ and $-1$. Since this is so, the sampling distribution of $r$ is usually not normal. Notice the (a) part of Figure 16.1. As $R$ increases in size, the sampling distribution becomes increasingly skewed. Suppose that the parameter value of $r$ is .90. As samples are drawn from this population, they will vary. However, there is a limit that these samples can take in an upward direction, 1. Downward there is practically no limit. So the sampling distribution of $r$ when the parameter value $R$ is .90 is negatively skewed. Similarly, the sampling distribution when $R$ is $-.90$ is positively skewed. The sampling distribution of $r$ is normal when the parameter value is zero. As this value approaches median values of $R$ (.50) the skewness begins to be considerable. The degree of skewness is also a function of the size of the sample; the smaller the sample, the greater the degree of skewness. Since in our work we are usually not interested in $r$'s of zero or thereabouts, this skewness in the sampling distribution has to be taken into account in all of our work in which $r$'s are manipulated.

## TESTING THE SIGNIFICANCE OF $r$

Suppose that we compute the Pearson $r$ for two sets of data based upon 82 cases and obtain an $r$ of .30. We wish to know if this $r$ indicates a real relationship between these two variables. We start with the usual null hypothesis that the population $R$ is zero. By this we are saying that our obtained $r$ is merely a chance deviation in the sampling distribution in which the population $R$ is 0 ($H_0$: $R = 0$). We test this $r$ in the usual way by making a $z$ or $t$ test. Again these are defined as the ratio of a deviation to a standard deviation. The deviation in this case is our obtained $r$; the standard deviation is the standard error of this $r$. We obtain the standard error of the $r$ by the use of the following formula:

$$s_{r0} = \frac{1}{\sqrt{N-1}} \qquad\qquad (16.1)$$

where $s_{r0}$ = standard error when $R$ is assumed to be 0
$N$ = number of pairs used in computing $r$

$$s_{r0} = \frac{1}{\sqrt{82-1}}$$

$$= \frac{1}{\sqrt{81}} = \frac{1}{9}$$

$$= .11$$

$$z = \frac{.30}{.11}$$

$$= 2.73$$

which is significant at the 1 percent level.

Let us next take a case in which the $N$ is small. Suppose that we have $N = 18$ and $r = .50$. The above technique is suitable when $N$ is large, but when $N$ is less than 30, the standard error of $r$ should be computed by the following formula:

$$t = \frac{r}{\sqrt{1 - r^2}} \sqrt{N - 2} \qquad (16.2)$$

Then

$$t = \frac{.50}{\sqrt{1 - .50^2}} \sqrt{18 - 2}$$

$$= \frac{.50}{\sqrt{.75}} \sqrt{16}$$

$$= \frac{.50(4)}{.866}$$

$$= \frac{2}{.866}$$

$$= 2.31$$

This value with $(N - 2)$ degrees of freedom, or 16 degrees of freedom, is significant at the 5 percent level ($t_{.05} = 2.12$ for a $df$ of 16).

A table has been developed which makes it unnecessary to go through either of these procedures. This table (Appendix F) is entered with $N - 2$ degrees of freedom, and the sizes of the $r$'s needed to be significant for various degrees of freedom are shown. Suppose that we solve the previous problem by the use of this table. $N$ is 18; hence $df$ is 16. For a $df$ of 16, we see that the $r$ must be at least equal to .468 to be significant at the 5 percent level and .590 at the 1 percent level. Since the obtained $r$ is .50 and falls between these two table values, we can reject the null hypothesis at the 5 percent level. In general the obtained $r$ is compared with the table values. If it is larger than both, the null hypothesis is rejected at the 1 percent level; if it falls between the two table values, the null hypothesis is rejected at the 5 percent level; and if it is smaller than both table values, the null hypothesis is not rejected.

The student should check the significance of any $r$ that he is going to use in further computations. For example, in testing the difference between means, the $r$ can rapidly be checked by the use of Appendix F, and the appropriate formula can be used in the calculation of the standard error of the difference between the two means.

## Fisher's Z Transformation

Fisher has developed a statistic which is a transformation of $r$. The sampling distribution of this statistic is approximately normal for all values. This is shown in the (b) part of Figure 16.1. This statistic is used in problems involving the sampling distribution of $r$.

The standard error of $Z$ is

$$s_Z = \frac{1}{\sqrt{N-3}} \qquad (16.3)$$

### Testing the Difference Between Two Correlation Coefficients for Uncorrelated Data

Suppose that we have the following data:

$r_{12} = .85 \qquad r_{34} = .75$

$N = 103 \qquad N = 147$

We wish to test the null hypothesis that the two population $R$'s do not differ ($H_0$: $R_{12} = R_{34}$). The first thing that we do is change each of these into Fisher's $Z$ statistic by the use of Appendix G. This gives us

$r_{12} = .85 \qquad r_{34} = .75$

$Z = 1.256 \qquad Z = .973$

The problem now is one of testing the difference between these two $Z$'s. The standard error of the difference between two $Z$'s is written as in equation 16.4, but we simplify it to the form of equation 16.5, before we use it.

$$s_{D_Z} = \sqrt{s_{Z_1}^2 + s_{Z_2}^2} \qquad (16.4)$$

$$= \sqrt{\left(\frac{1}{\sqrt{N_1 - 3}}\right)^2 + \left(\frac{1}{\sqrt{N_2 - 3}}\right)^2}$$

$$= \sqrt{\frac{1}{N_1 - 3} + \frac{1}{N_2 - 3}} \qquad (16.5)$$

$$= \sqrt{\frac{1}{103 - 3} + \frac{1}{147 - 3}}$$

$$= \sqrt{\frac{1}{100} + \frac{1}{144}}$$

$$= \sqrt{.01 + .0069}$$

$$= \sqrt{.0169}$$

$$= .13$$

Next we make the usual test of significance.

$$z = \frac{Z_1 - Z_2}{s_{D_z}}$$

$$= \frac{1.256 - .973}{.13}$$

$$= \frac{.283}{.13}$$

$$= 2.18$$

Since $z$ is more than 1.96, the difference is significant at the 5 percent level. Since there is a difference between the two $Z$ statistics, we conclude that there is significant difference between the two population correlation coefficients at the same level.

If the situation exists in which variable 1 is correlated with variable 2, and variable 1 is also correlated with variable 3, and all measurements are made on the same sample, this technique cannot be used, for here the relationship between the pairs of $r$'s has to be taken into account. Once again we are dealing with correlated data. A $t$ test that can be made when this situation occurs follows:

$$t = \frac{(r_{12} - r_{13})\sqrt{(N - 3)(1 + r_{23})}}{\sqrt{2(1 - r_{12}^2 - r_{13}^2 - r_{23}^2 + 2r_{12}r_{13}r_{23})}} \tag{16.6}$$

This is interpreted by going into the $t$ table with $N - 3$ degrees of freedom.

### Establishing the Confidence Interval for the Pearson $r$

Establishing the confidence interval for the Pearson $r$ is done by changing the $r$ to $Z$ and by setting up the confidence interval for $Z$.

$$Z_{95} = Z \pm 1.96 \, s_Z$$

After the confidence limits have been established for $Z$, these are translated to $r$'s by the use of Appendix G.

### TESTING THE SIGNIFICANCE OF OTHER CORRELATION COEFFICIENTS

THE POINT BISERIAL. Since the point biserial correlation coefficient is a product-moment correlation coefficient, the significance of such a statistic may be evaluated by using formula 16.2. In Chapter 8 we worked a problem in which we obtained a point biserial of .52 with $N = 100$. Using this information we compute the following and enter the $t$ table with $N - 2$ degrees of freedom.

$$t = \frac{r_{pb}\sqrt{N - 2}}{\sqrt{1 - r_{pb}^2}} \tag{16.7}$$

For these data:

$$t = \frac{.52\sqrt{98}}{\sqrt{1 - .52^2}}$$

$$= \frac{.52(9.8995)}{\sqrt{1 - .2704}}$$

$$= \frac{5.1477}{.854}$$

$$= 6.03$$

which with a *df* of 98 is significant.

This coefficient could be tested more easily by entering Appendix F with $N - 2$ degrees of freedom, which in this case is 98 degrees of freedom. From this we see that the coefficient is significant well beyond the .001 level.

THE BISERIAL *r*. Since the biserial *r* is only an estimate of a product-moment correlation coefficient, the preceding method of testing the significance of an *r* should not be used. Instead we use the following procedure. Again we shall use data from Chapter 8 from the solution of the biserial. There an $r_b$ of .65 was computed with $p_1 = .46$, $p_2 = .54$, and $N = 100$.

$$s_{r_b} = \frac{(\sqrt{p_1 p_2}/y) - r_b^2}{\sqrt{N}} \tag{16.8}$$

$$= \frac{[\sqrt{.46(.54)}/.3969] - .65^2}{\sqrt{100}}$$

$$= \frac{1.2547 - .4225}{10}$$

$$= .083$$

The significance of a biserial *r* can be tested using the null hypothesis of no difference between the estimated population biserial and zero with a formula similar to equation 16.1. This is done by dropping the $r_b^2$ in formula 16.8. Thus

$$s_{r_b} = \frac{\sqrt{p_1 p_2}}{y} \left(\frac{1}{\sqrt{N}}\right) \tag{16.9}$$

$$= \frac{1.255}{10}$$

$$= .1255$$

Then $t = .65/.1255 = 5.17$, which is significant beyond the .001 level.

From formula 16.8 it can be demonstrated that the standard error of the biserial *r* is large when $p_1$ and $p_2$ are extreme splits, for example, .90–.10 and

.95–.05. In such cases one of the means is computed on a small number of cases and is less reliable than the other mean. Both this statistic and the one previously discussed are thus most reliable when the point of dichotomy is close to the center of the distribution.

THE TETRACHORIC $r$. The sampling distribution of the tetrachoric $r$ is not very well understood. At the present time formulas exist for estimating the standard error of $r_t$, from which a test of significance may be made. See Ferguson (1966, p. 246) or Guilford (1965, p. 330).

THE PHI COEFFICIENT. The test of significance for the phi coefficient involves the use of chi-square. A study of formula 8.4

$$\Phi = \frac{ad - bc}{\sqrt{klmn}}$$

and 14.3

$$\chi^2 = \frac{N(ad - bc)^2}{klmn}$$

shows that the two equations are similar in respect to the terms involved. From these two equations it can be deduced that

$$\chi^2 = N\Phi^2 \qquad\qquad (16.10)$$

Because of this relationship, testing the significance of one or more phi coefficients is very simple. Suppose that we have a series of phi coefficients based upon an $N$ of 200. For this $N$ it is possible to establish the 1 and 5 percent values at which the phi coefficient is significant. From Appendix D we obtain the 1 and 5 percent values of chi-square for 1 degree of freedom. We substitute each in equation 16.10 and solve each for phi.

$$\chi^2 = N\Phi^2 \qquad\qquad \chi^2 = N\Phi^2$$
$$6.635 = 200\Phi^2 \qquad 3.841 = 200\Phi^2$$
$$\Phi^2 = .03318 \qquad\quad \Phi^2 = .019205$$
$$\Phi = .18 \qquad\qquad\quad \Phi = .14$$

Our computed phi coefficients can then be compared with this scale. For example, a phi coefficient of .16 is significant at the 5 percent level.

THE CORRELATION RATIO, ETA. To demonstrate the testing of the significance of eta we shall use the data in Table 8.5. In that scatterplot the data were tabulated in 12 columns and the number of cases was 200. Also we found the $SS_t$ to be 1718.48 and the $SS_b$ to be 869.19.

The significance of the correlation ratio may be evaluated by means of the $F$ test, in which $F$ is the ratio of the "between" mean-square divided by the "within" mean-square. The mean-square between columns is obtained by dividing the sum of the squares for the between columns by the number of columns in the scatterplot, minus one.

$$\frac{869.19}{c - 1} = \frac{869.19}{11} = 79.02$$

The within sum-of-squares is obtained as follows:

$$\Sigma y_w^2 = \Sigma y_t^2 - \Sigma y_b^2$$
$$= 1718.48 - 869.19$$
$$= 849.29$$

The mean-square for the within columns is obtained by dividing the sum of the squares for within columns by the number of pairs in the sample less the number of columns.

$$\frac{849.29}{N - c} = \frac{849.29}{200 - 12} = \frac{849.29}{188} = 4.52$$

Then the $F$ ratio is computed,

$$F = \frac{79.02}{4.52} = 17.48$$

This $F$ is interpreted by going into the $F$ table with $(c - 1)$ degrees of freedom for the between mean-square and $(N - c)$ degrees of freedom for the within mean-square. For these data, these degrees of freedom are 11 and 188. For this number of degrees of freedom, the obtained $F$ is shown to be significant beyond the 1 percent point, and hence the correlation ratio computed is highly significant.

THE SPEARMAN RANK ORDER CORRELATION COEFFICIENT. When $N$ is small a special table is available for testing the significance of this statistic (Edwards, 1967, p. 343). If the size of the sample is over 10, the $t$ test can be made using the following formula:

$$t = \frac{\rho}{\sqrt{1 - \rho^2}} \sqrt{N - 2} \tag{16.11}$$

This is the same formula used for testing the significance of an $r$. The student may recall that a table can be used in lieu of this formula for $r$. This same table (Appendix F) may be used when the number of pairs exceeds 10.

THE TAU COEFFICIENT.

$$z = \frac{T}{s_T} = \frac{T}{\sqrt{2(2N + 5)/[9N(N - 1)]}} \tag{16.12}$$

where $N \geq 10$. When $N < 10$ tables for testing the significance of tau will be found in Siegel (1956). Although rho and tau differ in size, they are of equal power, since a null hypothesis about rho or tau for the same data will be rejected at the same level of significance if there is a relationship between the two variables being correlated. Both rho and tau for the data of Table 8.8 are significant at the .05 level. Siegel (1956) noted that the effects of ties were slight.

THE COEFFICIENT OF CONCORDANCE, $W$. In Chapter 8 a coefficient of concordance was computed when five judges ($m$) ranked 10 individuals ($N$). It was found to be .82. The significance of $W$ may be evaluated by Appendix M. Entering this table with $m = 5$ and $N = 10$, we find the 1 and 5 percent values to be .43 and .35, respectively. Since the computed $W$ of .82 exceeds the 1 percent value, we conclude that this $W$ statistic is highly significant.

## Averaging Pearson $r$'s

Since Pearson $r$'s are not equal units of measurement, they should not be added and averaged. To find an average correlation coefficient, change each $r$ to its $Z$ using Appendix G. Then average these $Z$'s, and, using the same table, convert this $Z$ back to an $r$. This is the average $r$. For example, see the table below. The sum and mean in the $r$ column are inserted to show that there is a difference between the two techniques. If the $r$'s were all fairly similar in size, little would be gained by using this technique. Also if they were all small, they could be averaged as-is, for an inspection of Appendix G reveals that up to an $r$ of .25, the values of $r$ and $Z$ are quite similar.

| $r$ | $Z$ |
|---|---|
| .68 | .83 |
| .77 | 1.02 |
| .86 | 1.29 |
| .79 | 1.07 |
| .92 | 1.59 |
| $\Sigma = 4.02$ | $\Sigma = 5.80$ |
| $\overline{X} = .80$ | $\overline{X} = 1.16$   which converts to an $r$ of .82 |

A more precise procedure is to weight each $Z$ by multiplying each by $N - 3$. These products are summed, and if we were averaging four $Z$'s, this sum would be divided by $(N_1 - 3) + (N_2 - 3) + (N_3 - 3) + (N_4 - 3)$. This weighted $Z$ is then translated to an $r$ using Appendix G.

## EXERCISES

1. A student computes an $r$ of .32 with 22 cases. By use of the appropriate formula test the significance of this $r$. Check your results by using a table for testing the significance of an $r$.

**2.** By the use of Appendix F test the significance of each of the following Pearson $r$'s.

| | $r$ | $N$ |
|---|---|---|
| **a.** | .39 | 37 |
| **b.** | .38 | 15 |
| **c.** | .60 | 8 |
| **d.** | −.72 | 12 |

**3.** We find $r$'s of −.60 and −.51, $N$ being 85 in both sets of data. Use the formula for uncorrelated $r$'s to test the hypothesis of no significant difference between these two $r$'s.

**4.** The following correlations are found among three variables in the same sample of 103 individuals. Is the correlation between 1 and 2 significantly different from the correlation between 1 and 3?

| | 1 | 2 | 3 |
|---|---|---|---|
| 1 | — | .76 | .93 |
| 2 | — | — | .85 |
| 3 | — | — | — |

**5.** On five independent random samples of equal size the following $r$'s were obtained: .93, .81, .63, .78, .68. What is the best estimate of the population $R$?

**6.** Test the significance of the following $r_{p_b}$'s.

| | $r_{pb}$ | $N$ |
|---|---|---|
| **a.** | .40 | 16 |
| **b.** | .30 | 47 |
| **c.** | .20 | 14 |

**7.** Given $N = 100$, $r_b = .40$, $p_1 = .23$, and $p_2 = .77$, test the significance of this $r_b$.

**8.** In making an item analysis a test constructor obtains the following phi coefficients, $N = 150$. Test the significance of each.

| Item | Phi |
|---|---|
| 1 | .36 |
| 2 | .18 |
| 3 | .14 |
| 4 | .42 |
| 5 | .28 |
| 6 | .15 |

**9.** In computing an eta coefficient the data were set up in 15 columns and the following sums were obtained. Test the significance of the correlation ratio.

$SS_t = 1808$
$SS_b = 908$
$N = 80$

**10.** Test the significance of the following Spearman rho's.

| rho | N |
|-----|-----|
| .53 | 19 |
| .38 | 32 |
| .62 | 17 |
| .62 | 27 |

**11. a.** Seven judges ranked 8 objects. From the rankings a coefficient of concordance of .29 was obtained. Is this $W$ significant?

   **b.** In another situation 3 judges ranked 6 objects. From the rankings a $W$ of .53 was obtained. Test the significance of this $W$.

# Reliability, Validity, and Item Analysis
## Chapter 17

One of the major uses of correlation coefficients is in the computation of reliability, validity, and item statistics. In this chapter we shall first consider a bit of measurement theory, then the various types of reliability coefficients, the standard error of measurement, the statistical aspects of validity, and finally, a brief survey of item analysis techniques.

### MEASUREMENT THEORY

In modern educational and psychological measurement, each individual score or measurement is considered as being made up of two parts, a true score component and an error score component. This can be expressed as

$$X = X_t + X_e \tag{17.1}$$

where $X$ = any raw score or unit of measurement
$\quad X_t$ = true score component
$\quad X_e$ = error score component

The error component, or error score as it is sometimes called, is held to be random. Suppose we were to weigh a number of objects. At times there would be slight errors in the way we read the scale. These deviations would not always be in the same direction. Sometimes our readings would be too high and sometimes they would be too low. In the long run, they would tend to cancel out, and the mean of all of these errors of measurement would be zero. Such variations as these are described as *random* errors and their mean is always zero. But suppose that something was operating in our method or in our apparatus which caused us always to have an error which was on the high side. Errors such as these are described as *systematic*. Their mean would not be zero.

Whenever we give a test many factors enter into the error component of an individual's score. Some of these are guessing, misreading an item, daily fluctuations in an individual's health and emotional status, and many physical factors. The size of this error component is related to the reliability of any measuring device. The smaller the error component or error score, the more reliable the measuring instrument. In its simplest form, reliability means consistency. A reliable instrument leads to measurement units which are fairly similar from time to time.

Since any score can be broken into two parts, a true component and an error component, the variance of a test can also be so treated. We can write this as follows:

$$s^2 = s_t^2 + s_e^2 \tag{17.2}$$

which reads that the variance of a test is made up of the variance associated with true scores (true variance) and the variance associated with error scores (error variance). It is assumed that the correlation between $X_t$ and $X_e$ is equal to zero.

Next we shall divide equation 17.2 through by a constant, $s^2$:

$$\frac{s^2}{s^2} = \frac{s_t^2}{s^2} + \frac{s_e^2}{s^2} = 1$$

and by transposing

$$\frac{s_t^2}{s^2} = 1 - \frac{s_e^2}{s^2}$$

In modern test theory, reliability is defined as that part of the variance which is true variance:

$$r_{tt} = \frac{s_t^2}{s^2} \tag{17.3}$$

or by substituting an identity for the right-hand part of equation 17.3 we have

$$r_{tt} = 1 - \frac{s_e^2}{s^2} \tag{17.4}$$

which reads that the reliability is 1 minus that part of the total variance which is error variance.

## METHODS OF COMPUTING RELIABILITY COEFFICIENTS

There are a number of techniques used in the computation of reliability coefficients. In this book we shall limit ourselves to a consideration of the basic and simple techniques. These are the methods of test-retest, parallel forms, split halves, and Kuder-Richardson formula 20.

## Test-Retest Method

With the test-retest method a test is administered and then at a later date the same test is readministered to the same individuals. A Pearson product-moment correlation coefficient is computed between the two sets of scores. There are various conditions which affect this technique, limiting its effectiveness. The length of the time between the two administrations of the test is important in determining the size of the reliability coefficient. In general, the longer the time between the two administrations of the test, the lower the correlation. The research literature contains many studies involving intelligence tests, tests of special abilities, and interest inventories which bear this out. If the period between the administrations is very short, such as the second test immediately following the first, individuals may remember their answers and put the same responses down the second time without making a new effort to react to the test item. Such behavior tends to make reliability coefficients artificially high. When the period between testings is short, memory may be an important factor affecting the results. As the period increases in length, learning, maturation, senescence, and many other variables may enter the situation to lower the correlation coefficient. At the present time, this method is infrequently used when determining the reliability of paper-and-pencil tests. Coefficients computed by this method are frequently called *coefficients of stability*.

## Parallel Forms Method

This technique is also referred to as the method of equivalent forms. With this technique, we administer form A to a group of individuals and follow this immediately or fairly soon with form B of the same test. These two forms of the same test are said to be parallel or equivalent, because they are made up of the same types of item covering the same materials; they have the same means and variances; and if one form correlates to a certain extent with some other measure, then the other form correlates to the same degree.

As in the previously discussed method, a Pearson product-moment correlation coefficient is computed between the two sets of scores. This coefficient is sometimes referred to as a *coefficient of equivalence*. This method is widely used at the present time. It may be called the technique with the most universal applicability. When other methods cannot be used, we can usually fall back on this one. This is especially true when we wish to determine the reliability of speed tests. The memory factor, which was important with the test-retest method, is ruled out here. But such factors as learning, growth, and change are still present, and here again, the longer the period between the two test administrations, the lower the reliability coefficient tends to be.

## The Split-Half Method

An advantage of this method is that only one test is needed for the computation of the reliability coefficient. The test papers are scored so that from every single paper we have two scores. This is usually done by counting the number of odd-numbered items answered correctly and the number of even-numbered items answered correctly. Sometimes other splits are made, such as items 1 and 2 go into the first score, 3 and 4 into the second score, and 5 and 6 into the first score. Almost any split will be acceptable, except one taking the first half of the items against the second half; because tests are usually made with the easier items first and because students tend to complete the first half of a test and not the second, a division of this sort will result in two tests of a different nature.

Each paper now has an even and an odd score on it, or two other types of score, depending on the type of split made. Again a Pearson product-moment correlation coefficient is computed between the two sets of scores. A reliability coefficient of this type is often called a *coefficient of internal consistency*.

It so happens that the reliability of a test is directly related to the length of the test. When we scored our test on an odd-even basis, we actually cut the length of our original test in half. The reliability coefficient we have computed is then the equivalent of one for a test of half of the size of our original test. We make a correction for this effect by using what is known as the Spearman-Brown formula as follows:

$$r_{tt} = \frac{2r_{oe}}{1 + r_{oe}} \qquad (17.5)$$

where $r_{tt}$ = the reliability of the original test

$r_{oe}$ = the reliability coefficient obtained by correlating the scores on the odd items with the scores of the even items

Suppose that we have a 100-item test. As a result of computing the correlation coefficient between the odd scores and the even scores, we obtain a coefficient of .84 ($r_{oe} = .84$). By substitution, we find

$$r_{tt} = \frac{2(.84)}{1 + .84}$$

$$= \frac{1.68}{1.84}$$

$$= .91$$

This Spearman-Brown formula is frequently referred to as the Spearman-Brown prophecy formula, since it is used to predict reliability coefficients. For example, suppose we have a 20-item test, and we know its reliability. By the use of this formula we can predict what the reliability of this test would be if 60 additional similar items were added. Or suppose that we have a 100-item test of known reliability, and because of time limitations, we wish to reduce it to a 50-item test. Again, by the use of this Spearman-Brown formula we can estimate the reliability of this shortened test. In its general form, the Spearman-Brown prophecy formula is written as follows:

$$r_{tt} = \frac{N'r}{1 + (N' - 1)r} \tag{17.6}$$

where $N'$ = the number of times the test is longer or shorter than the original test

$r$ = the reliability of the test which is being lengthened or shortened

Here is how this formula is used. Suppose we have a 20-item test with a reliability coefficient of .60. We wish to know what the reliability of this test would be if 80 similar items were added to make it a 100-item test. $N'$ for this problem is 5, since the new test is five times the length of the original one. The solution is

$$r_{tt} = \frac{5(.60)}{1 + 4(.60)}$$

$$= \frac{3}{1 + 2.40}$$

$$= \frac{3}{3.40}$$

$$= .88$$

To illustrate further, suppose we have a 110-item test, the length of which is reduced to 55 items. The reliability of the original test is .88. What is the reliability of the shortened test? In this case $N'$ is .5, since the new test is one-half the length of the original one. The solution follows:

$$r_{tt} = \frac{.5(.88)}{1 + (.5 - 1)(.88)}$$

$$= \frac{.44}{1 + (-.5)(.88)}$$

$$= \frac{.44}{1 + (-.44)}$$

$$= \frac{.44}{.56}$$

$$= .786$$

Results obtained by the use of this formula agree very well with those obtained when tests are changed in length and the coefficient is computed for the new test.

Internal consistency coefficients are very suitable for use in computing the reliability of academic tests. However, they must not be used with tests in which speed is an important factor. A speed test is defined as one in which the items are so easy that, given enough time, there is no reason why any individual cannot respond correctly to all of the items. On such a test, the scores on the odd-numbered items are likely to be very similar to those obtained from the even-numbered ones, and hence the resulting correlation coefficient will be exaggerated.

### Kuder-Richardson Formula 20

The Kuder-Richardson formula 20 also yields a *coefficient of internal consistency*, and as such has some of the limitations of the split-half method. If an item analysis has been made for a test (this will be discussed later in this chapter), the Kuder-Richardson formula is easily applied to the data to obtain the reliability coefficient. One of the outcomes of an item analysis is a *difficulty measure* for each item on the test. Difficulty is defined as the proportion or percentage of those responding to an item who answered it correctly. The symbol $p$ is used to represent difficulty. An item with $p = .89$ was answered correctly by 89 percent of those who responded to it.

In working a Kuder-Richardson solution, we first set up a worksheet like that shown below. Column 1 consists of the number of the item. In column 2

| (1) | (2) | (3) | (4) |
|---|---|---|---|
| | | $q$ | |
| Item | $p$ | $(1 - p)$ | $pq$ |
| 1 | .60 | .40 | .2400 |
| 2 | .30 | .70 | .2100 |
| 3 | .71 | .29 | .2059 |
| etc. | | | |

$$\Sigma pq =$$

the difficulty value ($p$) of the item is recorded from the item analysis work. Column 3 is labeled $q$, which always means $1 - p$. In all statistical work, $p + q = 1$. Column 4 is labeled $pq$, which is the product of columns 2 and 3. This column should be carried to four decimal places. Then this last column is summed.

The formula for Kuder-Richardson No. 20 is

$$r_{tt} = \frac{k}{k - 1}\left[1 - \frac{\Sigma pq}{s^2}\right] \tag{17.7}$$

where  $k$ = number of items on the test
$s^2$ = the variance of the test
$pq$ = the quantity obtained from the worksheet

### The Size of Reliability Coefficients

In general, reliability coefficients of well-made standardized tests tend to be high, .9 or above. There is no hard and fast rule that says that any reliability has to be of a certain size before any test or measuring instrument can be useful. Today we look upon reliability as a relative thing, and there are certain areas and certain techniques where reliability coefficients fall well below this

.9, and the techniques are still used and found to be very useful. Rating scales are examples of this.

As noted previously, the length of any test influences the size of the reliability coefficient for that instrument. Since reliability coefficients are correlation coefficients, they too are greatly affected by the range of scores in the sample on which the reliability correlation is computed. The more homogeneous the sample, the lower the reliability coefficient. The size of the reliability coefficient will differ when computations are based on different samples. Thus no test has a single, characteristic reliability coefficient.

## STANDARD ERROR OF MEASUREMENT

Rather than use correlation coefficients to express the reliability of a test, some prefer to use a statistic called the *standard error of measurement*. This statistic, unlike the reliability coefficient, is not affected by the range of scores of the sample tested. It tends to be about the same for samples with different variances. This standard error of measurement is the standard deviation of a sample of scores of an individual about his true score. Suppose that it were possible to administer the same test repeatedly to the same individual, without any changes occurring in the individual. These scores would be different. If we took the mean of all of them, we could use this as an estimate of the individual's true score, and the standard deviation of these scores about this true score would be referred to as the standard error of measurement.

It is not reasonable to readminister the same test over and over again to the same individual. The best we can do is estimate the standard error of measurement. This is done by the following formula:

$$s_e = s\sqrt{1 - r_{tt}} \qquad (17.8)$$

where  $s_e$ = standard error of measurement
$s$ = standard deviation of the test
$r_{tt}$ = reliability of the test

This equation is very easily derived from equation 17.4 in the following manner:

$$r_{tt} = 1 - \frac{s_e^2}{s^2}$$

First, clear fractions:

$$r_{tt}s^2 = s^2 - s_e^2$$

Then transpose terms

$$s_e^2 = s^2 - r_{tt}s^2$$

and factor the right-hand side of the equation

$$s_e^2 = s^2(1 - r_{tt})$$

by taking the square root of both sides

$$s_e = s\sqrt{1 - r_{tt}}$$

Lord (1959) found that a good estimate of the standard error of measurement may be obtained directly by the use of the following formula:

$$s_e = .432\sqrt{k} \qquad\qquad (17.9)$$

where $k$ is the number of items on the test.

## Interpretation of the Standard Error of Measurement

Suppose that we administer a test with a reliability of .91 to a group. We find that the standard deviation is 11. A student obtains a score of 77 on this test. First, we calculate the standard error of measurement.

$$s_e = s\sqrt{1 - r_{tt}}$$
$$= 11\sqrt{1 - .91}$$
$$= 11\sqrt{.09}$$
$$= 11(.3)$$
$$= 3.3$$

Since this standard error of measurement is a standard deviation, it is interpreted as such. We can now make the statement that the chances are 2 out of 3 that this individual's obtained score of 77 is not more than 3.3 units from his true score. Or we could make other statements, such as the chances are 95 in 100 that this obtained score is not more than 6.6 units from his true score, and the chances are 99 in 100 that this obtained score is not more than 9.9 units from his true score. Each of these probability statements is based upon the different areas cut off when various standard deviation units are measured from the mean. In this case the true score is taken as the mean.

The smaller this standard error of measurement, the more reliable the test and the more confidence that we can place in any score obtained by using the test. An inspection of equation 17.8 shows that if the test were perfectly reliable, that is, the reliability coefficient was equal to 1, the right-hand part of the equation would reduce to zero, and there would be no standard error of measurement. A perfectly reliable instrument would always yield true scores.

This has been but a brief summary of reliability. The student who wishes more information on reliability is referred to Magnusson (1967) or, at a more advanced level, to Nunnally (1967) and Stanley (1971). Most elementary texts on educational and psychological measurement contain considerable material on this topic. The student is also referred to the *Standards for Educational and*

*Psychological Tests and Manuals* published by the American Psychological Association (1966) for an adequate discussion of both reliability and validity.

## VALIDITY

An interested person investigating the educational and psychological literature will find that there are many names used to describe the different kinds of validity. However, in the monograph of the American Psychological Association just mentioned the many types of validity have been reduced to three general forms: content, construct, and criterion-related.

### Content Validity
Content validity is a nonstatistical type of validity that is usually associated with achievement tests. When a test is so constructed that it adequately covers both the content and the objectives of a course or part of a course of learning, it is said to have content validity. An adequate job of sampling items on the part of the test constructor is usually enough to assure that a test has content validity.

### Construct Validity
Construct validity is determined by investigating the psychological qualities, traits, or factors measured by a test. Factorial validity is an example of this. Test constructors who build tests to measure abilities and adjustment have demonstrated that the traits studied can be reduced to statistical elements called factors. For example, mental ability has been reduced to verbal, numerical, spatial, and memory factors. A test that correlates significantly with such factors as these would be said to have factorial validity, a type of construct validity.

Construct validity can be logical as well as statistical. For example, it might be hypothesized that differences in academic achievement are related to a construct called study-habit skills, other things being equal. If a study-habit inventory is made and administered, construct validity is demonstrated if the inventory scores are correlated with other evidence of study habits.

### Criterion-Related Validity
Criterion-related validity is a very common type of validity, and it is primarily statistical. It is the correlation between a set of scores or some other predictor with an external measure. This external measure is referred to as a criterion. For example, we validate intelligence tests by obtaining a set of test scores on a group such as college freshmen, and then we later obtain the grade-point averages that these freshmen made during their first semester of college. A correlation coefficient is then run between our two sets of measurements. In this illustration the grade-point averages are the criterion. Some other

variables used as criteria are ratings of performance, ratings of adjustment, units produced in a certain period of time, amount of sales, and number of errors made.

These validity coefficients tend to be much lower than reliability coefficients. An examination of the research over the years will show that they tend to fall within the band of .4–.6, with a median value of about .5. A little reflection will show why this is so. If we consider the relationship between intelligence test scores and grades, we find that more is involved than intelligence in obtaining grades. Such factors as motivation and interests of the student, grading practices of the teachers, emotional adjustment, health of the student, and time available for study all influence the grades that a student obtains. An examination of the intelligence test scores for any classroom will show that there are one or two individuals who have high measured ability and low achievement or lower ability and high achievement. Individuals like these reduce the size of the computed correlation coefficients.

When ratings are used as criteria, we find that the ratings tend to be unreliable because of the nature of the trait being rated, lack of knowledge of the ratees by the judges, and other errors associated with ratings. Again, the validity coefficients tend to be in the middle of the range. In industrial situations where tests of special ability are widely used, the situation is no better. Here the number of units produced in a given period of time is used as the criterion. Many factors affect the number of units turned out by each worker, and we again have to consider such factors as motivation of the workers, speed of the machines, availability of raw materials, and such physical aspects of the environment as noise, light, heat, and presence or absence of other workers.

As was pointed out in Chapter 9 the logical step after computing a validity coefficient is to set up the regression equation for the data, so that predictions can be made when future individuals take the same test. For example, in predicting freshmen grades in a particular school a regression equation was set up. When new students or applicants now take the admission tests, their scores can be referred to the regression equation and a probability statement made for their success in that particular school. This would assume that both the type of freshmen and the grading practices of the school stay about the same. In actual practice, several predictors are used. Freshmen grade-point indexes might be predicted using a test as above together with high school rank. Then a multiple $R$ (Chapter 8) would be computed between freshmen grades and the combined effects of the predictor test and high school rank. Additional predictors might also be added, such as scores on a mathematics test or scores on interest and personality scales. The use of more than one variable raises the correlation coefficient and thus leads to better predictions.

As we have previously learned, when we make predictions, we attach a probability statement to our prediction. This was done using the standard error of estimate which is as follows (Chapter 9):

$$s_{yx} = s_y\sqrt{1 - r_{xy}^2}$$

$R$ is used instead of $r$ if we have multiple regression. To review the use of this statistic, suppose that in a given situation we find the standard error of estimate to be .2 in predicting freshmen first-semester index from a test of academic ability. For a given student whose score on the predictor $(X)$ is 120, we find that the predicted grade-point average $(Y)$ as read from the regression line is 5. We can then state that the chances are 2 out of 3 that the grade-point average for this student, whose score on a test of mental ability is 120, will be between 4.8 and 5.2. If we were to add and subtract 2 standard errors of estimate $(2 \times .2)$, we could state that the chances are 95 in 100 that this individual's index will fall between 4.6 and 5.4; that is, his predicted index plus and minus .4.

An examination of the formula for the standard error of estimate shows that if $r$ is perfect (equal to 1), the right-hand part of the equation reduces to 0, and there is no standard error of estimate. From this it follows that the higher our validity, the better we can predict. Let us look at the efficiency of these validity coefficients. We can start by noting the relationship between the coefficient of determination (a name given to $r^2$) and the coefficient of nondetermination $(k^2)$:

$$r^2 + k^2 = 1 \tag{17.10}$$

then

$$k^2 = 1 - r^2$$

and

$$k = \sqrt{1 - r^2} \tag{17.11}$$

Instead of $r + k$ equaling 1, the sum of their squares equals 1. When equation 17.11 is solved for various values of $r$, the corresponding values of $k$ obtained are shown in Table 17.1. From this it will be noted that as $r$ increases, $k$ decreases but at a much slower rate. When $r$ has increased to .50, $k$ is equal to .87. When $r$ is equal to .70, the value of $k$ is very close to it. Even when the value of $r$ is in the .90's, the $k$ values are considerable.

This $k$ statistic is of little value in itself. You may have noticed that it is one of the terms of the formula for the standard error of estimate. The formula for the standard error of estimate could be written

$$s_{xy} = sk$$

We also use this statistic in making a statement about the predictive efficiency of our validity coefficient. Suppose that we have a situation where the validity

**Table 17.1** Relation-
ship
Between
$r$ and $k$

| $r$ | $k$ |
| --- | --- |
| 1.00 | .000 |
| .95 | .310 |
| .90 | .440 |
| .80 | .600 |
| .70 | .710 |
| .60 | .800 |
| .50 | .870 |
| .40 | .920 |
| .30 | .950 |
| .20 | .980 |
| .10 | .995 |
| .00 | 1.000 |

coefficient between predictor and predicted variable is .45. First we solve for $k$:

$$k = \sqrt{1 - r^2}$$
$$= \sqrt{1 - .45^2}$$
$$= \sqrt{1 - .2025}$$
$$= \sqrt{.7975}$$
$$= .89$$

Then we compute our index of forecasting efficiency by subtracting the obtained $k$ from 1 and multiplying the remainder by 100:

$$E = (1 - k)100 \qquad\qquad (17.12)$$
$$= (1 - .89)100$$
$$= (.11)100$$
$$= 11\%$$

We can interpret this by saying that with a validity coefficient of .45 we can predict 11 percent better than we would have been able to had there been no relationship between the two variables. In other words, we have decreased the size of the error of prediction by 11 percent.

This discussion may make it appear that tests are not much good in predicting future behavior. But the point to remember is that we are usually not interested in an individual's exact grade-point average or score on another test. We are more concerned about whether he will pass or fail as a student, for instance, and be successful or eliminated as a trainee. Rather than knowing

**Table 17.2**    Data with a Dichotomized
Criterion and a Predictor
with a Cut-Off Point

|  | Predictor $X$ | | |
|  | High | Low |  |
| Criterion $Y$ High | $a$ 60 | $b$ 20 | 80 |
| Low | $c$ 40 | $d$ 80 | 120 |
|  | 100 | 100 | 200 |

his specific score, we are interested in his approximate location on the other variable. Our tests usually do an efficient job in placing individuals in this manner.

## An Index of Predictive Association

Goodman and Kruskal have proposed a statistic, *lambda,* $\lambda$, which gives an indication of the reduction of errors made in a prediction scheme. Lambda is defined as follows:

$$\lambda = \frac{\Sigma f_c - f_r}{N - f_r} \tag{17.13}$$

where $f_c$ = sum of the largest cell frequencies in the columns
  $f_r$ = largest row total

To illustrate the use of lambda we shall use the data in Table 17.2. For these data the criterion is dichotomized and a cut-off appears in the predictor, making it possible to set up the data in a 2 × 2 table. With this type of prediction scheme two types of error are possible: (1) some individuals who are low on the predictor are high on the criterion (false positives, cell *b*); and (2) some who are high on the predictor are low on the criterion (false acceptances or false negatives, cell *c*).

Computing lambda for these data we have

$$\lambda = \frac{(60 + 80) - 120}{200 - 120}$$

$$= \frac{20}{80}$$

$$= .25$$

This lambda of .25 indicates that, given the information about the prediction of $X$, the probability of error in predicting $Y$ from $X$ is reduced on the average 25 percent. For the same data the phi coefficient is

$$\Phi = \frac{ad - bc}{\sqrt{(a + d)(a + c)(b + c)(c + d)}}$$

$$= \frac{4800 - 800}{\sqrt{96,000,000}}$$

$$= \frac{4000}{9800}$$

$$= .41$$

This phi of .41 tells us that there is a high correlation between the predictor and the criterion, but nothing more. A test of significance can be applied to this phi of .41 ($X^2 = N\phi^2$). This results in $X^2 = 33.62$, which is significant beyond the .001 level. There are many times when a significant phi is obtained and lambda equals zero (Curtis, 1971). In these cases the data have no predictive value. Lambda and phi are equal when both predictor and criterion have 50-50 splits. Lambda may also be applied to contingency tables larger than a 2 × 2 table. In this case it would be compared with Cramér's statistic, $\phi'$, rather than phi (Hays, 1965).

## Attenuation

Since in any given situation both variables are unreliabale to a certain extent, any correlation coefficient computed between the two would tend to be lower than the true, or theoretical, relationship between the two variables. This lowered $r$ is referred to as *attenuated*. Formulas have been developed to correct for this attenuation. The most general one is

$$r_c = \frac{r_{xy}}{\sqrt{r_{xx}r_{yy}}}$$  (17.14)

where     $r_c$ = the correlation between $X$ and $Y$ corrected for attenuation

$r_{xx}$, $r_{yy}$ = the reliability coefficient of test $X$ and criterion $Y$ respectively

$r_{xy}$ = the computed validity coefficient

In practice, it is best not to use this formula, for the reliability of our tests is at present quite high, and it is doubtful if it will ever be considerably improved. We have to use these tests in our daily work, and we must be aware that they are not perfect and that any prediction made with them is likely to be inaccurate to a certain extent. However, many of the criterion measures that we use are low in reliability because of inaccuracies inherent in our measurement techniques. Some research workers therefore feel justified in correcting for attenuation in the criterion scores only. Equation 17.14 then becomes

$$r_c = \frac{r_{xy}}{\sqrt{r_{yy}}}$$  (17.15)

It should be emphasized at this point that no test in itself contains validity or reliability. After all, a test is merely a piece of paper with symbols printed on it. Whether or not any test has validity or reliability depends on the manner in which the test is used. For a given situation, a particular test may have high validity and reliability; in another situation both may be low. It might be added that any test or scale has any number of validity or reliability coefficients, depending upon how, when, where, and by whom it was used.

## ITEM ANALYSIS

To determine the merit of any test item, test results must be subjected to an item analysis. As a result of this item analysis, three kinds of information are obtained concerning the item: (1) the difficulty of the item, (2) the discrimination index of the item, and (3) the effectiveness of the distractors. The first of these, the difficulty of the item, is the proportion of individuals who answer the item correctly. The second, the discrimination index, is a measure of how well the item separates two groups. The purpose of most tests is to "spread out" the individuals taking it. The item which separates good students from poor ones, adjusted individuals from maladjusted ones, or those with artistic ability from those without such ability is said to discriminate. The third result, the effectiveness of the distractors, applies only to multiple-choice items. But since most of our standardized tests and many of the teacher-made tests use this type of item almost exclusively, a thorough item analysis is made to see how the incorrect responses in the multiple-choice item are working. We shall now discuss each of these results of an item analysis in detail.

### Item Difficulty

As previously defined in this chapter, the difficulty of an item ($p$) is the proportion of individuals who answer an item correctly. In a power test, that is, a test in which everyone has had a chance to read every item, the calculation of the item difficulty offers no problems. But when speed is a factor in a test—and many of our tests are of this nature—adjustments have to made in the calculation of the item difficulty. Basically, item difficulty is the result of dividing the number who answered the item correctly by the number who took the test. On a speed test many people never get to the items near the end of the test. In that case the number of correct answers should be divided by the number of individuals who reached the item. The usual procedure in determining how far an individual went on a test is to assume that when the test period was over the individual was working on the item following the last one that he marked on his answer sheet. All items beyond this one are not assumed to have been reached. On the first half of any test such corrections make little difference; but unless something like this is done on the items near the end of the test, both the item difficulty values and the indexes of discrimination will be spuriously large. It also follows that as a result of corrections like this the size of the sample upon which the item statistics depend becomes much smaller, and correspondingly the reliability of the item statistics is reduced. Perhaps in constructing tests, even speed tests, it is a good idea to give the examinees enough time to answer all of the items. One way to do this is to attach a group of buffer items at the end of the test to keep those who finish first busy. These items are not scored.

With most tests we wish to discriminate throughout the range. We wish to be able to grade students on the conventional letter system or on some other classification scheme using the results of the test. To have a test which discriminates over the entire range, items are selected which range from very easy to very difficult and which average, in the long run, to a difficulty value of 50 percent. From the viewpoint of item difficulty, a well-made test starts with a few very easy items, continues with the items of increasing difficulty, and ends with a few items which only a very few of the examinees answer correctly. There would be more difficulty values clustering about the center than at either extreme, but there would be a balance so that the average item difficulty is approximately 50 percent.

Sometimes standardized tests contain directions that the obtained scores must be corrected for guessing. Or a teacher may feel that the scores on an objective test are improved if a penalty is provided for those who guess on objective tests. The usual formula for correcting for guessing is

$$\text{Score} = R - \frac{W}{k-1} \tag{17.16}$$

where $R$ = the number of right answers on a student's paper

$W$ = the number of wrong answers on a student's paper

$k$ = the number of possible choices in the test item

For the ordinary true-false item, $k = 2$. For five-response multiple-choice items, $k = 5$. It should be noted that there is far from universal agreement on the use of correction formulas. There are some who argue that they should not be used at all because in many cases they do not correct for guessing but rather for marking the incorrect answer because of misinformation or on account of errors in calculation. When such a formula is used, all mistakes are assumed to be the result of blind guessing and in many cases this cannot be proved. If a test is well made, appropriate for those who are taking it, and possesses reasonable time limits, there is little justification for the use of a formula to correct for guessing.

## Item Discrimination

There are two general ways of demonstrating item discrimination: (1) a test of the significance of the difference between two proportions, and (2) correlational techniques. In the first method the percentage or proportion of individuals who answer the item correctly in the high group is tested against the proportion in the low group. If the difference is a significant one, the item is accepted as being one which discriminates. A disadvantage of this technique is that although it selects those items that discriminate, it does not reveal how well each one discriminates (see Chapter 13).

In much item analysis work, the student will find that it is customary to compare the responses in the top 27 percent of the papers with those in the lowest 27 percent of the papers. This goes back to the early item analysis work of T. L. Kelley, who demonstrated that when the high and low groups were made up of the top and bottom 50 percent, those papers clustering about the median had little influence on the discrimination index. A comparison of the upper 40 percent with the lower 40 percent produced more clear-cut results. This difference becomes increasingly sharp as the papers in the high and low groups become more extreme. Also as the number of papers in the two groups decreases, the standard errors increase and the item statistics become more unreliable. Kelley demonstrated that when the responses of the individuals in the upper 27 percent were compared with those in the lower 27 percent, the ratio of the difference between the means of the two groups over the probable error[1] of the difference between the means was at a maximum. Since that time many workers have accepted his findings and used this value of 27 percent in selecting the high and low groups. A major reason for this is that a number of useful computing devices have been made which are based on high and low groups consisting of the upper and lower 27 percent. That other percentages will work just as well has been suggested by research which compared results based on groups using 27 percent with those using a series of other percentages.

In the correlational approach to item analysis, a correlation coefficient is computed that shows the relationship of the responses to the total test score. In other words, it shows how well the item is doing what the test itself is doing. Suppose that we have the responses of 200 individuals to a 56-item test. To begin, we are going to employ a method that uses all of the papers. We set up a distribution of scores on the $y$ axis of our scatterplot, and since the test item is usually scored on a right or wrong basis, we have only two categories on the $x$ axis (see Table 17.3).

We take the first paper, note the total score and whether or not the item was answered correctly, and tally the item at the appropriate place on the frequency distributions. Then we take the next paper and tally the response. We continue this process for all of the papers. When the tallying is finished, we may compute either a point-biserial or biserial $r$. If the distribution of test scores is dichotomized as the response to the item is, we may use either the tetrachoric $r$ or the phi coefficient. We would have to repeat this same process for the other 55 items on this test.

---

[1] Probable error = .6745 standard error. This is an outmoded statistic.

**Table 17.3**  Responses of 200
Students to One Item
of a 56-Item Test

| Test Score | Right | Wrong |
|---|---|---|
| 51–53 | 6 | 1 |
| 48–50 | 18 | 2 |
| 45–47 | 22 | 4 |
| 42–44 | 18 | 7 |
| 39–41 | 28 | 6 |
| 36–38 | 15 | 8 |
| 33–35 | 12 | 12 |
| 30–32 | 8 | 10 |
| 27–29 | 5 | 4 |
| 24–26 |  | 4 |
| 21–23 | 1 | 4 |
| 18–20 |  | 3 |
| 15–17 |  | 1 |
| 12–14 |  | 1 |
|  | 133 | 67 |

Admittedly, this is a tremendous amount of clerical work and if it were
not for a large number of computing and data processing devices that have
been set up, many would think twice before getting involved in item analysis.
Flanagan was one of the first to produce one of these shortcuts. From his
chart correlation, coefficients are read directly when one enters the chart with
the percent answering the item correctly in the upper or lower 27 percent of
the papers.

**Table 17.4**  Item Analysis Worksheet

| (1) Item | (2) Upper 27% Correct | (3) Lower 27% Correct | (4) Difficulty $p$ | (5) Discrimination $r$ |
|---|---|---|---|---|
| 1 | .90 | .50 | .70 | .48 |
| 2 | .70 | .25 | .48 | .46 |
| 3 | .60 | .10 | .35 | .56 |
| 4 | .40 | .60 | .50 | −.20 |
| 5 | .28 | .02 | .15 | .55 |
| 6 | .40 | .10 | .25 | .40 |
| 7 | .70 | .70 | .70 | .00 |
| 8 | . | . |  |  |
| ⋮ | . | . |  |  |
| 50 | .45 | .40 | .42 | .05 |

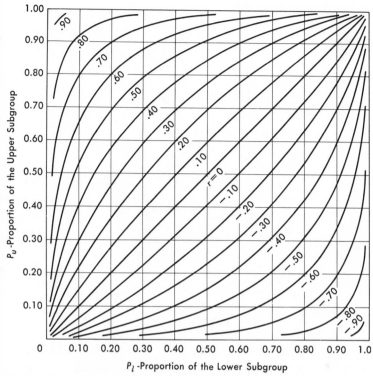

**Figure 17.1**  An abac for estimating biserial coefficients of correlation between item and total score when the sample has been restricted to the highest and lowest 27 percent of the total-score distribution.

Some data are presented in Table 17.4 which will be used to illustrate how Flanagan's method is used. Suppose that we have a large number of papers. From these we take 370 at random from the large group, and we separate the highest and lowest 27 percent of these for study. Since 27 percent of 370 is 100, we put 100 papers in the high group and 100 in the low group. Next we tally the responses of the individuals in each group and change these obtained frequencies to percentages or proportions. These values appear in columns 2 and 3 of Table 17.4. In column 4 are the difficulty values ($p$) of the items or, better, an estimate of the difficulty values obtained by averaging the proportions in columns 2 and 3. The correlation coefficients in column 5 are obtained for example from Figure 17.1 by entering the figure for item 1 with .90 on the $y$ axis and .50 on the $x$ axis. We then go up to .90 and then over to .50. The point where .90 and .50 intersect is the estimated value of $r$. It is a point estimated between two of the arcs of the figure. In this case we find $r$ to be .48. In this manner we read all the discrimination indices from the figure.

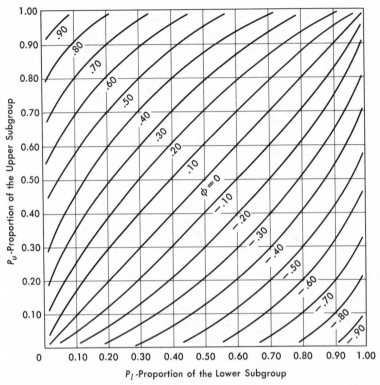

**Figure 17.2** An abac for graphic estimates of the phi coefficient when one variable has an even division of cases in two categories.

In Figure 17.2 another abac is provided for estimating phi coefficients using a worksheet similar to Table 17.4. Taking the first item, where the proportions are .90 and .50, we go up to .90 and across to .50 and read the value at the point of intersection. In this case it is approximately .43. These values will not be the same as those obtained with Figure 17.1; but if we apply each method separately to the same data, almost the same items will be selected as being discriminating ones. We may also use Guilford's abac with groups of sizes different from the upper and lower 27 percent. The only requirement is that there be an equal number in each group. That is, we could use it if we contrast the upper 50 percent with the lower 50 percent or the upper 10 percent with the lower 10 percent. In Chapter 8 a method was presented for estimating point-biserial correlation coefficients by the use of an abac (Figure 8.1). In addition a short method was given for obtaining tetrachoric correlation coefficients. We may use both of these statistics as indices of item discrimination.

To test the significance of these various indices of discrimination, we use one of the methods described in Chapter 16. Flanagan's r is often taken as an estimate of the Pearson r. Appendix F, constructed for estimating the significance of the Pearson r, may be used in testing the significance of Flanagan's r. In selecting items for use in the next edition of a teacher's test or for a standardized test, items are chosen on the basis of both their difficulty and discrimination indices. Items are selected to obtain a spread of difficulty values, and items are chosen that demonstrate significant differences between the two criterion groups. For example, items like 4 and 7 in Table 17.4 contribute nothing to the functioning of a test and would not be chosen.

### Item Analysis When the Number of Papers Is Small
Item analysis statistics are essentially large sample statistics, and probably the foregoing methodology should not be used when the total number of papers is less than 100. Data obtained from item analyses of small samples will produce item statistics that are unreliable. However, in many situations, all that one has is about 30 papers. With such samples a crude sort of item analysis can be made by using the upper ten papers versus the lower ten. The percentage who answer the item correctly in each group can then be determined, and from these an estimate of the item difficulty can be made. Estimates as to whether or not an item is discriminating can be made by an inspection of the number of individuals in the high and low groups for each item. Those items with no discrimination or a negative discrimination can be spotted at once. To determine the discriminating items, we might adopt a standard such as there has to be a difference of at least 2 between the frequency of the high and low group before the item is considered to be discriminating. While this technique is crude it is better than nothing and will be helpful in making better classroom tests.[2]

### The Computer and Item Analyses
In recent years item analysis work has been greatly simplified and speeded up by the use of computers. When the computer is used for scoring and analysis, students use a special pencil to make their responses on small cards. In addition to scoring the cards, the computer provides a printout of the number of rights, wrongs, and omissions for each student; it also changes each examinee's score to a standard score and puts these standard scores into a frequency distribution; further, it computes the mean and standard deviation and obtains the proportion responding to each part of each item. Obviously the proportion corresponding to the keyed correct response is the difficulty of the item. The computer then provides the correlation between responses

---

[2] Item analysis techniques are discussed in detail in educational or psychological measurement texts (Downie, 1967, Ebel, 1972, Henryssen, 1971).

to each part of each item and total test scores. Again the correlation for the response keyed as being correct is the discrimination index of the item. It might be added that the computer also produces the reliability coefficient (*KR* 20) and the standard error of measurement as part of a routine test analysis. The computer does all of this in a very short time, but time on a computer is expensive. Many people using tests—probably the majority of us —do not have access to a computer. Nevertheless, tests can be improved, and the methods described in this book are provided in an attempt to show how this may be done.

### The Effectiveness of Distractors

In estimating the effectiveness of the different distractors of the multiple-choice item, two procedures may be used. With the first procedure the responses to the different distractors of each test item can be counted for the high and low groups. Good distractors are those which are selected more frequently by members of the low-scoring group. When a reversal is found, that is, a particular distractor is more popular in the high group than in the low group, the test maker may revise the distractor, trying to make it less appealing to the better students; or he may rewrite the entire item. Analysis of distractors may also reveal options used by no one in either the high or the low group. Such distractors contribute nothing to the test. When such nonfunctioning distractors appear, they should be revised in an attempt to make them useful.

The other method of studying the distractors is to obtain the mean score of the individuals who respond to each of the distractors as well as to the correct answer. When this technique is used, it is not unusual to find that the mean score for individuals who selected a certain distractor is higher than the mean score for those who selected the correct answer. Here again we have evidence of a distractor lowering the discriminatory power of an item. Good distractors will yield lower mean scores than the papers associated with the correct answer. Also, with this technique certain distractors may appear that are selected by no one.

### EXERCISES

**1.** Below are tabulated the responses of 25 individuals to each of the items on a 15-item test.

    **a.** Obtain an odd and even score for each individual and then compute the reliability of the test.

    **b.** Use Kuder-Richardson formula 20 to compute the reliability of the test.

    **c.** Calculate the standard deviation of the test scores.

    **d.** What is the standard error of measurement for this test?

    **e.** Select an individual with a score of 10 on this test and interpret the standard error of measurement.

| Individual | 1 | 2 | 3 | 4 | 5 | 6 | 7 | 8 | 9 | 10 | 11 | 12 | 13 | 14 | 15 | Score |
|---|---|---|---|---|---|---|---|---|---|---|---|---|---|---|---|---|
| A | + | + | + | 0 | + | + | 0 | 0 | 0 | 0 | + | 0 | 0 | 0 | 0 | 6 |
| B | + | + | + | + | + | + | 0 | + | + | + | 0 | + | + | + | + | 13 |
| C | 0 | + | + | + | + | + | 0 | 0 | 0 | + | 0 | + | + | + | 0 | 9 |
| D | + | 0 | + | 0 | 0 | + | + | + | 0 | 0 | + | 0 | 0 | + | 0 | 7 |
| E | + | + | + | + | + | + | + | + | + | 0 | + | + | + | + | + | 14 |
| F | + | 0 | 0 | + | + | + | + | + | + | + | + | + | + | 0 | + | 12 |
| G | + | + | + | 0 | 0 | + | + | + | 0 | 0 | + | + | + | + | + | 11 |
| H | 0 | 0 | + | + | 0 | + | + | 0 | 0 | 0 | 0 | 0 | 0 | 0 | 0 | 4 |
| I | + | + | 0 | + | + | 0 | + | + | + | + | + | 0 | + | + | + | 12 |
| J | + | + | + | + | 0 | + | + | + | + | 0 | + | + | + | + | + | 13 |
| K | + | 0 | 0 | + | 0 | + | 0 | 0 | + | 0 | 0 | + | 0 | + | 0 | 6 |
| L | 0 | + | + | + | + | + | 0 | 0 | 0 | 0 | 0 | + | 0 | 0 | + | 8 |
| M | + | + | + | 0 | 0 | + | 0 | 0 | 0 | 0 | + | 0 | + | + | + | 8 |
| N | + | + | + | + | + | + | + | + | 0 | + | + | 0 | + | + | + | 13 |
| O | + | + | + | + | 0 | + | 0 | 0 | 0 | + | 0 | + | + | + | 0 | 9 |
| P | + | + | + | + | + | 0 | + | + | + | 0 | 0 | 0 | 0 | 0 | 0 | 8 |
| Q | + | + | + | + | + | + | 0 | 0 | 0 | + | 0 | 0 | + | 0 | 0 | 8 |
| R | + | + | + | + | + | + | 0 | + | + | + | 0 | + | 0 | 0 | 0 | 10 |
| S | 0 | + | 0 | 0 | 0 | + | 0 | + | 0 | 0 | 0 | 0 | 0 | 0 | 0 | 3 |
| T | + | + | + | + | + | + | + | + | + | + | + | + | + | + | + | 15 |
| U | + | + | 0 | + | + | + | 0 | + | 0 | + | 0 | 0 | 0 | + | + | 9 |
| V | 0 | + | + | + | 0 | + | 0 | + | 0 | 0 | + | 0 | 0 | 0 | 0 | 6 |
| W | + | + | 0 | + | + | + | + | + | + | + | + | 0 | + | + | + | 13 |
| X | + | + | + | + | + | + | 0 | + | + | + | + | 0 | + | + | 0 | 12 |
| Y | + | + | + | + | 0 | + | + | + | + | + | + | + | + | + | + | 14 |
| Number correct | 20 | 21 | 19 | 20 | 16 | 22 | 12 | 18 | 11 | 13 | 16 | 10 | 17 | 15 | 13 | 243 |

**2. a.** A certain 25-item test has a reliability coefficient of .72. To this are added 75 similar and well-made items. What is the reliability of the new test?

   **b.** Suppose that a test made of 150 items takes 80 minutes to administer. The author of the test decides to shorten it to 50 items so that it will be a 25–30 minute test. The reliability of the original test is .93. What might you expect the reliability of the shortened test to be?

   **c.** Another test has a reliability coefficient of .60. This test is made up of 20 items. How many well-made and similar items will have to be added to increase the reliability to .90?

**3. a.** A certain test has a reliability coefficient of .92 and a validity coefficient of .48 with criterion $Y$. The standard deviation of this criterion is 10.2 and that of the test is 9.6. Calculate the standard error of estimate when predicting $Y$ from $X$.

   **b.** Suppose that from a regression equation for the above data you calculate a criterion score of 77 for an individual. How would you associate the standard error of estimate with this predicted score?

   **c.** Calculate the index of forecasting efficiency for this validity coefficient. What does this mean?

**4.** Sometimes we square the validity coefficient and call it a coefficient of determination. We then interpret this as being the amount of variance in $Y$ that is accounted for or determined by the variance in $X$. With this same line of reasoning, how might one best describe $k^2$?

**5.** Take the data in problem 3a and apply the attenuation formula for correcting test scores. Exactly what have you done? Can you defend the use of this statistic?

**6.** Investigate the pros and cons of the use of correction formulas in scoring test papers. Any educational and psychological measurement book may be consulted.

**7.** If test papers are available, apply an item analysis, calculating for each item the difficulty value and the discrimination index. Use any of the computing charts or devices discussed in this chapter. Summarize your results. Examine the items that are not good. Have you any ideas why they did not hold up in the item analysis?

**8.** Suppose that we have a dichotomized criterion and a predictor dichotomized as in the data below.

| Predictor | Criterion High | Low |
|---|---|---|
| High | 70 | 30 |
| Low | 50 | 50 |

    **a.** Compute the Goodman-Kruskal statistic for these data.
    **b.** Next compute phi for the same data.
    **c.** What does the lambda statistic tell you that the phi coefficient does not?

# Distribution-Free Statistical Tests
## Chapter 18

Although a few distribution-free or nonparametric statistics have been known and used for many decades, it was not until recently that statisticians have devoted much attention to them. The modern period, so to speak, began in the mid-1930s, and growth has continued at an accelerated rate up to the present. In distribution-free or nonparametric tests no assumptions are made about the precise form of the sampled population. Sometimes certain assumptions are made, such as that a distribution is continuous, or that the sampled populations have identical shapes or distributions symmetrical about the same point. However, never are the assumptions so elaborate as to imply a completely specified population distribution, as is the case with the normal curve. Most of the statistics studied so far—with the exception of centiles, chi-square, and several of the correlation coefficients—have implied a normal distribution of the parameter, and $z$, $t$, and $F$ are parametric tests.

Bradley (1968) makes a detailed listing of the advantages and disadvantages of distribution-free statistical tests. Some of the major points he makes are:

**1.** *Simplicity of derivation.* The derivation of classical tests requires a level of competence in mathematics far above that attained by the typical research worker, whereas most distribution-free statistics can be derived using simple combinational formulas.

**2.** *Ease of application.* Frequently the only mathematical operations needed for distribution-free statistical tests are ranking, counting, adding, and subtracting.

**3.** *Speed of application.* When sample size is small or moderate distribution-free models are generally faster than parametric techniques.

**4.** *Scope of application.* Because they are based on fewer and less elaborate assumptions than the classical tests, distribution-free statistical tests can be correctly applied to a much larger class of populations.

**5.** *Susceptibility to violation of assumptions.* Since the assumptions are fewer and less elaborate with nonparametric statistical tests, they are less susceptible to violation. These violations are easier to detect with nonparametric tests. The effect of the violation of assumptions is important with both types of statistics, but Bradley feels that the effects of violation of assumptions can be more readily and economically taken care of with distribution-free statistical tests.

**6.** *Type of measurement required.* Distribution-free statistical tests usually require at least ordinal data, though sometimes nominal data may be used. Parametric tests generally require measurements on an interval or ratio scale.

**7.** *Influence of sample size.* When sample sizes are $\leq 10$, distribution-free statistical tests are easier, quicker, and only slightly less efficient—even if all the assumptions of the parametric test have been met. At such sample size, violations of parametric assumptions are most devastating; hence in these cases nonparametric tests are most appropriate. As sample size increases the nonparametric tests become more laborious and time consuming and frequently become a much less efficient statistical test.

**8.** *Statistical efficiency.* In terms of practicality—the amount of human effort required to conduct and analyze an experiment—nonparametric statistics are frequently more convenient than their parametric counterparts. In terms of a mathematical criterion of statistical efficiency, distribution-free tests are often superior or equal to their parametric counterparts when the assumptions of the nonparametric test are met, but the assumptions of the parametric test are not. If both tests are applied when all assumptions of the parametric test can be met, distribution-free statistics are only slightly less efficient with an extremely small sample size; but they become increasingly less efficient as sample size increases.

Moses (1952) also noted that nonparametric tests were (1) easier to apply, (2) applicable to rank data, (3) usable when two sets of observations come from different populations, (4) the only alternative when sample size is small, and (5) useful at a specified significance level as stated (whatever happened to be the shape of the distribution from which the sample distribution was drawn). He also noted that among the disadvantages of nonparametric tests was their lower statistical efficiency.

All writers on this subject show that parametric statistical tests are more powerful than nonparametric tests. The "power" of a statistical test refers to the probability of rejecting the null when it is actually false. In other words, power is equal to 1 minus the probability of making a Type II error, or $1 - \beta$. One important factor contributing to the power of a statistical test is the sample size, $N$. As $N$ increases, the probability of making a Type II error decreases and hence the power of the test increases. It will be noted that some of the statistical tests discussed in this chapter are stated as being more powerful than others.

We will now examine a few of the more widely used distribution-free tests of significance, together with their application. The student who wishes to study this topic in more detail may refer to Bradley (1968), Kraft (1968), Siegal (1956), and Tate and Clelland (1957).

## TESTS FOR CORRELATED SAMPLES

### The Sign Test

The sign test is among the simplest of the tests for correlated samples. In discussing this test, we shall refer to Table 18.1. In this table data are found in pairs. Individuals are matched and assigned to one of the two groups at random at the beginning of the experiment. A treatment is applied and the resulting scores for the pairs of individuals are shown in the table. In using a test like this it is assumed that the variable under consideration is distributed continuously and that both members of each pair are treated similarly except for the experimental variable.

In analyzing data we consider each pair of scores separately, noting whether the second score is smaller or larger than the first. We then assign the appropriate sign as shown. Ties may be handled by dropping them from the computations or by giving one tie a plus sign and the next a minus sign, and so on.

The null hypothesis tested here is that the median change is zero. This hypothesis is rejected if there are too few of any one sign. Where there are ten or less pairs we apply a test of significance by the use of the binomial expansion, with $p = .5$ and $N = $ the number of pairs. For our data, $N = 10$. From Table 18.1, we note that we have 8 pluses and 2 minuses. By chance we would expect 5 of each. The question is, does this frequency of 8 pluses differ significantly from what we would expect by chance? With the use of the binomial expansion we find the probability of obtaining 8 or more plus signs is equal to .0009766 + .0097656 + .0439452, which adds up to .0547.

**Table 18.1**  The
Sign Test

| $X$ | $Y$ | $X - Y$ |
|-----|-----|---------|
| 16 | 4 | + |
| 12 | 18 | − |
| 22 | 10 | + |
| 16 | 14 | + |
| 14 | 12 | + |
| 10 | 14 | − |
| 20 | 10 | + |
| 18 | 12 | + |
| 10 | 4 | + |
| 22 | 12 | + |

The process of doubling this probability for a two-tailed test gives a $p$ of .1094, or .11. This forces us to let the null hypothesis stand.

When the number of pairs is greater than ten, the results may be evaluated using a $z$ score and the formula for the mean and standard deviation of the binomial (formulas 10.2 and 10.3). A correction for discontinuity must also be made (see description in Chapter 10).

Since this test uses only information on the direction of the differences between pairs, it is not the most efficient. The next test combines the sign of the difference with the relative size of the difference. It is a more powerful test.

### Wilcoxon's Matched-Pairs Signed-Ranks Test

To illustrate Wilcoxon's matched-pairs signed-ranks test of significance, we shall again use the data presented in Table 18.1, which have been recopied into Table 18.2. First we obtain the difference between each pair of scores. Then we rank the absolute values of these differences, "absolute" meaning to disregard signs in our ranking. For these data the two differences of 2 are the smallest. Since these occupy positions one and two of the ranking, we give the average of the two ranks, this being 1.5. The next smallest difference is $-4$. We give this a rank of 3. We continue in this manner until all the differences are ranked. Whenever a difference of zero appears we disregard it in the computations.

In the last two columns of Table 18.2, the ranks have been summed according to the sign of the differences. The smaller of these is taken as Wilcoxon's $T$ statistic. If there were no difference between the two groups, $T$ would equal $\bar{T}$, the mean sum of the ranks. This latter statistic can be obtained by the following formula:

$$\bar{T} = \frac{N(N + 1)}{4} \qquad\qquad (18.1)$$

For these data,

$$\bar{T} = \frac{10(10 + 1)}{4}$$

$$= \frac{10(11)}{4}$$

$$= \frac{110}{4}$$

$$= 27.5$$

If the obtained $T$ differs significantly from $\bar{T}$, we can reject the null hypothesis. For smaller sums of ranks, for $N$'s between 7 and 25, Wilcoxon has developed a table (reproduced as Appendix I) which may be entered directly to test the significance of $T$. For our problem, let us assume a two-tailed test with alpha equal to .05. From Appendix I, for $N = 10$ we see that a $T$ of 8 or less is significant at the .05 level for the two-tailed test. Our $T$ is exactly

**Table 18.2**   Data of Table 18.1 Used to Illustrate the
Wilcoxon Matched-Pairs Signed-Ranks Test

| X | Y | Difference | Absolute Rank of Difference | R(+) | R(−) |
|----|----|----|----|----|----|
| 16 | 4 | 12 | 9.5 | 9.5 | |
| 12 | 18 | −6 | 5.0 | | 5 |
| 22 | 10 | 12 | 9.5 | 9.5 | |
| 16 | 14 | 2 | 1.5 | 1.5 | |
| 14 | 12 | 2 | 1.5 | 1.5 | |
| 10 | 14 | −4 | 3.0 | | 3 |
| 20 | 10 | 10 | 7.5 | 7.5 | |
| 18 | 12 | 6 | 5.0 | 5.0 | |
| 10 | 4 | 6 | 5.0 | 5.0 | |
| 22 | 12 | 10 | 7.5 | 7.5 | |
| | | | | $\Sigma R(+) = 47.0$ | $\Sigma R(-) = 8$ |

8; hence we reject the null hypothesis at the 5 percent level. If we compare our findings obtained on the sign test with those for the Wilcoxon test, we find contradictory results. With the sign test we were unable to reject the null hypothesis, which we rejected at the 5 percent level for Wilcoxon's test. Thus the Wilcoxon test is a more powerful test than the sign test because it takes into account more information than the sign test. In this case the magnitudes of differences are used and lead to the different results.

When $N$ is greater than 25, the sum of the ranks may be taken as normally distributed with

$$T = \frac{N(N + 1)}{4}$$

$$s = \sqrt{\frac{(2N + 1)T}{6}} \quad \text{or} \quad \sqrt{\frac{N(N + 1)(2N + 1)}{24}} \tag{18.2}$$

and the sum of the $T$'s may be treated with the familiar $z$ test with $T - T$ as the deviation.

## TESTS FOR UNCORRELATED DATA

### The Median Test
The median test is applied to see if two groups come from populations which have the same median. In using this test the size of the two samples need not be the same. Table 18.3 shows arithmetic addition scores for 27 individuals. We shall test the hypothesis of no difference between these two sets of scores by using the median test. Group $X$ contains 13 scores and group $Y$, 14. We

**Table 18.3**
Scores on an
Addition Test

| X | Y |
|---|---|
| 12 | 7 |
| 16 | 12 |
| 18 | 14 |
| 7 | 18 |
| 6 | 5 |
| 4 | 16 |
| 11 | 9 |
| 12 | 10 |
| 8 | 14 |
| 20 | 3 |
| 18 | 18 |
| 16 | 9 |
| 10 | 7 |
|  | 4 |

first compute the median of the entire set of 27 scores. We find the median for these data to be 11.

If both of these distributions came from the same populations, half of the X values and half of the Y values would lie above this median of 11; and half of each of the two distributions would lie below it. A contingency table is set up as follows:

|  | X | Y |  |
|---|---|---|---|
| Above median | 7<br>a | 6<br>b | 13<br>k |
| Not above median | 6<br>c | 8<br>d | 14<br>l |
|  | 13<br>m | 14<br>n | 27<br>N |

We next make a test of significance, using chi-square as follows:

$$\chi^2 = \frac{N(ad - bc| - N/2)^2}{klmn}$$

$$= \frac{27[(56 - 36) - 27/2]^2}{(13)(14)(13)(14)}$$

$$= \frac{27(42.25)}{33124}$$

$$= .034$$

which with 1 degree of freedom is not significant.

When the cell frequencies in the contingency become very small (1 or 2) this chi-square solution should not be used—even when corrected by Yates' correction for continuity. A more complicated method, known as Fisher's exact method, must be used. This may be found in Siegel (1956, pp. 96–104).

The median test is also applicable when there are more than two groups. The technique is the same. First the median of all the scores taken together is obtained. Then a contingency table similar to the one above is set up containing the number of scores above the median and the number not above the median for each of the groups. Chi-square is then applied to the data in the contingency table, and the appropriate conclusion drawn. In this type of problem, the expected frequency for a cell is half of the number of cases in the group. It should be noted that when a score falls exactly at the median, this score is included in the group referred to as "not above the median."

## Mann-Whitney $U$ Test

A more powerful test than the median test for uncorrelated data is the Mann-Whitney $U$ test. The test is used with independently drawn random samples, the sizes of which need not be the same. When the sample sizes are very small, that is, when both $N_1$ and $N_2$ are made up of eight or less measures, the reader is referred to Siegel (1956) for a method and tables that apply to such sample sizes. When the larger of the two samples is nine or more, the procedure described below is used.

In Table 18.4 we have two samples, $N_X$ with seven cases and $N_Y$ with nine cases. In this case we are testing the hypothesis that both samples come from the same population. First we rank all the scores in one composite distribution in an increasing order of size (considering algebraic signs when negative measures are present). In Table 18.4 the score of 3 in column $Y$ is the lowest one; hence we give it a rank of 1. The score of 4 in column $X$ is the

**Table 18.4**   Two Small Independent Samples Illustrating the Use of the Mann-Whitney $U$ Test

| $X(N = 7)$ | $Y(N = 9)$ | $R_X$ | $R_Y$ |
|---|---|---|---|
| 14 | 18 | 11.5 | 15.0 |
| 12 | 16 | 9.0 | 14.0 |
| 13 | 15 | 10.0 | 13.0 |
| 10 | 14 | 8.0 | 11.5 |
| 7 | 19 | 5.5 | 16.0 |
| 6 | 7 | 3.5 | 5.5 |
| 4 | 8 | 2.0 | 7.0 |
| | 6 | | 3.5 |
| | 3 | | 1.0 |
| $N_X = 7$ | $N_Y = 9$ | $\Sigma R_X = 49.5$ | $\Sigma R_Y = 86.5$ |

second lowest, and we give it a rank of 2. We have a score of 6 in both the $X$ and $Y$ distribution. We handle such ties by giving them the average of the next two ranks, which for this distribution is 3.5. We continue the ranking until all scores receive a rank. Then we sum the columns of ranks.

At this point we can make an arithmetic check on the work. The sum of the two columns of ranks must be equal to $N(N + 1)/2$. For this problem

$$\Sigma R_X + \Sigma R_Y = \frac{N(N + 1)}{2}$$

$$49.5 + 86.5 = \frac{16(17)}{2}$$

$$136 = 136$$

We obtain $U$ as follows:

$$U_1 = N_1 N_2 + \frac{N_1(N_1 + 1)}{2} - \Sigma R_X \tag{18.3}$$

$$U_2 = N_1 N_2 + \frac{N_2(N_2 + 1)}{2} - \Sigma R_Y \tag{18.3a}$$

For our problem,

$$U_1 = 7(9) + \frac{7(8)}{2} - 49.5$$

$$= 63 + 28 - 49.5$$

$$= 41.5$$

Equations 18.3 and 18.3a give different results. To complete our problem we need the smaller of the two. By solving equation 18.3a we have

$$U_2 = 7(9) + \frac{9(10)}{2} - 86.5$$

$$= 63 + 45 - 86.5$$

$$= 21.5$$

We can check to see if we have the smaller value of $U$ by the following:

$$U_2 = N_1 N_2 - U_1$$

$$= 7(9) - 41.5$$

$$= 63 - 41.5$$

$$= 21.5$$

Refer to Appendix J, which we use to evaluate $U$. Notice that Appendix J is made up of four parts: $p$ values of .001, .01, .025, and .05; or if we were making a directional test, $p$'s of .002, .02, .05, and .10. For example, suppose that $N_1 = 10$, and $N_2 = 15$. From part (b) of Appendix J we see that a $U$

of 33 or less makes it possible to reject the null hypothesis at the 1 percent level for a two-tailed test or at the 2 percent level for a one-tailed test. Suppose that we set alpha equal to .01 for our problem. Examination of part (b) of Appendix J shows that when $N_1 = 7$ and $N_2 = 9$, $U$ must be 9 or less to be significant. Our obtained $U$ of 21.5 is, therefore, not significant, and the null hypothesis stands.

When either $N_1$ or $N_2$ is larger than 20, we solve for the $U$ statistic, which is considered to be normally distributed for samples of this size. To illustrate this technique, some fictitious data have been assembled in Table 18.5. In group $X$ ($N_1$) there are 21 cases and in group $Y$ ($N_2$) there are 10. The data are treated exactly as they were previously. First the two sets of scores are ranked together, with the smallest score being given the smallest rank. Then these two columns of ranks are summed. Again these can be checked. $N(N + 1)/2$ equals the sum of the two sets of ranks. Next $U_1$ is found by formula 18.3,

$$U_1 = N_1 N_2 + \frac{N_1(N_1 + 1)}{2} - \Sigma R_X$$

$$= 21(10) + \frac{21(21 + 1)}{2} - 272.5$$

$$= 210 + 231 - 272.5$$

$$= 168.5$$

The $z$ ratio is then computed by the use of the following equation:

$$z = \frac{U_1 - (N_1 N_2/2)}{\sqrt{[N_1 N_2(N_1 + N_2 + 1)]/12}} \tag{18.4}$$

where the numerator is $U_1$ minus the second term, which is the mean of $U$, and the denominator is the standard deviation of the $U$.

$$z = \frac{168.5 - 21(10)/2}{\sqrt{[21(10)(21 + 10 + 1)]/12}}$$

$$= \frac{168.5 - 105}{\sqrt{6720/12}}$$

$$= \frac{63.5}{\sqrt{560}}$$

$$= \frac{63.5}{23.7}$$

$$= 2.68$$

A $z$ value of this size leads to the rejection of the null hypothesis at the 1 percent level.

This problem could have also been solved for $U_2$.

$$U_2 = N_1 N_2 + \frac{N_2(N_2 + 1)}{2} - \Sigma R_Y$$

$$= 21(10) + \frac{10(10 + 1)}{2} - 223.5$$

$$= 210 + 55 - 223.5$$

$$= 41.5$$

If this value of 41.5 is placed in the numerator of formula 18.4 in place of $U_1$, it will be found that the value of the numerator of $z$ this time is $-63.5$. It does not matter, then, for which $U$ value we solve. The size of the $z$ is unaffected, but the sign is affected. In a two-tailed test this makes no difference in the interpretation of $z$.

The Mann-Whitney test is said by Siegel (1956) to be a powerful test and an excellent substitute for the $t$ test.

### Wald-Wolfowitz Runs Test

Another test applicable to unrelated samples is the Wald-Wolfowitz runs test. With this technique, we test the hypothesis of no difference in respect

**Table 18.5**   Illustration of the Mann-Whitney
Test for Larger Sets of Data

| X | Y | $R_X$ | $R_Y$ |
|---|---|---|---|
| 7 | 18 | 12.5 | 31.0 |
| 10 | 14 | 18.0 | 27.0 |
| 12 | 15 | 21.5 | 29.0 |
| 6 | 16 | 10.0 | 30.0 |
| 8 | 12 | 14.5 | 21.5 |
| 4 | 10 | 6.5 | 18.0 |
| 6 | 8 | 10.0 | 14.5 |
| 14 | 13 | 27.0 | 24.0 |
| 2 | 13 | 2.5 | 24.0 |
| 4 | 3 | 6.5 | 4.5 |
| 6 | | 10.0 | |
| 7 | | 12.5 | |
| 3 | | 4.5 | |
| 5 | | 8.0 | |
| 2 | | 2.5 | |
| 1 | | 1.0 | |
| 13 | | 24.0 | |
| 14 | | 27.0 | |
| 11 | | 20.0 | |
| 10 | | 18.0 | |
| 9 | | 16.0 | |
| $N_1 = 21$ | $N_2 = 10$ | $\Sigma R_X = 272.5$ | $\Sigma R_Y = 223.5$ |

**Table 18.6**   Two Sets of
of Independent
Data for the
Wald-Wolfowitz
Runs Test

| X | Y |
|---|---|
| 12 | 7 |
| 8 | 9 |
| 6 | 16 |
| 18 | 17 |
| 14 | 20 |
| 15 | 19 |
| 3 | 2 |
| 5 | 10 |
| 4 | 22 |
| 1 | 21 |
| $N_1 = 10$ | $N_2 = 10$ |

to average value, variability, and skewness, for example. In Table 18.6 are two sets of data taken on two independent samples. The scores are arranged in a continuous series from high to low, and the identity of the distribution from which the score came is recorded below it. The data in Table 18.6 are set up as follows:

| 22 | 21 | 20 | 19 | 18 | 17 | 16 | 15 | 14 | 12 | 10 | 9 | 8 | 7 | 6 | 5 | 4 | 3 | 2 | 1 |
|----|----|----|----|----|----|----|----|----|----|----|---|---|---|---|---|---|---|---|---|
| Y | Y | Y | Y | X | Y | Y | X | X | X | Y | Y | X | Y | X | X | X | X | Y | X |

| 1 | 2 | 3 | 4 | 5 | 6 | 7 | 8 | 9 | 10 |

After this is done the number of runs is determined. Starting with the score of 22, we have four scores in a row that came from the $Y$ distribution. These four scores constitute the first run. The score of 18 is from the $X$ distribution. This is run 2. All the runs are then marked as shown and the number of runs counted.

If the two sets of data are from the same population, we would expect the number of runs to be large. That is, there would be a good mixture of high, medium, and low scores in each distribution, and this would lead to a large number of short runs. When $N_1$ and $N_2$ (the numbers in each distribution) are 20 or less, the significance of the number of runs observed can be tested directly from tables (Appendix K). From these tables we see that for the values of $N$ in our sample, the number of runs is not significant, and hence there is no difference between the two samples.

When the size of the two samples is over 20, the significance of the difference is tested as follows:

$$z = \frac{[r - (2N_1 N_2/N_1 + N_2) + 1] - .5}{2N_1 N_2(2N_1 N_2 - N_1 - N_2)/[(N_1 + N_2)^2(N_1 + N_2 - 1)]} \tag{18.5}$$

where                          $r$ = number of runs

$(2N_1 N_2)/(N_1 + N_2)$ = mean of the runs

denominator = the standard deviation

$- .5$ = correction for continuity

With this technique the problem of ties is an important one. If the ties are all in one sample or the other, no problem arises. If the ties are in both samples, they should be mixed in all possible ways. If the results agree by all three ways, there is no problem. But if the results differ, the probability of each $z$ is determined and the average of these taken in the evaluation of the null hypothesis. As an example, suppose that in two series of scores a score of 14 appears three times, twice in the $X$ distribution and once in the $Y$ distribution. It could be arranged in the series of scores three ways:

| 14 14 14 | 14 14 14 | 14 14 14 |
|----------|----------|----------|
|    or    |    or    |          |
| X  X  Y  | X  Y  X  | Y  X  X  |

For this series of ties, then, the problem would have to be solved three times, determining the different number of runs which the various combinations cause.

This test is not as good as some of the others because of the general nature of the hypothesis that may be tested with it. Moses (1952) states that tests which test hypotheses against many alternatives simultaneously are not very good in the prevention of the acceptance of the null hypothesis erroneously with respect to one alternative. Since this test is designed to disclose differences of many types, it is not as efficient as the Mann-Whitney test or others that are set up to test specific differences.

### Kruskal-Wallis $H$ Test

The Kruskal-Wallis $H$ test is used to test whether or not a group of independent samples is from the same or different populations. Data to illustrate this test are presented in Table 18.7.

Table 18.7 shows three sets of data. The scores of the three sets are combined and ranked with the lowest score receiving a rank value of 1. Ties are treated in the usual fashion for ranking data. Then the ranks of the three columns are summed. If the sample proportions are similar, the total sum of the ranks would be divided proportionately among the various samples on the basis of sample size. The sums of the ranks of the various samples are then tested against these proportions of the total sum of ranks, by use of the following formula:

$$H = \frac{12}{N(N + 1)} \left[ \sum \frac{R_i^2}{N_i} \right] - 3(N + 1) \tag{18.6}$$

where $N$ = the number in all samples combined

$R_i$ = sum of ranks and $N_i$ the numbers in $i$ samples

**Table 18.7**   The Kruskal-Wallis $H$ Test

| $X$ | $R_x$ | $Y$ | $R_y$ | $Z$ | $R_z$ |
|---|---|---|---|---|---|
| 12 | 13.5 | 13 | 16.0 | 13 | 16.0 |
| 16 | 21.0 | 18 | 22.0 | 14 | 19.0 |
| 14 | 19.0 | 14 | 19.0 | 7 | 8.5 |
| 2 | 1.5 | 13 | 16.0 | 8 | 10.5 |
| 12 | 13.5 | 8 | 10.5 | 4 | 4.5 |
|  |  | 7 | 8.5 | 3 | 3.0 |
|  |  | 6 | 7.0 | 2 | 1.5 |
|  |  | 4 | 4.5 | 5 | 6.0 |
|  |  |  |  | 9 | 12.0 |
| $N_x = 5$ | $\Sigma R_x = 68.5$ | $N_y = 8$ | $\Sigma R_y = 103.5$ | $N_z = 9$ | $\Sigma R_z = 81.0$ |

For the data in Table 18.7,

$$H = \frac{12}{22(23)} \left[ \frac{68.5^2}{5} + \frac{103.5^2}{8} + \frac{81^2}{9} \right] - 3(23)$$

$$= \frac{12}{506} (938.45 + 1339.03 + 729) - 69$$

$$= .0237(3006.48) - 69$$

$$= 71.3 - 69$$

$$= 2.3$$

If the number of cases in the samples is from one to five, special tables are used in the interpretation of $H$. (Refer to Appendix L.) When the samples contain five or more cases, $H$ is interpreted as chi-square with the number of samples minus 1 degree of freedom. For this problem $df$ is equal to 2, and this $H$ is not significant.

The Kruskal-Wallis test is a powerful nonparametric test.

## EXERCISES

**1.** The following represent the scores of 9 individuals on a pretest and a post-test. Apply the sign test to the data and draw the appropriate conclusion.

| A | B |
|---|---|
| 16 | 22 |
| 8 | 12 |
| 8 | 13 |
| 12 | 14 |
| 10 | 11 |
| 9 | 7 |
| 22 | 24 |
| 10 | 17 |
| 11 | 18 |

**2.** In the data below, column $X$ represents 12 scores of members of a control group in an experiment. Column $Y$ represents the scores of 12 matched individuals who were given the same test after a period of stress. Use the sign test to test the hypothesis of no difference.

| X | Y |
|----|----|
| 46 | 36 |
| 68 | 50 |
| 60 | 58 |
| 58 | 40 |
| 42 | 44 |
| 43 | 43 |
| 40 | 29 |
| 56 | 36 |
| 38 | 46 |
| 58 | 48 |
| 42 | 38 |
| 48 | 42 |

**3.** With the data in problem 2 test the hypothesis of no difference using Wilcoxon's matched-pairs signed-ranks test.

**4.** Below are the scores of a group of normals and a group of psychotics on the picture completion scale of the Wechsler-Bellevue. Test the hypothesis of no difference using the median test.

| N | P |
|----|----|
| 6 | 7 |
| 6 | 2 |
| 14 | 12 |
| 13 | 13 |
| 15 | 8 |
| 6 | 6 |
| 8 | 4 |
| 7 | 2 |
| 10 | 2 |
| 14 | 12 |
| 10 | |
| 14 | |

**5.** In the data below are three samples of scores obtained on the arithmetic scale of the Wechsler-Bellevue. Apply the median test to all three samples.

| A | B | C |
|----|----|----|
| 8 | 2 | 12 |
| 7 | 4 | 3 |
| 14 | 6 | 10 |
| 10 | 14 | 4 |
| 8 | 10 | 14 |
| 6 | 8 | 11 |
| | 6 | 10 |
| | 2 | 8 |

**6.** Apply the Mann-Whitney $U$ test to the data in problem 4.

**7.** Assume the data in problem 2 to be uncorrelated. Apply the Mann-Whitney $U$ test.

**8.** Two groups of children made the following scores on the vocabulary test of the Stanford-Binet. By use of the Mann-Whitney $U$ test, test for differences in the two groups.

| Group I | | Group II |
|---|---|---|
| 10 | 33 | 14 |
| 18 | 36 | 18 |
| 36 | 26 | 22 |
| 22 | 24 | 16 |
| 28 | 31 | 38 |
| 29 | 13 | 26 |
| 32 | 19 | 28 |
| 15 | 23 | 12 |
| 18 | 25 | |
| 36 | 27 | 11 |
| 21 | 32 | 19 |
| 27 | | 16 |

**9.** Apply the Wald-Wolfowitz runs test to the data in problem 7 above.

**10.** By using the data in problem 4, test the hypothesis of no difference using the Kruskal-Wallis $H$ test.

**11.** The following data from Borislow[1] show the changed response scores of three groups on the Edwards Personal Preference Schedule. Test the differences among these three groups, using any appropriate nonparametric tests.

| Control Group | Social Desirability Group | Personal Desirability Group |
|---|---|---|
| 47 | 76 | 87 |
| 38 | 76 | 76 |
| 34 | 75 | 52 |
| 32 | 73 | 51 |
| 32 | 62 | 50 |
| 30 | 47 | 42 |
| | | 38 |

[1] *Journal of Applied Psychology*, 1958, *42*, 25.

# Appendixes

# Appendix A
## Squares, Square Roots, and Reciprocals of Integers from 1 to 1000

| $n$ | $n^2$ | $\sqrt{n}$ | $\dfrac{1}{n}$ | $\dfrac{1}{\sqrt{n}}$ |
|---|---|---|---|---|
| 1 | 1 | 1.0000 | 1.000000 | 1.0000 |
| 2 | 4 | 1.4142 | .500000 | .7071 |
| 3 | 9 | 1.7321 | .333333 | .5774 |
| 4 | 16 | 2.0000 | .250000 | .5000 |
| 5 | 25 | 2.2361 | .200000 | .4472 |
| 6 | 36 | 2.4495 | .166667 | .4082 |
| 7 | 49 | 2.6458 | .142857 | .3780 |
| 8 | 64 | 2.8284 | .125000 | .3536 |
| 9 | 81 | 3.0000 | .111111 | .3333 |
| 10 | 100 | 3.1623 | .100000 | .3162 |
| 11 | 121 | 3.3166 | .090909 | .3015 |
| 12 | 144 | 3.4641 | .083333 | .2887 |
| 13 | 169 | 3.6056 | .076923 | .2774 |
| 14 | 196 | 3.7417 | .071429 | .2673 |
| 15 | 225 | 3.8730 | .066667 | .2582 |
| 16 | 256 | 4.0000 | .062500 | .2500 |
| 17 | 289 | 4.1231 | .058824 | .2425 |
| 18 | 324 | 4.2426 | .055556 | .2357 |
| 19 | 361 | 4.3589 | .052632 | .2294 |
| 20 | 400 | 4.4721 | .050000 | .2236 |
| 21 | 441 | 4.5826 | .047619 | .2182 |
| 22 | 484 | 4.6904 | .045455 | .2132 |
| 23 | 529 | 4.7958 | .043478 | .2085 |
| 24 | 576 | 4.8990 | .041667 | .2041 |
| 25 | 625 | 5.0000 | .040000 | .2000 |
| 26 | 676 | 5.0990 | .038462 | .1961 |
| 27 | 729 | 5.1962 | .037037 | .1925 |
| 28 | 784 | 5.2915 | .035714 | .1890 |
| 29 | 841 | 5.3852 | .034483 | .1857 |
| 30 | 900 | 5.4772 | .033333 | .1826 |
| 31 | 961 | 5.5678 | .032258 | .1796 |
| 32 | 1024 | 5.6569 | .031250 | .1768 |
| 33 | 1089 | 5.7446 | .030303 | .1741 |
| 34 | 1156 | 5.8310 | .029412 | .1715 |
| 35 | 1225 | 5.9161 | .028571 | .1690 |

SOURCE: J. G. Peatman. *Descriptive and Sampling Statistics*. New York: Harper & Row, 1947. Reprinted by permission of the publisher.

Appendix A (*Continued*)

| $n$ | $n^2$ | $\sqrt{n}$ | $\dfrac{1}{n}$ | $\dfrac{1}{\sqrt{n}}$ |
|---|---|---|---|---|
| 36 | 1296 | 6.0000 | .027778 | .1667 |
| 37 | 1369 | 6.0828 | .027027 | .1644 |
| 38 | 1444 | 6.1644 | .026316 | .1622 |
| 39 | 1521 | 6.2450 | .025641 | .1601 |
| 40 | 1600 | 6.3246 | .025000 | .1581 |
| 41 | 1681 | 6.4031 | .024390 | .1562 |
| 42 | 1764 | 6.4807 | .023810 | .1543 |
| 43 | 1849 | 6.5574 | .023256 | .1525 |
| 44 | 1936 | 6.6332 | .022727 | .1508 |
| 45 | 2025 | 6.7082 | .022222 | .1491 |
| 46 | 2116 | 6.7823 | .021739 | .1474 |
| 47 | 2209 | 6.8557 | .021277 | .1459 |
| 48 | 2304 | 6.9282 | .020833 | .1443 |
| 49 | 2401 | 7.0000 | .020408 | .1429 |
| 50 | 2500 | 7.0711 | .020000 | .1414 |
| 51 | 2601 | 7.1414 | .019608 | .1400 |
| 52 | 2704 | 7.2111 | .019231 | .1387 |
| 53 | 2809 | 7.2801 | .018868 | .1374 |
| 54 | 2916 | 7.3485 | .018519 | .1361 |
| 55 | 3025 | 7.4162 | .018182 | .1348 |
| 56 | 3136 | 7.4833 | .017857 | .1336 |
| 57 | 3249 | 7.5498 | .017544 | .1325 |
| 58 | 3364 | 7.6158 | .017241 | .1313 |
| 59 | 3481 | 7.6811 | .016949 | .1302 |
| 60 | 3600 | 7.7460 | .016667 | .1291 |
| 61 | 3721 | 7.8102 | .016393 | .1280 |
| 62 | 3844 | 7.8740 | .016129 | .1270 |
| 63 | 3969 | 7.9373 | .015873 | .1260 |
| 64 | 4096 | 8.0000 | .015625 | .1250 |
| 65 | 4225 | 8.0623 | .015385 | .1240 |
| 66 | 4356 | 8.1240 | .015152 | .1231 |
| 67 | 4489 | 8.1854 | .014925 | .1222 |
| 68 | 4624 | 8.2462 | .014706 | .1213 |
| 69 | 4761 | 8.3066 | .014493 | .1204 |
| 70 | 4900 | 8.3666 | .014286 | .1195 |
| 71 | 5041 | 8.4261 | .014085 | .1187 |
| 72 | 5184 | 8.4853 | .013889 | .1179 |
| 73 | 5329 | 8.5440 | .013699 | .1170 |
| 74 | 5476 | 8.6023 | .013514 | .1162 |
| 75 | 5625 | 8.6603 | .013333 | .1155 |
| 76 | 5776 | 8.7178 | .013158 | .1147 |
| 77 | 5929 | 8.7750 | .012987 | .1140 |
| 78 | 6084 | 8.8318 | .012821 | .1132 |
| 79 | 6241 | 8.8882 | .012658 | .1125 |
| 80 | 6400 | 8.9443 | .012500 | .1118 |
| 81 | 6561 | 9.0000 | .012346 | .1111 |
| 82 | 6724 | 9.0554 | .012195 | .1104 |
| 83 | 6889 | 9.1104 | .012048 | .1098 |
| 84 | 7056 | 9.1652 | .011905 | .1091 |
| 85 | 7225 | 9.2195 | .011765 | .1085 |

APPENDIX A (*Continued*)

| $n$ | $n^2$ | $\sqrt{n}$ | $\dfrac{1}{n}$ | $\dfrac{1}{\sqrt{n}}$ |
|---|---|---|---|---|
| 86 | 7396 | 9.2736 | .011628 | .1078 |
| 87 | 7569 | 9.3274 | .011494 | .1072 |
| 88 | 7744 | 9.3808 | .011364 | .1066 |
| 89 | 7921 | 9.4340 | :011236 | .1060 |
| 90 | 8100 | 9.4868 | .011111 | .1054 |
| 91 | 8281 | 9.5394 | .010989 | .1048 |
| 92 | 8464 | 9.5917 | .010870 | .1043 |
| 93 | 8649 | 9.6437 | .010753 | .1037 |
| 94 | 8836 | 9.6954 | .010638 | .1031 |
| 95 | 9025 | 9.7468 | .010526 | .1026 |
| 96 | 9216 | 9.7980 | .010417 | .1021 |
| 97 | 9409 | 9.8489 | .010309 | .1015 |
| 98 | 9604 | 9.8995 | .010204 | .1010 |
| 99 | 9801 | 9.9499 | .010101 | .1005 |
| 100 | 10000 | 10.0000 | .010000 | .1000 |
| 101 | 10201 | 10.0499 | .009901 | .0995 |
| 102 | 10404 | 10.0995 | .009804 | .0990 |
| 103 | 10609 | 10.1489 | .009709 | .0985 |
| 104 | 10816 | 10.1980 | .009615 | .0981 |
| 105 | 11025 | 10.2470 | .009524 | .0976 |
| 106 | 11236 | 10.2956 | .009434 | .0971 |
| 107 | 11449 | 10.3441 | .009346 | .0967 |
| 108 | 11664 | 10.3923 | .009259 | .0962 |
| 109 | 11881 | 10.4403 | .009174 | .0958 |
| 110 | 12100 | 10.4881 | .009091 | .0953 |
| 111 | 12321 | 10.5357 | .009009 | .0949 |
| 112 | 12544 | 10.5830 | .008929 | .0945 |
| 113 | 12769 | 10.6301 | .008850 | .0941 |
| 114 | 12996 | 10.6771 | .008772 | .0937 |
| 115 | 13225 | 10.7238 | .008696 | .0933 |
| 116 | 13456 | 10.7703 | .008621 | .0928 |
| 117 | 13689 | 10.8167 | .008547 | .0925 |
| 118 | 13924 | 10.8628 | .008475 | .0921 |
| 119 | 14161 | 10.9087 | .008403 | .0917 |
| 120 | 14400 | 10.9545 | .008333 | .0913 |
| 121 | 14641 | 11.0000 | .008264 | .0909 |
| 122 | 14884 | 11.0454 | .008197 | .0905 |
| 123 | 15129 | 11.0905 | .008130 | .0902 |
| 124 | 15376 | 11.1355 | .008065 | .0898 |
| 125 | 15625 | 11.1803 | .008000 | .0894 |
| 126 | 15876 | 11.2250 | .007937 | .0891 |
| 127 | 16129 | 11.2694 | .007874 | .0887 |
| 128 | 16384 | 11.3137 | .007813 | .0884 |
| 129 | 16641 | 11.3578 | .007752 | .0880 |
| 130 | 16900 | 11.4018 | .007692 | .0877 |
| 131 | 17161 | 11.4455 | .007634 | .0874 |
| 132 | 17424 | 11.4891 | .007576 | .0870 |
| 133 | 17689 | 11.5326 | .007519 | .0867 |
| 134 | 17956 | 11.5758 | .007463 | .0864 |
| 135 | 18225 | 11.6190 | .007407 | .0861 |

APPENDIX A (*Continued*)

| $n$ | $n^2$ | $\sqrt{n}$ | $\dfrac{1}{n}$ | $\dfrac{1}{\sqrt{n}}$ |
|---|---|---|---|---|
| 136 | 18496 | 11.6619 | .007353 | .0857 |
| 137 | 18769 | 11.7047 | .007299 | .0854 |
| 138 | 19044 | 11.7473 | .007246 | .0851 |
| 139 | 19321 | 11.7898 | .007194 | .0848 |
| 140 | 19600 | 11.8322 | .007143 | .0845 |
| 141 | 19881 | 11.8743 | .007092 | .0842 |
| 142 | 20164 | 11.9164 | .007042 | .0839 |
| 143 | 20449 | 11.9583 | .006993 | .0836 |
| 144 | 20736 | 12.0000 | .006944 | .0833 |
| 145 | 21025 | 12.0416 | .006897 | .0830 |
| 146 | 21316 | 12.0830 | .006849 | .0828 |
| 147 | 21609 | 12.1244 | .006803 | .0825 |
| 148 | 21904 | 12.1655 | .006757 | .0822 |
| 149 | 22201 | 12.2066 | .006711 | .0819 |
| 150 | 22500 | 12.2474 | .006667 | .0816 |
| 151 | 22801 | 12.2882 | .006623 | .0814 |
| 152 | 23104 | 12.3288 | .006579 | .0811 |
| 153 | 23409 | 12.3693 | .006536 | .0808 |
| 154 | 23716 | 12.4097 | .006494 | .0806 |
| 155 | 24025 | 12.4499 | .006452 | .0803 |
| 156 | 24336 | 12.4900 | .006410 | .0801 |
| 157 | 24649 | 12.5300 | .006369 | .0798 |
| 158 | 24964 | 12.5698 | .006329 | .0796 |
| 159 | 25281 | 12.6095 | .006289 | .0793 |
| 160 | 25600 | 12.6491 | .006250 | .0791 |
| 161 | 25921 | 12.6886 | .006211 | .0788 |
| 162 | 26244 | 12.7279 | .006173 | .0786 |
| 163 | 26569 | 12.7671 | .006135 | .0783 |
| 164 | 26896 | 12.8062 | .006098 | .0781 |
| 165 | 27225 | 12.8452 | .006061 | .0778 |
| 166 | 27556 | 12.8841 | .006024 | .0776 |
| 167 | 27889 | 12.9228 | .005988 | .0774 |
| 168 | 28224 | 12.9615 | .005952 | .0772 |
| 169 | 28561 | 13.0000 | .005917 | .0769 |
| 170 | 28900 | 13.0384 | .005882 | .0767 |
| 171 | 29241 | 13.0767 | .005848 | .0765 |
| 172 | 29584 | 13.1149 | .005814 | .0762 |
| 173 | 29929 | 13.1529 | .005780 | .0760 |
| 174 | 30276 | 13.1909 | .005747 | .0758 |
| 175 | 30625 | 13.2288 | .005714 | .0756 |
| 176 | 30976 | 13.2665 | .005682 | .0754 |
| 177 | 31329 | 13.3041 | .005650 | .0752 |
| 178 | 31684 | 13.3417 | .005618 | .0750 |
| 179 | 32041 | 13.3791 | .005587 | .0747 |
| 180 | 32400 | 13.4164 | .005556 | .0745 |
| 181 | 32761 | 13.4536 | .005525 | .0743 |
| 182 | 33124 | 13.4907 | .005495 | .0741 |
| 183 | 33489 | 13.527 | .005464 | .0739 |
| 184 | 33856 | 13.5647 | .005435 | .0737 |
| 185 | 34225 | 13.6015 | .005405 | .0735 |

APPENDIX A (*Continued*)

| $n$ | $n^2$ | $\sqrt{n}$ | $\dfrac{1}{n}$ | $\dfrac{1}{\sqrt{n}}$ |
|---|---|---|---|---|
| 186 | 34596 | 13.6382 | .005376 | .0733 |
| 187 | 34969 | 13.6748 | .005348 | .0731 |
| 188 | 35344 | 13.7113 | .005319 | .0729 |
| 189 | 35721 | 13.7477 | .005291 | .0727 |
| 190 | 36100 | 13.7840 | .005263 | .0725 |
| 191 | 36481 | 13.8203 | .005236 | .0724 |
| 192 | 36864 | 13.8564 | .005208 | .0722 |
| 193 | 37249 | 13.8924 | .005181 | .0720 |
| 194 | 37636 | 13.9284 | .005155 | .0718 |
| 195 | 38025 | 13.9642 | .005128 | .0716 |
| 196 | 38416 | 14.0000 | .005102 | .0714 |
| 197 | 38809 | 14.0357 | .005076 | .0712 |
| 198 | 39204 | 14.0712 | .005051 | .0711 |
| 199 | 39601 | 14.1067 | .005025 | .0709 |
| 200 | 40000 | 14.1421 | .005000 | .0707 |
| 201 | 40401 | 14.1774 | .004975 | .0705 |
| 202 | 40804 | 14.2127 | .004950 | .0704 |
| 203 | 41209 | 14.2478 | .004926 | .0702 |
| 204 | 41616 | 14.2829 | .004902 | .0700 |
| 205 | 42025 | 14.3178 | .004878 | .0698 |
| 206 | 42436 | 14.3527 | .004854 | .0697 |
| 207 | 42849 | 14.3875 | .004831 | .0695 |
| 208 | 43264 | 14.4222 | .004808 | .0693 |
| 209 | 43681 | 14.4568 | .004785 | .0692 |
| 210 | 44100 | 14.4914 | .004762 | .0690 |
| 211 | 44521 | 14.5258 | .004739 | .0688 |
| 212 | 44944 | 14.5602 | .004717 | .0687 |
| 213 | 45369 | 14.5945 | .004695 | .0685 |
| 214 | 45796 | 14.6287 | .004673 | .0684 |
| 215 | 46225 | 14.6629 | .004651 | .0682 |
| 216 | 46656 | 14.6969 | .004630 | .0680 |
| 217 | 47089 | 14.7309 | .004608 | .0679 |
| 218 | 47524 | 14.7648 | .004587 | .0677 |
| 219 | 47961 | 14.7986 | .004566 | .0676 |
| 220 | 48400 | 14.8324 | .004545 | .0674 |
| 221 | 48841 | 14.8661 | .004525 | .0673 |
| 222 | 49284 | 14.8997 | .004505 | .0671 |
| 223 | 49729 | 14.9332 | .004484 | .0670 |
| 224 | 50176 | 14.9666 | .004464 | .0668 |
| 225 | 50625 | 15.0000 | .004444 | .0667 |
| 226 | 51076 | 15.0333 | .004425 | .0665 |
| 227 | 51529 | 15.0665 | .004405 | .0664 |
| 228 | 51984 | 15.0997 | .004386 | .0662 |
| 229 | 52441 | 15.1327 | .004367 | .0661 |
| 230 | 52900 | 15.1658 | .004348 | .0659 |
| 231 | 53361 | 15.1987 | .004329 | .0658 |
| 232 | 53824 | 15.2315 | .004310 | .0657 |
| 233 | 54289 | 15.2643 | .004292 | .0655 |
| 234 | 54756 | 15.2971 | .004274 | .0654 |
| 235 | 55225 | 15.3297 | .004255 | .0652 |

APPENDIX A (*Continued*)

| $n$ | $n^2$ | $\sqrt{n}$ | $\dfrac{1}{n}$ | $\dfrac{1}{\sqrt{n}}$ |
|-----|-------|------------|----------------|------------------------|
| 236 | 55696 | 15.3623 | .004237 | .0651 |
| 237 | 56169 | 15.3948 | .004219 | .0650 |
| 238 | 56644 | 15.4272 | .004202 | .0648 |
| 239 | 57121 | 15.4596 | .004184 | .0647 |
| 240 | 57600 | 15.4919 | .004167 | .0645 |
| 241 | 58081 | 15.5242 | .004149 | .0644 |
| 242 | 58564 | 15.5563 | .004132 | .0643 |
| 243 | 59049 | 15.5885 | .004115 | .0642 |
| 244 | 59536 | 15.6205 | .004098 | .0640 |
| 245 | 60025 | 15.6525 | .004082 | .0639 |
| 246 | 60516 | 15.6844 | .004065 | .0638 |
| 247 | 61009 | 15.7162 | .004049 | .0636 |
| 248 | 61504 | 15.7480 | .004032 | .0635 |
| 249 | 62001 | 15.7797 | .004016 | .0634 |
| 250 | 62500 | 15.8114 | .004000 | .0632 |
| 251 | 63001 | 15.8430 | .003984 | .0631 |
| 252 | 63504 | 15.8745 | .003968 | .0630 |
| 253 | 64009 | 15.9060 | .003953 | .0629 |
| 254 | 64516 | 15.9374 | .003937 | .0627 |
| 255 | 65025 | 15.9687 | .003922 | .0626 |
| 256 | 65536 | 16.0000 | .003906 | .0625 |
| 257 | 66049 | 16.0312 | .003891 | .0624 |
| 258 | 66564 | 16.0624 | .003876 | .0623 |
| 259 | 67081 | 16.0935 | .003861 | .0621 |
| 260 | 67600 | 16.1245 | .003846 | .0620 |
| 261 | 68121 | 16.1555 | .003831 | .0619 |
| 262 | 68644 | 16.1864 | .003817 | .0618 |
| 263 | 69169 | 16.2173 | .003802 | .0617 |
| 264 | 69696 | 16.2481 | .003788 | .0615 |
| 265 | 70225 | 16.2788 | .003774 | .0614 |
| 266 | 70756 | 16.3095 | .003759 | .0613 |
| 267 | 71289 | 16.3401 | .003745 | .0612 |
| 268 | 71824 | 16.3707 | .003731 | .0611 |
| 269 | 72361 | 16.4012 | .003717 | .0610 |
| 270 | 72900 | 16.4317 | .003704 | .0609 |
| 271 | 73441 | 16.4621 | .003690 | .0607 |
| 272 | 73984 | 16.4924 | .003676 | .0606 |
| 273 | 74529 | 16.5227 | .003663 | .0605 |
| 274 | 75076 | 16.5529 | .003650 | .0604 |
| 275 | 75625 | 16.5831 | .003636 | .0603 |
| 276 | 76176 | 16.6132 | .003623 | .0602 |
| 277 | 76729 | 16.6433 | .003610 | .0601 |
| 278 | 77284 | 16.6733 | .003597 | .0600 |
| 279 | 77841 | 16.7033 | .003584 | .0599 |
| 280 | 78400 | 16.7332 | .003571 | .0598 |
| 281 | 78961 | 16.7631 | .003559 | .0597 |
| 282 | 79524 | 16.7929 | .003546 | .0595 |
| 283 | 80089 | 16.8226 | .003534 | .0594 |
| 284 | 80656 | 16.8523 | .003521 | .0593 |
| 285 | 81225 | 16.8819 | .003509 | .0592 |

APPENDIX A (*Continued*)

| $n$ | $n^2$ | $\sqrt{n}$ | $\dfrac{1}{n}$ | $\dfrac{1}{\sqrt{n}}$ |
|---|---|---|---|---|
| 286 | 81796 | 16.9115 | .003497 | .0591 |
| 287 | 82369 | 16.9411 | .003484 | .0590 |
| 288 | 82944 | 16.9706 | .003472 | .0589 |
| 289 | 83521 | 17.0000 | .003460 | .0588 |
| 290 | 84100 | 17.0294 | .003448 | .0587 |
| 291 | 84681 | 17.0587 | .003436 | .0586 |
| 292 | 85264 | 17.0880 | .003425 | .0585 |
| 293 | 85849 | 17.1172 | .003413 | .0584 |
| 294 | 86436 | 17.1464 | .003401 | .0583 |
| 295 | 87025 | 17.1756 | .003390 | .0582 |
| 296 | 87616 | 17.2047 | .003378 | .0581 |
| 297 | 88209 | 17.2337 | .003367 | .0580 |
| 298 | 88804 | 17.2627 | .003356 | .0579 |
| 299 | 89401 | 17.2916 | .003344 | .0578 |
| 300 | 90000 | 17.3205 | .003333 | .0577 |
| 301 | 90601 | 17.3494 | .003322 | .0576 |
| 302 | 91204 | 17.3781 | .003311 | .0575 |
| 303 | 91809 | 17.4069 | .003300 | .0574 |
| 304 | 92416 | 17.4356 | .003289 | .0574 |
| 305 | 93025 | 17.4642 | .003279 | .0573 |
| 306 | 93636 | 17.4929 | .003268 | .0572 |
| 307 | 94249 | 17.5214 | .003257 | .0571 |
| 308 | 94864 | 17.5499 | .003247 | .0570 |
| 309 | 95481 | 17.5784 | .003236 | .0569 |
| 310 | 96100 | 17.6068 | .003226 | .0568 |
| 311 | 96721 | 17.6352 | .003215 | .0567 |
| 312 | 97344 | 17.6635 | .003205 | .0566 |
| 313 | 97969 | 17.6918 | .003195 | .0565 |
| 314 | 98596 | 17.7200 | .003185 | .0564 |
| 315 | 99225 | 17.7482 | .003175 | .0563 |
| 316 | 99856 | 17.7764 | .003165 | .0563 |
| 317 | 100489 | 17.8045 | .003155 | .0562 |
| 318 | 101124 | 17.8326 | .003145 | .0561 |
| 319 | 101761 | 17.8606 | .003135 | .0560 |
| 320 | 102400 | 17.8885 | .003125 | .0559 |
| 321 | 103041 | 17.9165 | .003115 | .0558 |
| 322 | 103684 | 17.9444 | .003106 | .0557 |
| 323 | 104329 | 17.9722 | .003096 | .0556 |
| 324 | 104976 | 18.0000 | .003086 | .0556 |
| 325 | 105625 | 18.0278 | .003077 | .0555 |
| 326 | 106276 | 18.0555 | .003067 | .0554 |
| 327 | 106929 | 18.0831 | .003058 | .0553 |
| 328 | 107584 | 18.1108 | .003049 | .0552 |
| 329 | 108241 | 18.1384 | .003040 | .0551 |
| 330 | 108900 | 18.1659 | .003030 | .0550 |
| 331 | 109561 | 18.1934 | .003021 | .0550 |
| 332 | 110224 | 18.2209 | .003012 | .0549 |
| 333 | 110889 | 18.2483 | .003003 | .0548 |
| 334 | 111556 | 18.2757 | .002994 | .0547 |
| 335 | 112225 | 18.3030 | .002985 | .0546 |

## APPENDIX A (*Continued*)

| $n$ | $n^2$ | $\sqrt{n}$ | $\dfrac{1}{n}$ | $\dfrac{1}{\sqrt{n}}$ |
|---|---|---|---|---|
| 336 | 112896 | 18.3303 | .002976 | .0546 |
| 337 | 113569 | 18.3576 | .002967 | .0545 |
| 338 | 114244 | 18.3848 | .002959 | .0544 |
| 339 | 114921 | 18.4120 | .002950 | .0543 |
| 340 | 115600 | 18.4391 | .002941 | .0542 |
| 341 | 116281 | 18.4662 | .002933 | .0542 |
| 342 | 116964 | 18.4932 | .002924 | .0541 |
| 343 | 117649 | 18.5203 | .002915 | .0540 |
| 344 | 118336 | 18.5472 | .002907 | .0539 |
| 345 | 119025 | 18.5742 | .002899 | .0538 |
| 346 | 119716 | 18.6011 | .002890 | .0538 |
| 347 | 120409 | 18.6279 | .002882 | .0537 |
| 348 | 121104 | 18.6548 | .002874 | .0536 |
| 349 | 121801 | 18.6815 | .002865 | .0535 |
| 350 | 122500 | 18.7083 | .002857 | .0535 |
| 351 | 123201 | 18.7350 | .002849 | .0534 |
| 352 | 123904 | 18.7617 | .002841 | .0533 |
| 353 | 124609 | 18.7883 | .002833 | .0532 |
| 354 | 125316 | 18.8149 | .002825 | .0531 |
| 355 | 126025 | 18.8414 | .002817 | .0531 |
| 356 | 126736 | 18.8680 | .002809 | .0530 |
| 357 | 127449 | 18.8944 | .002801 | .0529 |
| 358 | 128164 | 18.9209 | .002793 | .0529 |
| 359 | 128881 | 18.9473 | .002786 | .0528 |
| 360 | 129600 | 18.9737 | .002778 | .0527 |
| 361 | 130321 | 19.0000 | .002770 | .0526 |
| 362 | 131044 | 19.0263 | .002762 | .0526 |
| 363 | 131769 | 19.0526 | .002755 | .0525 |
| 364 | 132496 | 19.0788 | .002747 | .0524 |
| 365 | 133225 | 19.1050 | .002740 | .0523 |
| 366 | 133956 | 19.1311 | .002732 | .0523 |
| 367 | 134689 | 19.1572 | .002725 | .0522 |
| 368 | 135424 | 19.1833 | .002717 | .0521 |
| 369 | 136161 | 19.2094 | .002710 | .0521 |
| 370 | 136900 | 19.2354 | .002703 | .0520 |
| 371 | 137641 | 19.2614 | .002695 | .0519 |
| 372 | 138384 | 19.2873 | .002688 | .0518 |
| 373 | 139129 | 19.3132 | .002681 | .0518 |
| 374 | 139876 | 19.3391 | .002674 | .0517 |
| 375 | 140625 | 19.3649 | .002667 | .0516 |
| 376 | 141376 | 19.3907 | .002660 | .0516 |
| 377 | 142129 | 19.4165 | .002653 | .0515 |
| 378 | 142884 | 19.4422 | .002646 | .0514 |
| 379 | 143641 | 19.4679 | .002639 | .0514 |
| 380 | 144400 | 19.4936 | .002632 | .0513 |
| 381 | 145161 | 19.5192 | .002625 | .0512 |
| 382 | 145924 | 19.5448 | .002618 | .0512 |
| 383 | 146689 | 19.5704 | .002611 | .0511 |
| 384 | 147456 | 19.5959 | .002604 | .0510 |
| 385 | 148225 | 19.6214 | .002597 | .0510 |

APPENDIX A (*Continued*)

| $n$ | $n^2$ | $\sqrt{n}$ | $\dfrac{1}{n}$ | $\dfrac{1}{\sqrt{n}}$ |
|---|---|---|---|---|
| 386 | 148996 | 19.6469 | .002591 | .0509 |
| 387 | 149769 | 19.6723 | .002584 | .0508 |
| 388 | 150544 | 19.6977 | .002577 | .0508 |
| 389 | 151321 | 19.7231 | .002571 | .0507 |
| 390 | 152100 | 19.7484 | .002564 | .0506 |
| 391 | 152881 | 19.7737 | .002558 | .0506 |
| 392 | 153664 | 19.7990 | .002551 | .0505 |
| 393 | 154449 | 19.8242 | .002545 | .0504 |
| 394 | 155236 | 19.8494 | .002538 | .0504 |
| 395 | 156025 | 19.8746 | .002532 | .0503 |
| 396 | 156816 | 19.8997 | .002525 | .0503 |
| 397 | 157609 | 19.9249 | .002519 | .0502 |
| 398 | 158404 | 19.9499 | .002513 | .0501 |
| 399 | 159201 | 19.9750 | .002506 | .0501 |
| 400 | 160000 | 20.0000 | .002500 | .0500 |
| 401 | 160801 | 20.0250 | .002494 | .0499 |
| 402 | 161604 | 20.0499 | .002488 | .0499 |
| 403 | 162409 | 20.0749 | .002481 | .0498 |
| 404 | 163216 | 20.0998 | .002475 | .0498 |
| 405 | 164025 | 20.1246 | .002469 | .0497 |
| 406 | 164836 | 20.1494 | .002463 | .0496 |
| 407 | 165649 | 20.1742 | .002457 | .0496 |
| 408 | 166464 | 20.1990 | .002451 | .0495 |
| 409 | 167281 | 20.2237 | .002445 | .0494 |
| 410 | 168100 | 20.2485 | .002439 | .0494 |
| 411 | 168921 | 20.2731 | .002433 | .0493 |
| 412 | 169744 | 20.2978 | .002427 | .0493 |
| 413 | 170569 | 20.3224 | .002421 | .0492 |
| 414 | 171396 | 20.3470 | .002415 | .0491 |
| 415 | 172225 | 20.3715 | .002410 | .0491 |
| 416 | 173056 | 20.3961 | .002404 | .0490 |
| 417 | 173889 | 20.4206 | .002398 | .0490 |
| 418 | 174724 | 20.4450 | .002392 | .0489 |
| 419 | 175561 | 20.4695 | .002387 | .0489 |
| 420 | 176400 | 20.4939 | .002381 | .0488 |
| 421 | 177241 | 20.5183 | .002375 | .0487 |
| 422 | 178084 | 20.5426 | .002370 | .0487 |
| 423 | 178929 | 20.5670 | .002364 | .0486 |
| 424 | 179776 | 20.5913 | .002358 | .0486 |
| 425 | 180625 | 20.6155 | .002353 | .0485 |
| 426 | 181476 | 20.6398 | .002347 | .0485 |
| 427 | 182329 | 20.6640 | .002342 | .0484 |
| 428 | 183184 | 20.6882 | .002336 | .0483 |
| 429 | 184041 | 20.7123 | .002331 | .0483 |
| 430 | 184900 | 20.7364 | .002326 | .0482 |
| 431 | 185761 | 20.7605 | .002320 | .0482 |
| 432 | 186624 | 20.7846 | .002315 | .0481 |
| 433 | 187489 | 20.8087 | .002309 | .0481 |
| 434 | 188356 | 20.8327 | .002304 | .0480 |
| 435 | 189225 | 20.8567 | .002299 | .0479 |

## APPENDIX A (*Continued*)

| $n$ | $n^2$ | $\sqrt{n}$ | $\dfrac{1}{n}$ | $\dfrac{1}{\sqrt{n}}$ |
|---|---|---|---|---|
| 436 | 190096 | 20.8806 | .002294 | .0479 |
| 437 | 190969 | 20.9045 | .002288 | .0478 |
| 438 | 191844 | 20.9284 | .002283 | .0478 |
| 439 | 192721 | 20.9523 | .002278 | .0477 |
| 440 | 193600 | 20.9762 | .002273 | .0477 |
| 441 | 194481 | 21.0000 | .002268 | .0476 |
| 442 | 195364 | 21.0238 | .002262 | .0476 |
| 443 | 196249 | 21.0476 | .002257 | .0475 |
| 444 | 197136 | 21.0713 | .002252 | .0475 |
| 445 | 198025 | 21.0950 | .002247 | .0474 |
| 446 | 198916 | 21.1187 | .002242 | .0474 |
| 447 | 199809 | 21.1424 | .002237 | .0473 |
| 448 | 200704 | 21.1660 | .002232 | .0472 |
| 449 | 201601 | 21.1896 | .002227 | .0472 |
| 450 | 202500 | 21.2132 | .002222 | .0471 |
| 451 | 203401 | 21.2368 | .002217 | .0471 |
| 452 | 204304 | 21.2603 | .002212 | .0470 |
| 453 | 205209 | 21.2838 | .002208 | .0470 |
| 454 | 206116 | 21.3073 | .002203 | .0469 |
| 455 | 207025 | 21.3307 | .002198 | .0469 |
| 456 | 207936 | 21.3542 | .002193 | .0468 |
| 457 | 208849 | 21.3776 | .022188 | .0468 |
| 458 | 209764 | 21.4009 | .002183 | .0467 |
| 459 | 210681 | 21.4243 | .002179 | .0467 |
| 460 | 211600 | 21.4476 | .002174 | .0466 |
| 461 | 212521 | 21.4709 | .002169 | .0466 |
| 462 | 213444 | 21.4942 | .002165 | .0465 |
| 463 | 214369 | 21.5174 | .002160 | .0465 |
| 464 | 215296 | 21.5407 | .002155 | .0464 |
| 465 | 216225 | 21.5639 | .002151 | .0464 |
| 466 | 217156 | 21.5870 | .002146 | .0463 |
| 467 | 218089 | 21.6102 | .002141 | .0463 |
| 468 | 219024 | 21.6333 | .002137 | .0462 |
| 469 | 219961 | 21.6564 | .002132 | .0462 |
| 470 | 220900 | 21.6795 | .002128 | .0461 |
| 471 | 221841 | 21.7025 | .002123 | .0461 |
| 472 | 222784 | 21.7256 | .002119 | .0460 |
| 473 | 223729 | 21.7486 | .002114 | .0460 |
| 474 | 224676 | 21.7715 | .002110 | .0459 |
| 475 | 225625 | 21.7945 | .002105 | .0459 |
| 476 | 226576 | 21.8174 | .002101 | .0458 |
| 477 | 227529 | 21.8403 | .002096 | .0458 |
| 478 | 228484 | 21.8632 | .002092 | .0457 |
| 479 | 229441 | 21.8861 | .002088 | .0457 |
| 480 | 230400 | 21.9089 | .002083 | .0456 |
| 481 | 231361 | 21.9317 | .002079 | .0456 |
| 482 | 232324 | 21.9545 | .002075 | .0455 |
| 483 | 233289 | 21.9773 | .002070 | .0455 |
| 484 | 234256 | 22.0000 | .002066 | .0455 |
| 485 | 235225 | 22.0227 | .002062 | .0454 |

APPENDIX A (*Continued*)

| $n$ | $n^2$ | $\sqrt{n}$ | $\dfrac{1}{n}$ | $\dfrac{1}{\sqrt{n}}$ |
|---|---|---|---|---|
| 486 | 236196 | 22.0454 | .002058 | .0454 |
| 487 | 237169 | 22.0681 | .002053 | .0453 |
| 488 | 238144 | 22.0907 | .002049 | .0453 |
| 489 | 239121 | 22.1133 | .002045 | .0452 |
| 490 | 240100 | 22.1359 | .002041 | .0452 |
| 491 | 241081 | 22.1585 | .002037 | .0451 |
| 492 | 242064 | 22.1811 | .002033 | .0451 |
| 493 | 243049 | 22.2036 | .002028 | .0450 |
| 494 | 244036 | 22.2261 | .002024 | .0450 |
| 495 | 245025 | 22.2486 | .002020 | .0449 |
| 496 | 246016 | 22.2711 | .002016 | .0448 |
| 497 | 247009 | 22.2935 | .002012 | .0449 |
| 498 | 248004 | 22.3159 | .002008 | .0449 |
| 499 | 249001 | 22.3383 | .002004 | .0448 |
| 500 | 250000 | 22.3607 | .002000 | .0447 |
| 501 | 251001 | 22.3830 | .001996 | .0447 |
| 502 | 252004 | 22.4054 | .001992 | .0446 |
| 503 | 253009 | 22.4277 | .001988 | .0446 |
| 504 | 254016 | 22.4499 | .001984 | .0445 |
| 505 | 255025 | 22.4722 | .001980 | .0445 |
| 506 | 256036 | 22.4944 | .001976 | .0445 |
| 507 | 257049 | 22.5167 | .001972 | .0444 |
| 508 | 258064 | 22.5389 | .001969 | .0444 |
| 509 | 259081 | 22.5610 | .001965 | .0443 |
| 510 | 260100 | 22.5832 | .001961 | .0443 |
| 511 | 261121 | 22.6053 | .001957 | .0442 |
| 512 | 262144 | 22.6274 | .001953 | .0442 |
| 513 | 263169 | 22.6495 | .001949 | .0442 |
| 514 | 264196 | 22.6716 | .001946 | .0441 |
| 515 | 265225 | 22.6936 | .001942 | .0441 |
| 516 | 266256 | 22.7156 | .001938 | .0440 |
| 517 | 267289 | 22.7376 | .001934 | .0440 |
| 518 | 268324 | 22.7596 | .001931 | .0439 |
| 519 | 269361 | 22.7816 | .001927 | .0439 |
| 520 | 270400 | 22.8035 | .001923 | .0439 |
| 521 | 271441 | 22.8254 | .001919 | .0438 |
| 522 | 272484 | 22.8473 | .001916 | .0438 |
| 523 | 273529 | 22.8692 | .001912 | .0437 |
| 524 | 274576 | 22.8910 | .001908 | .0437 |
| 525 | 275625 | 22.9129 | .001905 | .0436 |
| 526 | 276676 | 22.9347 | .001901 | .0436 |
| 527 | 277729 | 22.9565 | .001898 | .0436 |
| 528 | 278784 | 22.9783 | .001894 | .0435 |
| 529 | 279841 | 23.0000 | .001890 | .0435 |
| 530 | 280900 | 23.0217 | .001887 | .0434 |
| 531 | 281961 | 23.0434 | .001883 | .0434 |
| 532 | 283024 | 23.0651 | .001880 | .0434 |
| 533 | 284089 | 23.0868 | .001876 | .0433 |
| 534 | 285156 | 23.1084 | .001873 | .0433 |
| 535 | 286225 | 23.1301 | .001869 | .0432 |

APPENDIX A (*Continued*)

| $n$ | $n^2$ | $\sqrt{n}$ | $\dfrac{1}{n}$ | $\dfrac{1}{\sqrt{n}}$ |
|---|---|---|---|---|
| 536 | 287296 | 23.1517 | .001866 | .0432 |
| 537 | 288369 | 23.1733 | .001862 | .0432 |
| 538 | 289444 | 23.1948 | .001859 | .0431 |
| 539 | 290521 | 23.2164 | .001855 | .0431 |
| 540 | 291600 | 23.2379 | .001852 | .0430 |
| 541 | 292681 | 23.2594 | .001848 | .0430 |
| 542 | 293764 | 23.2809 | .001845 | .0430 |
| 543 | 294849 | 23.3024 | .001842 | .0429 |
| 544 | 295936 | 23.3238 | .001838 | .0429 |
| 545 | 297025 | 23.3452 | .001835 | .0428 |
| 546 | 298116 | 23.3666 | .001832 | .0428 |
| 547 | 299209 | 23.3880 | .001828 | .0428 |
| 548 | 300304 | 23.4094 | .001825 | .0427 |
| 549 | 301401 | 23.4307 | .001821 | .0427 |
| 550 | 302500 | 23.4521 | .001818 | .0426 |
| 551 | 303601 | 23.4734 | .001815 | .0426 |
| 552 | 304704 | 23.4947 | .001812 | .0426 |
| 553 | 305809 | 23.5160 | .001808 | .0425 |
| 554 | 306916 | 23.5372 | .001805 | .0425 |
| 555 | 308025 | 23.5584 | .001802 | .0424 |
| 556 | 309136 | 23.5797 | .001799 | .0424 |
| 557 | 310249 | 23.6008 | .001795 | .0424 |
| 558 | 311364 | 23.6220 | .001792 | .0423 |
| 559 | 312481 | 23.6432 | .001789 | .0423 |
| 560 | 313600 | 23.6643 | .001786 | .0423 |
| 561 | 314721 | 23.6854 | .001783 | .0422 |
| 562 | 315844 | 23.7065 | .001779 | .0422 |
| 563 | 316969 | 23.7276 | .001776 | .0421 |
| 564 | 318096 | 23.7487 | .001773 | .0421 |
| 565 | 319225 | 23.7697 | .001770 | .0421 |
| 566 | 320356 | 23.7908 | .001767 | .0420 |
| 567 | 321489 | 23.8118 | .001764 | .0420 |
| 568 | 322624 | 23.8328 | .001761 | .0420 |
| 569 | 323761 | 23.8537 | .001757 | .0419 |
| 570 | 324900 | 23.8747 | .001754 | .0419 |
| 571 | 326041 | 23.8956 | .001751 | .0418 |
| 572 | 327184 | 23.9165 | .001748 | .0418 |
| 573 | 328329 | 23.9374 | .001745 | .0418 |
| 574 | 329476 | 23.9583 | .001742 | .0417 |
| 575 | 330625 | 23.9792 | .001739 | .0417 |
| 576 | 331776 | 24.0000 | .001736 | .0417 |
| 577 | 332929 | 24.0208 | .001733 | .0416 |
| 578 | 334084 | 24.0416 | .001730 | .0416 |
| 579 | 335241 | 24.0624 | .001727 | .0416 |
| 580 | 336400 | 24.0832 | .001724 | .0415 |
| 581 | 337561 | 24.1039 | .001721 | .0415 |
| 582 | 338724 | 24.1247 | .001718 | .0415 |
| 583 | 339889 | 24.1454 | .001715 | .0414 |
| 584 | 341056 | 24.1661 | .001712 | .0414 |
| 585 | 342225 | 24.1868 | .001709 | .0413 |

APPENDIX A (*Continued*)

| $n$ | $n^2$ | $\sqrt{n}$ | $\dfrac{1}{n}$ | $\dfrac{1}{\sqrt{n}}$ |
|---|---|---|---|---|
| 586 | 343396 | 24.2074 | .001706 | .0413 |
| 587 | 344569 | 24.2281 | .001704 | .0413 |
| 588 | 345744 | 24.2487 | .001701 | .0412 |
| 589 | 346921 | 24.2693 | .001698 | .0412 |
| 590 | 348100 | 24.2899 | .001695 | .0412 |
| 591 | 349281 | 24.3105 | .001692 | .0411 |
| 592 | 350464 | 24.3311 | .001689 | .0411 |
| 593 | 351649 | 24.3516 | .001686 | .0411 |
| 594 | 352836 | 24.3721 | .001684 | .0410 |
| 595 | 354025 | 24.3926 | .001681 | .0410 |
| 596 | 355216 | 24.4131 | .001678 | .0410 |
| 597 | 356409 | 24.4336 | .001675 | .0409 |
| 598 | 357604 | 24.4540 | .001672 | .0409 |
| 599 | 358801 | 24.4745 | .001669 | .0409 |
| 600 | 360000 | 24.4949 | .001667 | .0408 |
| 601 | 361201 | 24.5153 | .001664 | .0408 |
| 602 | 362404 | 24.5357 | .001661 | .0408 |
| 603 | 363609 | 24.5561 | .001658 | .0407 |
| 604 | 364816 | 24.5764 | .001656 | .0407 |
| 605 | 366025 | 24.5967 | .001653 | .0407 |
| 606 | 367236 | 24.6171 | .001650 | .0406 |
| 607 | 368449 | 24.6374 | .001647 | .0406 |
| 608 | 369664 | 24.6577 | .001645 | .0406 |
| 609 | 370881 | 24.6779 | .001642 | .0405 |
| 610 | 372100 | 24.6982 | .001639 | .0405 |
| 611 | 373321 | 24.7184 | .001637 | .0405 |
| 612 | 374544 | 24.7386 | .001634 | .0404 |
| 613 | 375769 | 24.7588 | .001631 | .0404 |
| 614 | 376996 | 24.7790 | .001629 | .0404 |
| 615 | 378225 | 24.7992 | .001626 | .0403 |
| 616 | 379456 | 24.8193 | .001623 | .0403 |
| 617 | 380689 | 24.8395 | .001621 | .0403 |
| 618 | 381924 | 24.8596 | .001618 | .0402 |
| 619 | 383161 | 24.8797 | .001616 | .0402 |
| 620 | 384400 | 24.8998 | .001613 | .0402 |
| 621 | 385641 | 24.9199 | .001610 | .0401 |
| 622 | 386884 | 24.9399 | .001608 | .0401 |
| 623 | 388129 | 24.9600 | .001605 | .0401 |
| 624 | 389376 | 24.9800 | .001603 | .0400 |
| 625 | 390625 | 25.0000 | .001600 | .0400 |
| 626 | 391876 | 25.0200 | .001597 | .0400 |
| 627 | 393129 | 25.0400 | .001595 | .0399 |
| 628 | 394384 | 25.0599 | .001592 | .0399 |
| 629 | 395641 | 25.0799 | .001590 | .0399 |
| 630 | 396900 | 25.0998 | .001587 | .0398 |
| 631 | 398161 | 25.1197 | .001585 | .0398 |
| 632 | 399424 | 25.1396 | .001582 | .0398 |
| 633 | 400689 | 25.1595 | .001580 | .0397 |
| 634 | 401956 | 25.1794 | .001577 | .0397 |
| 635 | 403225 | 25.1992 | .001575 | .0397 |

APPENDIX A (*Continued*)

| $n$ | $n^2$ | $\sqrt{n}$ | $\dfrac{1}{n}$ | $\dfrac{1}{\sqrt{n}}$ |
|---|---|---|---|---|
| 636 | 404496 | 25.2190 | .001572 | .0397 |
| 637 | 405769 | 25.2389 | .001570 | .0396 |
| 638 | 407044 | 25.2587 | .001567 | .0396 |
| 639 | 408321 | 25.2784 | .001565 | .0396 |
| 640 | 409600 | 25.2982 | .001563 | .0395 |
| 641 | 410881 | 25.3180 | .001560 | .0395 |
| 642 | 412164 | 25.3377 | .001558 | .0395 |
| 643 | 413449 | 25.3574 | .001555 | .0394 |
| 644 | 414736 | 25.3772 | .001553 | .0394 |
| 645 | 416025 | 25.3969 | .001550 | .0394 |
| 646 | 417316 | 25.4165 | .001548 | .0393 |
| 647 | 418609 | 25.4362 | .001546 | .0393 |
| 648 | 419904 | 25.4558 | .001543 | .0393 |
| 649 | 421201 | 25.4755 | .001541 | .0393 |
| 650 | 422500 | 25.4951 | .001538 | .0392 |
| 651 | 423801 | 25.5147 | .001536 | .0392 |
| 652 | 425104 | 25.5343 | .001534 | .0392 |
| 653 | 426409 | 25.5539 | .001531 | .0391 |
| 654 | 427716 | 25.5734 | .001529 | .0391 |
| 655 | 429025 | 25.5930 | .001527 | .0391 |
| 656 | 430336 | 25.6125 | .001524 | .0390 |
| 657 | 431649 | 25.6320 | .001522 | .0390 |
| 658 | 432964 | 25.6515 | .001520 | .0390 |
| 659 | 434281 | 25.6710 | .001517 | .0390 |
| 660 | 435600 | 25.6905 | .001515 | .0389 |
| 661 | 436921 | 25.7099 | .001513 | .0389 |
| 662 | 438244 | 25.7294 | .001511 | .0389 |
| 663 | 439569 | 25.7488 | .001508 | .0388 |
| 664 | 440896 | 25.7682 | .001506 | .0388 |
| 665 | 442225 | 25.7876 | .001504 | .0388 |
| 666 | 443556 | 25.8070 | .001502 | .0387 |
| 667 | 444889 | 25.8263 | .001499 | .0387 |
| 668 | 446224 | 25.8457 | .001497 | .0387 |
| 669 | 447561 | 25.8650 | .001495 | .0387 |
| 670 | 448900 | 25.8844 | .001493 | .0386 |
| 671 | 450241 | 25.9037 | .001490 | .0386 |
| 672 | 451584 | 25.9230 | .001488 | .0386 |
| 673 | 452929 | 25.9422 | .001486 | .0385 |
| 674 | 454276 | 25.9615 | .001484 | .0385 |
| 675 | 455625 | 25.9808 | .001481 | .0385 |
| 676 | 456976 | 26.0000 | .001479 | .0385 |
| 677 | 458329 | 26.0192 | .001477 | .0384 |
| 678 | 459684 | 26.0384 | .001475 | .0384 |
| 679 | 461041 | 26.0576 | .001473 | .0384 |
| 680 | 462400 | 26.0768 | .001471 | .0383 |
| 681 | 463761 | 26.0960 | .001468 | .0383 |
| 682 | 465124 | 26.1151 | .001466 | .0383 |
| 683 | 466489 | 26.1343 | .001464 | .0383 |
| 684 | 467856 | 26.1534 | .001462 | .0382 |
| 685 | 469225 | 26.1725 | .001460 | .0382 |

APPENDIX A (*Continued*)

| n | $n^2$ | $\sqrt{n}$ | $\dfrac{1}{n}$ | $\dfrac{1}{\sqrt{n}}$ |
|---|---|---|---|---|
| 686 | 470596 | 26.1916 | .001458 | .0382 |
| 687 | 471969 | 26.2107 | .001456 | .0382 |
| 688 | 473344 | 26.2298 | .001453 | .0381 |
| 689 | 474721 | 26.2488 | .001451 | .0381 |
| 690 | 476100 | 26.2679 | .001449 | .0381 |
| 691 | 477481 | 26.2869 | .001447 | .0380 |
| 692 | 478864 | 26.3059 | .001445 | .0380 |
| 693 | 480249 | 26.3249 | .001443 | .0380 |
| 694 | 481636 | 26.3439 | .001441 | .0380 |
| 695 | 483025 | 26.3629 | .001439 | .0379 |
| 696 | 484416 | 26.3818 | .001437 | .0379 |
| 697 | 485809 | 26.4008 | .001435 | .0379 |
| 698 | 487204 | 26.4197 | .001433 | .0379 |
| 699 | 488601 | 26.4386 | .001431 | .0378 |
| 700 | 490000 | 26.4575 | .001429 | .0378 |
| 701 | 491401 | 26.4764 | .001427 | .0378 |
| 702 | 492804 | 26.4953 | .001425 | .0377 |
| 703 | 494209 | 26.5141 | .001422 | .0377 |
| 704 | 495616 | 26.5330 | .001420 | .0377 |
| 705 | 497025 | 26.5518 | .001418 | .0377 |
| 706 | 498436 | 26.5707 | .001416 | .0376 |
| 707 | 499849 | 26.5895 | .001414 | .0376 |
| 708 | 501264 | 26.6083 | .001412 | .0376 |
| 709 | 502681 | 26.6271 | .001410 | .0376 |
| 710 | 504100 | 26.6458 | .001408 | .0375 |
| 711 | 505521 | 26.6646 | .001406 | .0375 |
| 712 | 506944 | 26.6833 | .001404 | .0375 |
| 713 | 508369 | 26.7021 | .001403 | .0375 |
| 714 | 509796 | 26.7208 | .001401 | .0374 |
| 715 | 511225 | 26.7395 | .001399 | .0374 |
| 716 | 512656 | 26.7582 | .001397 | .0374 |
| 717 | 514089 | 26.7769 | .001395 | .0373 |
| 718 | 515524 | 26.7955 | .001393 | .0373 |
| 719 | 516961 | 26.8142 | .001391 | .0373 |
| 720 | 518400 | 26.8328 | .001389 | .0373 |
| 721 | 519841 | 26.8514 | .001387 | .0372 |
| 722 | 521284 | 26.8701 | .001385 | .0372 |
| 723 | 522729 | 26.8887 | .001383 | .0372 |
| 724 | 524176 | 26.9072 | .001381 | .0372 |
| 725 | 525625 | 26.9258 | .001379 | .0371 |
| 726 | 527076 | 26.9444 | .001377 | .0371 |
| 727 | 528529 | 26.9629 | .001376 | .0371 |
| 728 | 529984 | 26.9815 | .001374 | .0371 |
| 729 | 531441 | 27.0000 | .001372 | .0370 |
| 730 | 532900 | 27.0185 | .001370 | .0370 |
| 731 | 534361 | 27.0370 | .001368 | .0370 |
| 732 | 535824 | 27.0555 | .001366 | .0370 |
| 733 | 537289 | 27.0740 | .001364 | .0369 |
| 734 | 538756 | 27.0924 | .001362 | .0369 |
| 735 | 540225 | 27.1109 | .001361 | .0369 |

APPENDIX A (*Continued*)

| $n$ | $n^2$ | $\sqrt{n}$ | $\dfrac{1}{n}$ | $\dfrac{1}{\sqrt{n}}$ |
|---|---|---|---|---|
| 736 | 541696 | 27.1293 | .001359 | .0369 |
| 737 | 543169 | 27.1477 | .001357 | .0368 |
| 738 | 544644 | 27.1662 | .001355 | .0368 |
| 739 | 546121 | 27.1846 | .001353 | .0368 |
| 740 | 547600 | 27.2029 | .001351 | .0368 |
| 741 | 549081 | 27.2213 | .001350 | .0367 |
| 742 | 550564 | 27.2397 | .001348 | .0367 |
| 743 | 552049 | 27.2580 | .001346 | .0367 |
| 744 | 553536 | 27.2764 | .001344 | .0367 |
| 745 | 555025 | 27.2947 | .001342 | .0366 |
| 746 | 556516 | 27.3130 | .001340 | .0366 |
| 747 | 558009 | 27.3313 | .001339 | .0366 |
| 748 | 559504 | 27.3496 | .001337 | .0366 |
| 749 | 561001 | 27.3679 | .001335 | .0365 |
| 750 | 562500 | 27.3861 | .001333 | .0365 |
| 751 | 564001 | 27.4044 | .001332 | .0365 |
| 752 | 565504 | 27.4226 | .001330 | .0365 |
| 753 | 567009 | 27.4408 | .001328 | .0364 |
| 754 | 568516 | 27.4591 | .001326 | .0364 |
| 755 | 570025 | 27.4773 | .001325 | .0364 |
| 756 | 571536 | 27.4955 | .001323 | .0364 |
| 757 | 573049 | 27.5136 | .001321 | .0363 |
| 758 | 574564 | 27.5318 | .001319 | .0363 |
| 759 | 576081 | 27.5500 | .001318 | .0363 |
| 760 | 577600 | 27.5681 | .001316 | .0363 |
| 761 | 579121 | 27.5862 | .001314 | .0363 |
| 762 | 580644 | 27.6043 | .001312 | .0362 |
| 763 | 582169 | 27.6225 | .001311 | .0362 |
| 764 | 583696 | 27.6405 | .001309 | .0362 |
| 765 | 585225 | 27.6586 | .001307 | .0362 |
| 766 | 586756 | 27.6767 | .001305 | .0361 |
| 767 | 588289 | 27.6948 | .001304 | .0361 |
| 768 | 589824 | 27.7128 | .001302 | .0361 |
| 769 | 591361 | 27.7308 | .001300 | .0361 |
| 770 | 592900 | 27.7489 | .001299 | .0360 |
| 771 | 594441 | 27.7669 | .001297 | .0360 |
| 772 | 595984 | 27.7849 | .001295 | .0360 |
| 773 | 597529 | 27.8029 | .001294 | .0360 |
| 774 | 599076 | 27.8209 | .001292 | .0359 |
| 775 | 600625 | 27.8388 | .001290 | .0359 |
| 776 | 602176 | 27.8568 | .001289 | .0359 |
| 777 | 603729 | 27.8747 | .001287 | .0359 |
| 778 | 605284 | 27.8927 | .001285 | .0359 |
| 779 | 606841 | 27.9106 | .001284 | .0358 |
| 780 | 608400 | 27.9285 | .001282 | .0358 |
| 781 | 609961 | 27.9464 | .001280 | .0358 |
| 782 | 611524 | 27.9643 | .001279 | .0358 |
| 783 | 613089 | 27.9821 | .001277 | .0357 |
| 784 | 614656 | 28.0000 | .001276 | .0357 |
| 785 | 616225 | 28.0179 | .001274 | .0357 |

APPENDIX A (*Continued*)

| $n$ | $n^2$ | $\sqrt{n}$ | $\dfrac{1}{n}$ | $\dfrac{1}{\sqrt{n}}$ |
|---|---|---|---|---|
| 786 | 617796 | 28.0357 | .001272 | .0357 |
| 787 | 619369 | 28.0535 | .001271 | .0356 |
| 788 | 620944 | 28.0713 | .001269 | .0356 |
| 789 | 622521 | 28.0891 | .001267 | .0356 |
| 790 | 624100 | 28.1069 | .001266 | .0356 |
| 791 | 625681 | 28.1247 | .001264 | .0356 |
| 792 | 627264 | 28.1425 | .001263 | .0355 |
| 793 | 628849 | 28.1603 | .001261 | .0355 |
| 794 | 630436 | 28.1780 | .001259 | .0355 |
| 795 | 632025 | 28.1957 | .001258 | .0355 |
| 796 | 633616 | 28.2135 | .001256 | .0354 |
| 797 | 635209 | 28.2312 | .001255 | .0354 |
| 798 | 636804 | 28.2489 | .001253 | .0354 |
| 799 | 638401 | 28.2666 | .001252 | .0354 |
| 800 | 640000 | 28.2843 | .001250 | .0354 |
| 801 | 641601 | 28.3019 | .001248 | .0353 |
| 802 | 643204 | 28.3196 | .001247 | .0353 |
| 803 | 644809 | 28.3373 | .001245 | .0353 |
| 804 | 646416 | 28.3549 | .001244 | .0353 |
| 805 | 648025 | 28.3725 | .001242 | .0352 |
| 806 | 649636 | 28.3901 | .001241 | .0352 |
| 807 | 651249 | 28.4077 | .001239 | .0352 |
| 808 | 652864 | 28.4253 | .001238 | .0352 |
| 809 | 654481 | 28.4429 | .001236 | .0352 |
| 810 | 656100 | 28.4605 | .001235 | .0351 |
| 811 | 657721 | 28.4781 | .001233 | .0351 |
| 812 | 659344 | 28.4956 | .001232 | .0351 |
| 813 | 660969 | 28.5132 | .001230 | .0351 |
| 814 | 662596 | 28.5307 | .001229 | .0351 |
| 815 | 664225 | 28.5482 | .001227 | .0350 |
| 816 | 665856 | 28.5657 | .001225 | .0350 |
| 817 | 667489 | 28.5832 | .001224 | .0350 |
| 818 | 669124 | 28.6007 | .001222 | .0350 |
| 819 | 670761 | 28.6182 | .001221 | .0349 |
| 820 | 672400 | 28.6356 | .001220 | .0349 |
| 821 | 674041 | 28.6531 | .001218 | .0349 |
| 822 | 675684 | 28.6705 | .001217 | .0349 |
| 823 | 677329 | 28.6880 | .001215 | .0349 |
| 824 | 678976 | 28.7054 | .001214 | .0348 |
| 825 | 680625 | 28.7228 | .001212 | .0348 |
| 826 | 682276 | 28.7402 | .001211 | .0348 |
| 827 | 683929 | 28.7576 | .001209 | .0348 |
| 828 | 685584 | 28.7750 | .001208 | .0348 |
| 829 | 687241 | 28.7924 | .001206 | .0347 |
| 830 | 688900 | 28.8097 | .001205 | .0347 |
| 831 | 690561 | 28.8271 | .001203 | .0347 |
| 832 | 692224 | 28.8444 | .001202 | .0347 |
| 833 | 693889 | 28.8617 | .001200 | .0346 |
| 834 | 695556 | 28.8791 | .001199 | .0346 |
| 835 | 697225 | 28.8964 | .001198 | .0346 |

## APPENDIX A (*Continued*)

| $n$ | $n^2$ | $\sqrt{n}$ | $\dfrac{1}{n}$ | $\dfrac{1}{\sqrt{n}}$ |
|---|---|---|---|---|
| 836 | 698896 | 28.9137 | .001196 | .0346 |
| 837 | 700569 | 28.9310 | .001195 | .0346 |
| 838 | 702244 | 28.9482 | .001193 | .0345 |
| 839 | 703921 | 28.9655 | .001192 | .0345 |
| 840 | 705600 | 28.9828 | .001190 | .0345 |
| 841 | 707281 | 29.0000 | .001189 | .0345 |
| 842 | 708964 | 29.0172 | .001188 | .0345 |
| 843 | 710649 | 29.0345 | .001186 | .0344 |
| 844 | 712336 | 29.0517 | .001185 | .0344 |
| 845 | 714025 | 29.0689 | .001183 | .0344 |
| 846 | 715716 | 29.0861 | .001182 | .0344 |
| 847 | 717409 | 29.1033 | .001181 | .0344 |
| 848 | 719104 | 29.1204 | .001179 | .0343 |
| 849 | 720801 | 29.1376 | .001178 | .0343 |
| 850 | 722500 | 29.1548 | .001176 | .0343 |
| 851 | 724201 | 29.1719 | .001175 | .0343 |
| 852 | 725904 | 29.1890 | .001174 | .0343 |
| 853 | 727609 | 29.2062 | .001172 | .0342 |
| 854 | 729316 | 29.2233 | .001171 | .0342 |
| 855 | 731025 | 29.2404 | .001170 | .0342 |
| 856 | 732736 | 29.2575 | .001168 | .0342 |
| 857 | 734449 | 29.2746 | .001167 | .0342 |
| 858 | 736164 | 29.2916 | .001166 | .0341 |
| 859 | 737881 | 29.3087 | .001164 | .0341 |
| 860 | 739600 | 29.3258 | .001163 | .0341 |
| 861 | 741321 | 29.3428 | .001161 | .0341 |
| 862 | 743044 | 29.3598 | .001160 | .0341 |
| 863 | 744769 | 29.3769 | .001159 | .0340 |
| 864 | 746496 | 29.3939 | .001157 | .0340 |
| 865 | 748225 | 29.4109 | .001156 | .0340 |
| 866 | 749956 | 29.4279 | .001155 | .0340 |
| 867 | 751689 | 29.4449 | .001153 | .0340 |
| 868 | 753424 | 29.4618 | .001152 | .0339 |
| 869 | 755161 | 29.4788 | .001151 | .0339 |
| 870 | 756900 | 29.4958 | .001149 | .0339 |
| 871 | 758641 | 29.5127 | .001148 | .0339 |
| 872 | 760384 | 29.5296 | .001147 | .0339 |
| 873 | 762129 | 29.5466 | .001145 | .0338 |
| 874 | 763876 | 29.5635 | .001144 | .0338 |
| 875 | 765625 | 29.5804 | .001143 | .0338 |
| 876 | 767376 | 29.5973 | .001142 | .0338 |
| 877 | 769129 | 29.6142 | .001140 | .0338 |
| 878 | 770884 | 29.6311 | .001139 | .0337 |
| 879 | 772641 | 29.6479 | .001138 | .0337 |
| 880 | 774400 | 29.6648 | .001136 | .0337 |
| 881 | 776161 | 29.6816 | .001135 | .0337 |
| 882 | 777924 | 29.6985 | .001134 | .0337 |
| 883 | 779689 | 29.7153 | .001133 | .0337 |
| 884 | 781456 | 29.7321 | .001131 | .0336 |
| 885 | 783225 | 29.7489 | .001130 | .0336 |

APPENDIX A (*Continued*)

| $n$ | $n^2$ | $\sqrt{n}$ | $\dfrac{1}{n}$ | $\dfrac{1}{\sqrt{n}}$ |
|---|---|---|---|---|
| 886 | 784996 | 29.7658 | .001129 | .0336 |
| 887 | 786769 | 29.7825 | .001127 | .0336 |
| 888 | 788544 | 29.7993 | .001126 | .0336 |
| 889 | 790321 | 29.8161 | .001125 | .0335 |
| 890 | 792100 | 29.8329 | .001124 | .0335 |
| 891 | 793881 | 29.8496 | .001122 | .0335 |
| 892 | 795664 | 29.8664 | .001121 | .0335 |
| 893 | 797449 | 29.8831 | .001120 | .0335 |
| 894 | 799236 | 29.8998 | .001119 | .0334 |
| 895 | 801025 | 29.9166 | .001117 | .0334 |
| 896 | 802816 | 29.9333 | .001116 | .0334 |
| 897 | 804609 | 29.9500 | .001115 | .0334 |
| 898 | 806404 | 29.9666 | .001114 | .0334 |
| 899 | 808201 | 29.9833 | .001112 | .0334 |
| 900 | 810000 | 30.0000 | .001111 | .0333 |
| 901 | 811801 | 30.0167 | .001110 | .0333 |
| 902 | 813604 | 30.0333 | .001109 | .0333 |
| 903 | 815409 | 30.0500 | .001107 | .0333 |
| 904 | 817216 | 30.0666 | .001106 | .0333 |
| 905 | 819025 | 30.0832 | .001105 | .0332 |
| 906 | 820836 | 30.0998 | .001104 | .0332 |
| 907 | 822649 | 30.1164 | .001103 | .0332 |
| 908 | 824464 | 30.1330 | .001101 | .0332 |
| 909 | 826281 | 30.1496 | .001100 | .0332 |
| 910 | 828100 | 30.1662 | .001099 | .0331 |
| 911 | 829921 | 30.1828 | .001098 | .0331 |
| 912 | 831744 | 30.1993 | .001096 | .0331 |
| 913 | 833569 | 30.2159 | .001095 | .0331 |
| 914 | 835396 | 30.2324 | .001094 | .0331 |
| 915 | 837225 | 30.2490 | .001093 | .0331 |
| 916 | 839056 | 30.2655 | .001092 | .0330 |
| 917 | 840889 | 30.2820 | .001091 | .0330 |
| 918 | 842724 | 30.2985 | .001089 | .0330 |
| 919 | 844561 | 30.3150 | .001088 | .0330 |
| 920 | 846400 | 30.3315 | .001087 | .0330 |
| 921 | 848241 | 30.3480 | .001086 | .0330 |
| 922 | 850084 | 30.3645 | .001085 | .0329 |
| 923 | 851929 | 30.3809 | .001083 | .0329 |
| 924 | 853776 | 30.3974 | .001082 | .0329 |
| 925 | 855625 | 30.4138 | .001081 | .0329 |
| 926 | 857476 | 30.4302 | .001080 | .0329 |
| 927 | 859329 | 30.4467 | .001079 | .0328 |
| 928 | 861184 | 30.4631 | .001078 | .0328 |
| 929 | 863041 | 30.4795 | .001076 | .0328 |
| 930 | 864900 | 30.4959 | .001075 | .0328 |
| 931 | 866761 | 30.5123 | .001074 | .0328 |
| 932 | 868624 | 30.5287 | .001073 | .0328 |
| 933 | 870489 | 30.5450 | .001072 | .0327 |
| 934 | 872356 | 30.5614 | .001071 | .0327 |
| 935 | 874225 | 30.5778 | .001070 | .0327 |

APPENDIX A (*Continued*)

| $n$ | $n^2$ | $\sqrt{n}$ | $\dfrac{1}{n}$ | $\dfrac{1}{\sqrt{n}}$ |
|---|---|---|---|---|
| 936 | 876096 | 30.5941 | .001068 | .0327 |
| 937 | 877969 | 30.6105 | .001067 | .0327 |
| 938 | 879844 | 30.6268 | .001066 | .0327 |
| 939 | 881721 | 30.6431 | .001065 | .0326 |
| 940 | 883600 | 30.6594 | .001064 | .0326 |
| 941 | 885481 | 30.6757 | .001063 | .0326 |
| 942 | 887364 | 30.6920 | .001062 | .0326 |
| 943 | 889249 | 30.7083 | .001060 | .0326 |
| 944 | 891136 | 30.7246 | .001059 | .0325 |
| 945 | 893025 | 30.7409 | .001058 | .0325 |
| 946 | 894916 | 30.7571 | .001057 | .0325 |
| 947 | 896809 | 30.7734 | .001056 | .0325 |
| 948 | 898704 | 30.7896 | .001055 | .0325 |
| 949 | 900601 | 30.8058 | .001054 | .0325 |
| 950 | 902500 | 30.8221 | .001053 | .0324 |
| 951 | 904401 | 30.8383 | .001052 | .0324 |
| 952 | 906304 | 30.8545 | .001050 | .0324 |
| 953 | 908209 | 30.8707 | .001049 | .0324 |
| 954 | 910116 | 30.8869 | .001048 | .0324 |
| 955 | 912025 | 30.9031 | .001047 | .0324 |
| 956 | 913936 | 30.9192 | .001046 | .0323 |
| 957 | 915849 | 30.9354 | .001045 | .0323 |
| 958 | 917764 | 30.9516 | .001044 | .0323 |
| 959 | 919681 | 30.9677 | .001043 | .0323 |
| 960 | 921600 | 30.9839 | .001042 | .0323 |
| 961 | 923521 | 31.0000 | .001041 | .0323 |
| 962 | 925444 | 31.0161 | .001040 | .0322 |
| 963 | 927369 | 31.0322 | .001038 | .0322 |
| 964 | 929296 | 31.0483 | .001037 | .0322 |
| 965 | 931225 | 31.0644 | .001036 | .0322 |
| 966 | 933156 | 31.0805 | .001035 | .0322 |
| 967 | 935089 | 31.0966 | .001034 | .0322 |
| 968 | 937024 | 31.1127 | .001033 | .0321 |
| 969 | 938961 | 31.1288 | .001032 | .0321 |
| 970 | 940900 | 31.1448 | .001031 | .0321 |
| 971 | 942841 | 31.1609 | .001030 | .0321 |
| 972 | 944784 | 31.1769 | .001029 | .0321 |
| 973 | 946729 | 31.1929 | .001028 | .0321 |
| 974 | 948676 | 31.2090 | .001027 | .0320 |
| 975 | 950625 | 31.2250 | .001026 | .0320 |
| 976 | 952576 | 31.2410 | .001025 | .0320 |
| 977 | 954529 | 31.2570 | .001024 | .0320 |
| 978 | 956484 | 31.2730 | .001022 | .0320 |
| 979 | 958441 | 31.2890 | .001021 | .0320 |
| 980 | 960400 | 31.3050 | .001020 | .0319 |
| 981 | 962361 | 31.3209 | .001019 | .0319 |
| 982 | 964324 | 31.3369 | .001018 | .0319 |
| 983 | 966289 | 31.3528 | .001017 | .0319 |
| 984 | 968256 | 31.3688 | .001016 | .0319 |
| 985 | 970225 | 31.3847 | .001015 | .0319 |

APPENDIX A (*Continued*)

| $n$ | $n^2$ | $\sqrt{n}$ | $\dfrac{1}{n}$ | $\dfrac{1}{\sqrt{n}}$ |
|---|---|---|---|---|
| 986 | 972196 | 31.4006 | .001014 | .0318 |
| 987 | 974169 | 31.4166 | .001013 | .0318 |
| 988 | 976144 | 31.4325 | .001012 | .0318 |
| 989 | 978121 | 31.4484 | .001011 | .0318 |
| 990 | 980100 | 31.4643 | .001010 | .0318 |
| 991 | 982081 | 31.4802 | .001009 | .0318 |
| 992 | 984064 | 31.4960 | .001008 | .0318 |
| 993 | 986049 | 31.5119 | .001007 | .0317 |
| 994 | 988036 | 31.5278 | .001006 | .0317 |
| 995 | 990025 | 31.5436 | .001005 | .0317 |
| 996 | 992016 | 31.5595 | .001004 | .0317 |
| 997 | 994009 | 31.5753 | .001003 | .0317 |
| 998 | 996004 | 31.5911 | .001002 | .0317 |
| 999 | 998001 | 31.6070 | .001001 | .0316 |
| 1000 | 1000000 | 31.6228 | .001000 | .0316 |

# Appendix B
## Areas and Ordinates of the Normal Curve in Terms of $x/\sigma$

| (1)<br>z<br>Standard<br>Score | (2)<br><br>Area from<br>Mean to $x/\sigma$ | (3)<br>Area in<br>Larger<br>Portion | (4)<br>Area in<br>Smaller<br>Portion | (5)<br>y<br>Ordinate<br>at $x/\sigma$ |
|---|---|---|---|---|
| 0.00 | .0000 | .5000 | .5000 | .3989 |
| 0.01 | .0040 | .5040 | .4960 | .3989 |
| 0.02 | .0080 | .5080 | .4920 | .3989 |
| 0.03 | .0120 | .5120 | .4880 | .3988 |
| 0.04 | .0160 | .5160 | .4840 | .3986 |
| 0.05 | .0199 | .5199 | .4801 | .3984 |
| 0.06 | .0239 | .5239 | .4761 | .3982 |
| 0.07 | .0279 | .5279 | .4721 | .3980 |
| 0.08 | .0319 | .5319 | .4681 | .3977 |
| 0.09 | .0359 | .5359 | .4641 | .3973 |
| 0.10 | .0398 | .5398 | .4602 | .3970 |
| 0.11 | .0438 | .5438 | .4562 | .3965 |
| 0.12 | .0478 | .5478 | .4522 | .3961 |
| 0.13 | .0517 | .5517 | .4483 | .3956 |
| 0.14 | .0557 | .5557 | .4443 | .3951 |
| 0.15 | .0596 | .5596 | .4404 | .3945 |
| 0.16 | .0636 | .5636 | .4364 | .3939 |
| 0.17 | .0675 | .5675 | .4325 | .3932 |
| 0.18 | .0714 | .5714 | .4286 | .3925 |
| 0.19 | .0753 | .5753 | .4247 | .3918 |
| 0.20 | .0793 | .5793 | .4207 | .3910 |
| 0.21 | .0832 | .5832 | .4168 | .3902 |
| 0.22 | .0871 | .5871 | .4129 | .3894 |
| 0.23 | .0910 | .5910 | .4090 | .3885 |
| 0.24 | .0948 | .5948 | .4052 | .3876 |
| 0.25 | .0987 | .5987 | .4013 | .3867 |
| 0.26 | .1026 | .6026 | .3974 | .3857 |
| 0.27 | .1064 | .6064 | .3936 | .3847 |
| 0.28 | .1103 | .6103 | .3897 | .3836 |
| 0.29 | .1141 | .6141 | .3859 | .3825 |
| 0.30 | .1179 | .6179 | .3821 | .3814 |
| 0.31 | .1217 | .6217 | .3783 | .3802 |
| 0.32 | .1255 | .6255 | .3745 | .3790 |
| 0.33 | .1293 | .6293 | .3707 | .3778 |
| 0.34 | .1331 | .6331 | .3669 | .3765 |
| 0.35 | .1368 | .6368 | .3632 | .3752 |
| 0.36 | .1406 | .6406 | .3594 | .3739 |
| 0.37 | .1443 | .6443 | .3557 | .3725 |
| 0.38 | .1480 | .6480 | .3520 | .3712 |
| 0.39 | .1517 | .6517 | .3483 | .3697 |
| 0.40 | .1554 | .6554 | .3446 | .3683 |
| 0.41 | .1591 | .6591 | .3409 | .3668 |
| 0.42 | .1628 | .6628 | .3372 | .3653 |
| 0.43 | .1664 | .6664 | .3336 | .3637 |
| 0.44 | .1700 | .6700 | .3300 | .3621 |

Source: A. L. Edwards. *Statistical Methods for the Behavioral Sciences*. New York: Holt, Rinehart and Winston, 1954. Reprinted by permission of the publisher.

APPENDIX B (*Continued*)

| (1)<br>z<br>Standard<br>Score | (2)<br>Area from<br>Mean to $x/\sigma$ | (3)<br>Area in<br>Larger<br>Portion | (4)<br>Area in<br>Smaller<br>Portion | (5)<br>y<br>Ordinate<br>at $x/\sigma$ |
|---|---|---|---|---|
| 0.45 | .1736 | .6736 | .3264 | .3605 |
| 0.46 | .1772 | .6772 | .3228 | .3589 |
| 0.47 | .1808 | .6808 | .3192 | .3572 |
| 0.48 | .1844 | .6844 | .3156 | .3555 |
| 0.49 | .1879 | .6879 | .3121 | .3538 |
| 0.50 | .1915 | .6915 | .3085 | .3521 |
| 0.51 | .1950 | .6950 | .3050 | .3503 |
| 0.52 | .1985 | .6985 | .3015 | .3485 |
| 0.53 | .2019 | .7019 | .2981 | .3467 |
| 0.54 | .2054 | .7054 | .2946 | .3448 |
| 0.55 | .2088 | .7088 | .2912 | .3429 |
| 0.56 | .2123 | .7123 | .2877 | .3410 |
| 0.57 | .2157 | .7157 | .2843 | .3391 |
| 0.58 | .2190 | .7190 | .2810 | .3372 |
| 0.59 | .2224 | .7224 | .2776 | .3352 |
| 0.60 | .2257 | .7257 | .2743 | .3332 |
| 0.61 | .2291 | .7291 | .2709 | .3312 |
| 0.62 | .2324 | .7324 | .2676 | .3292 |
| 0.63 | .2357 | .7357 | .2643 | .3271 |
| 0.64 | .2389 | .7389 | .2611 | .3251 |
| 0.65 | .2422 | .7422 | .2578 | .3230 |
| 0.66 | .2454 | .7454 | .2546 | .3209 |
| 0.67 | .2486 | .7486 | .2514 | .3187 |
| 0.68 | .2517 | .7517 | .2483 | .3166 |
| 0.69 | .2549 | .7549 | .2451 | .3144 |
| 0.70 | .2580 | .7580 | .2420 | .3123 |
| 0.71 | .2611 | .7611 | .2389 | .3101 |
| 0.72 | .2642 | .7642 | .2358 | .3079 |
| 0.73 | .2673 | .7673 | .2327 | .3056 |
| 0.74 | .2704 | .7704 | .2296 | .3034 |
| 0.75 | .2734 | .7734 | .2266 | .3011 |
| 0.76 | .2764 | .7764 | .2236 | .2989 |
| 0.77 | .2794 | .7794 | .2206 | .2966 |
| 0.78 | .2823 | .7823 | .2177 | .2943 |
| 0.79 | .2852 | .7852 | .2148 | .2920 |
| 0.80 | .2881 | .7881 | .2119 | .2897 |
| 0.81 | .2910 | .7910 | .2090 | .2874 |
| 0.82 | .2939 | .7939 | .2061 | .2850 |
| 0.83 | .2967 | .7967 | .2033 | .2827 |
| 0.84 | .2995 | .7995 | .2005 | .2803 |
| 0.85 | .3023 | .8023 | .1977 | .2780 |
| 0.86 | .3051 | .8051 | .1949 | .2756 |
| 0.87 | .3078 | .8078 | .1922 | .2732 |
| 0.88 | .3106 | .8106 | .1894 | .2709 |
| 0.89 | .3133 | .8133 | .1867 | .2685 |

APPENDIX B (*Continued*)

| (1)<br>z<br>Standard<br>Score | (2)<br><br>Area from<br>Mean to $x/\sigma$ | (3)<br>Area in<br>Larger<br>Portion | (4)<br>Area in<br>Smaller<br>Portion | (5)<br>y<br>Ordinate<br>at $x/\sigma$ |
|---|---|---|---|---|
| 0.90 | .3159 | .8159 | .1841 | .2661 |
| 0.91 | .3186 | .8186 | .1814 | .2637 |
| 0.92 | .3212 | .8212 | .1788 | .2613 |
| 0.93 | .3238 | .8238 | .1762 | .2589 |
| 0.94 | .3264 | .8264 | .1736 | .2565 |
| 0.95 | .3289 | .8289 | .1711 | .2541 |
| 0.96 | .3315 | .8315 | .1685 | .2516 |
| 0.97 | .3340 | .8340 | .1660 | .2492 |
| 0.98 | .3365 | .8365 | .1635 | .2468 |
| 0.99 | .3389 | .8389 | .1611 | .2444 |
| 1.00 | .3413 | .8413 | .1587 | .2420 |
| 1.01 | .3438 | .8438 | .1562 | .2396 |
| 1.02 | .3461 | .8461 | .1539 | .2371 |
| 1.03 | .3485 | .8485 | .1515 | .2347 |
| 1.04 | .3508 | .8508 | .1492 | .2323 |
| 1.05 | .3531 | .8531 | .1469 | .2299 |
| 1.06 | .3554 | .8554 | .1446 | .2275 |
| 1.07 | .3577 | .8577 | .1423 | .2251 |
| 1.08 | .3599 | .8599 | .1401 | .2227 |
| 1.09 | .3621 | .8621 | .1379 | .2203 |
| 1.10 | .3643 | .8643 | .1357 | .2179 |
| 1.11 | .3665 | .8665 | .1335 | .2155 |
| 1.12 | .3686 | .8686 | .1314 | .2131 |
| 1.13 | .3708 | .8708 | .1292 | .2107 |
| 1.14 | .3729 | .8729 | .1271 | .2083 |
| 1.15 | .3749 | .8749 | .1251 | .2059 |
| 1.16 | .3770 | .8770 | .1230 | .2036 |
| 1.17 | .3790 | .8790 | .1210 | .2012 |
| 1.18 | .3810 | .8810 | .1190 | .1989 |
| 1.19 | .3830 | .8830 | .1170 | .1965 |
| 1.20 | .3849 | .8849 | .1151 | .1942 |
| 1.21 | .3869 | .8869 | .1131 | .1919 |
| 1.22 | .3888 | .8888 | .1112 | .1895 |
| 1.23 | .3907 | .8907 | .1093 | .1872 |
| 1.24 | .3925 | .8925 | .1075 | .1849 |
| 1.25 | .3944 | .8944 | .1056 | .1826 |
| 1.26 | .3962 | .8962 | .1038 | .1804 |
| 1.27 | .3980 | .8980 | .1020 | .1781 |
| 1.28 | .3997 | .8997 | .1003 | .1758 |
| 1.29 | .4015 | .9015 | .0985 | .1736 |
| 1.30 | .4032 | .9032 | .0968 | .1714 |
| 1.31 | .4049 | .9049 | .0951 | .1691 |
| 1.32 | .4066 | .9066 | .0934 | .1669 |
| 1.33 | .4082 | .9082 | .0918 | .1647 |
| 1.34 | .4099 | .9099 | .0901 | .1626 |

APPENDIX B (*Continued*)

| (1) z Standard Score | (2) Area from Mean to $x/\sigma$ | (3) Area in Larger Portion | (4) Area in Smaller Portion | (5) y Ordinate at $x/\sigma$ |
|---|---|---|---|---|
| 1.35 | .4115 | .9115 | .0885 | .1604 |
| 1.36 | .4131 | .9131 | .0869 | .1582 |
| 1.37 | .4147 | .9147 | .0853 | .1561 |
| 1.38 | .4162 | .9162 | .0838 | .1539 |
| 1.39 | .4177 | .9177 | .0823 | .1518 |
| 1.40 | .4192 | .9192 | .0808 | .1497 |
| 1.41 | .4207 | .9207 | .0793 | .1476 |
| 1.42 | .4222 | .9222 | .0778 | .1456 |
| 1.43 | .4236 | .9236 | .0764 | .1435 |
| 1.44 | .4251 | .9251 | .0749 | .1415 |
| 1.45 | .4265 | .9265 | .0735 | .1394 |
| 1.46 | .4279 | .9279 | .0721 | .1374 |
| 1.47 | .4292 | .9292 | .0708 | .1354 |
| 1.48 | .4306 | .9306 | .0694 | .1334 |
| 1.49 | .4319 | .9319 | .0681 | .1315 |
| 1.50 | .4332 | .9332 | .0668 | .1295 |
| 1.51 | .4345 | .9345 | .0655 | .1276 |
| 1.52 | .4357 | .9357 | .0643 | .1257 |
| 1.53 | .4370 | 9370 | .0630 | .1238 |
| 1.54 | .4382 | .9382 | .0618 | .1219 |
| 1.55 | .4394 | .9394 | .0606 | .1200 |
| 1.56 | .4406 | .9406 | .0594 | .1182 |
| 1.57 | .4418 | .9418 | .0582 | .1163 |
| 1.58 | .4429 | .9429 | .0571 | .1145 |
| 1.59 | .4441 | .9441 | .0559 | .1127 |
| 1.60 | .4452 | .9452 | .0548 | .1109 |
| 1.61 | .4463 | .9463 | .0537 | .1092 |
| 1.62 | .4474 | .9474 | .0526 | .1074 |
| 1.63 | .4484 | .9484 | .0516 | .1057 |
| 1.64 | .4495 | .9495 | .0505 | .1040 |
| 1.65 | .4505 | .9505 | .0495 | .1023 |
| 1.66 | .4515 | .9515 | .0485 | .1006 |
| 1.67 | .4525 | .9525 | .0475 | .0989 |
| 1.68 | .4535 | .9535 | .0465 | .0973 |
| 1.69 | .4545 | .9545 | .0455 | .0957 |
| 1.70 | .4554 | .9554 | .0446 | .0940 |
| 1.71 | .4564 | .9564 | .0436 | .0925 |
| 1.72 | .4573 | .9573 | .0427 | .0909 |
| 1.73 | .4582 | .9582 | .0418 | .0893 |
| 1.74 | .4591 | .9591 | .0409 | .0878 |
| 1.75 | .4599 | .9599 | .0401 | .0863 |
| 1.76 | .4608 | .9608 | .0392 | .0848 |
| 1.77 | .4616 | .9616 | .0384 | .0833 |
| 1.78 | .4625 | .9625 | .0375 | .0818 |
| 1.79 | .4633 | .9633 | .0367 | .0804 |

APPENDIX B (*Continued*)

| (1)<br>z<br>Standard<br>Score | (2)<br><br>Area from<br>Mean to $x/\sigma$ | (3)<br>Area in<br>Larger<br>Portion | (4)<br>Area in<br>Smaller<br>Portion | (5)<br>y<br>Ordinate<br>at $x/\sigma$ |
|---|---|---|---|---|
| 1.80 | .4641 | .9641 | .0359 | .0790 |
| 1.81 | .4649 | .9649 | .0351 | .0775 |
| 1.82 | .4656 | .9656 | .0344 | .0761 |
| 1.83 | .4664 | .9664 | .0336 | .0748 |
| 1.84 | .4671 | .9671 | .0329 | .0734 |
| 1.85 | .4648 | .9678 | .0322 | .0721 |
| 1.86 | .4686 | .9686 | .0314 | .0707 |
| 1.87 | .4693 | .9693 | .0307 | .0694 |
| 1.88 | .4699 | .9699 | .0301 | .0681 |
| 1.89 | .4706 | .9706 | .0294 | .0669 |
| 1.90 | .4713 | .9713 | .0287 | .0656 |
| 1.91 | .4719 | .9719 | .0281 | .0644 |
| 1.92 | .4726 | .9726 | .0274 | .0632 |
| 1.93 | .4732 | .9732 | .0268 | .0620 |
| 1.94 | .4738 | .9738 | .0262 | .0608 |
| 1.95 | .4744 | .9744 | .0256 | .0596 |
| 1.96 | .4750 | .9750 | .0250 | .0584 |
| 1.97 | .4756 | .9756 | .0244 | .0573 |
| 1.98 | .4761 | .9761 | .0239 | .0562 |
| 1.99 | .4767 | .9767 | .0233 | .0551 |
| 2.00 | .4772 | .9772 | .0228 | .0540 |
| 2.01 | .4778 | .9778 | .0222 | .0529 |
| 2.02 | .4783 | .9783 | .0217 | .0519 |
| 2.03 | .4788 | .9788 | .0212 | .0508 |
| 2.04 | .4793 | .9793 | .0207 | .0498 |
| 2.05 | .4798 | .9798 | .0202 | .0488 |
| 2.06 | .4803 | .9803 | .0197 | .0478 |
| 2.07 | .4808 | .9808 | .0192 | .0468 |
| 2.08 | .4812 | .9812 | .0188 | .0459 |
| 2.09 | .4817 | .9817 | .0183 | .0449 |
| 2.10 | .4821 | .9821 | .0179 | .0440 |
| 2.11 | .4826 | .9826 | .0174 | .0431 |
| 2.12 | .4830 | .9830 | .0170 | .0422 |
| 2.13 | .4834 | .9834 | .0166 | .0413 |
| 2.14 | .4838 | .9838 | .0162 | .0404 |
| 2.15 | .4842 | .9842 | .0158 | .0396 |
| 2.16 | .4846 | .9846 | .0154 | .0387 |
| 2.17 | .4850 | .9850 | .0150 | .0379 |
| 2.18 | .4854 | .9854 | .0146 | .0371 |
| 2.19 | .4857 | .9857 | .0143 | .0363 |
| 2.20 | .4861 | .9861 | .0139 | .0355 |
| 2.21 | .4864 | .9864 | .0136 | .0347 |
| 2.22 | .4868 | .9868 | .0132 | .0339 |
| 2.23 | .4871 | .9871 | .0129 | .0332 |
| 2.24 | .4875 | .9875 | .0125 | .0325 |

APPENDIX B *(Continued)*

| (1)<br>z<br>Standard<br>Score | (2)<br><br>Area from<br>Mean to $x/\sigma$ | (3)<br>Area in<br>Larger<br>Portion | (4)<br>Area in<br>Smaller<br>Portion | (5)<br>y<br>Ordinate<br>at $x/\sigma$ |
|---|---|---|---|---|
| 2.25 | .4878 | .9878 | .0122 | .0317 |
| 2.26 | .4881 | .9881 | .0119 | .0310 |
| 2.27 | .4884 | .9884 | .0116 | .0303 |
| 2.28 | .4887 | .9887 | .0113 | .0297 |
| 2.29 | .4890 | .9890 | .0110 | .0290 |
| 2.30 | .4893 | .9893 | .0107 | .0283 |
| 2.31 | .4896 | .9896 | .0104 | .0277 |
| 2.32 | .4898 | .9898 | .0102 | .0270 |
| 2.33 | .4901 | .9901 | .0099 | .0264 |
| 2.34 | .4904 | .9904 | .0096 | .0258 |
| 2.35 | .4906 | .9906 | .0094 | .0252 |
| 2.36 | .4909 | .9909 | .0091 | .0246 |
| 2.37 | .4911 | .9911 | .0089 | .0241 |
| 2.38 | .4913 | .9913 | .0087 | .0235 |
| 2.39 | .4916 | .9916 | .0084 | .0229 |
| 2.40 | .4918 | .9918 | .0082 | .0224 |
| 2.41 | .4920 | .9920 | .0080 | .0219 |
| 2.42 | .4922 | .9922 | .0078 | .0213 |
| 2.43 | .4925 | .9925 | .0075 | .0208 |
| 2.44 | .4927 | .9927 | .0073 | .0203 |
| 2.45 | .4929 | .9929 | .0071 | .0198 |
| 2.46 | .4931 | .9931 | .0069 | .0194 |
| 2.47 | .4932 | .9932 | .0068 | .0189 |
| 2.48 | .4934 | .9934 | .0066 | .0184 |
| 2.49 | .4936 | .9936 | .0064 | .0180 |
| 2.50 | .4938 | .9938 | .0062 | .0175 |
| 2.51 | .4940 | .9940 | .0060 | .0171 |
| 2.52 | .4941 | .9941 | .0059 | .0167 |
| 2.53 | .4943 | .9943 | .0057 | .0163 |
| 2.54 | .4945 | .9945 | .0055 | .0158 |
| 2.55 | .4946 | .9946 | .0054 | .0154 |
| 2.56 | .4948 | .9948 | .0052 | .0151 |
| 2.57 | .4949 | .9949 | .0051 | .0147 |
| 2.58 | .4951 | .9951 | .0049 | .0143 |
| 2.59 | .4952 | .9952 | .0048 | .0139 |
| 2.60 | .4953 | .9953 | .0047 | .0136 |
| 2.61 | .4955 | .9955 | .0045 | .0132 |
| 2.62 | .4956 | .9956 | .0044 | .0129 |
| 2.63 | .4957 | .9957 | .0043 | .0126 |
| 2.64 | .4959 | .9959 | .0041 | .0122 |
| 2.65 | .4960 | .9960 | .0040 | .0119 |
| 2.66 | .4961 | .9961 | .0039 | .0116 |
| 2.67 | .4962 | .9962 | .0038 | .0113 |
| 2.68 | .4963 | .9963 | .0037 | .0110 |
| 2.69 | .4964 | .9964 | .0036 | .0107 |

## APPENDIX B (*Continued*)

| (1)<br>z<br>Standard<br>Score | (2)<br>Area from<br>Mean to $x/\sigma$ | (3)<br>Area in<br>Larger<br>Portion | (4)<br>Area in<br>Smaller<br>Portion | (5)<br>y<br>Ordinate<br>at $x/\sigma$ |
|---|---|---|---|---|
| 2.70 | .4965 | .9965 | .0035 | .0104 |
| 2.71 | .4966 | .9966 | .0034 | .0101 |
| 2.72 | .4967 | .9967 | .0033 | .0099 |
| 2.73 | .4968 | .9968 | .0032 | .0096 |
| 2.74 | .4969 | .9969 | .0031 | .0093 |
| 2.75 | .4970 | .9970 | .0030 | .0091 |
| 2.76 | .4971 | .9971 | .0029 | .0088 |
| 2.77 | .4972 | .9972 | .0028 | .0086 |
| 2.78 | .4973 | .9973 | .0027 | .0084 |
| 2.79 | .4974 | .9974 | .0026 | .0081 |
| 2.80 | .4974 | .9974 | .0026 | .0079 |
| 2.81 | .4975 | .9975 | .0025 | .0077 |
| 2.82 | .4976 | .9976 | .0024 | .0075 |
| 2.83 | .4977 | .9977 | .0023 | .0073 |
| 2.84 | .4977 | .9977 | .0023 | .0071 |
| 2.85 | .4978 | .9978 | .0022 | .0069 |
| 2.86 | .4979 | .9979 | .0021 | .0067 |
| 2.87 | .4979 | .9979 | .0021 | .0065 |
| 2.88 | .4980 | .9980 | .0020 | .0063 |
| 2.89 | .4981 | .9981 | .0019 | .0061 |
| 2.90 | .4981 | .9981 | .0019 | .0060 |
| 2.91 | .4982 | .9982 | .0018 | .0058 |
| 2.92 | .4982 | .9982 | .0018 | .0056 |
| 2.93 | .4983 | .9983 | .0017 | .0055 |
| 2.94 | .4984 | .9984 | .0016 | .0053 |
| 2.95 | .4984 | .9984 | .0016 | .0051 |
| 2.96 | .4985 | .9985 | .0015 | .0050 |
| 2.97 | .4985 | .9985 | .0015 | .0048 |
| 2.98 | .4986 | .9986 | .0014 | .0047 |
| 2.99 | .4986 | .9986 | .0014 | .0046 |
| 3.00 | .4987 | .9987 | .0013 | .0044 |
| 3.01 | .4987 | .9987 | .0013 | .0043 |
| 3.02 | .4987 | .9987 | .0013 | .0042 |
| 3.03 | .4988 | .9988 | .0012 | .0040 |
| 3.04 | .4988 | .9988 | .0012 | .0039 |
| 3.05 | .4989 | .9989 | .0011 | .0038 |
| 3.06 | .4989 | .9989 | .0011 | .0037 |
| 3.07 | .4989 | .9989 | .0011 | .0036 |
| 3.08 | .4990 | .9990 | .0010 | .0035 |
| 3.09 | .4990 | .9990 | .0010 | .0034 |
| 3.10 | .4990 | .9990 | .0010 | .0033 |
| 3.11 | .4991 | .9991 | .0009 | .0032 |
| 3.12 | .4991 | .9991 | .0009 | .0031 |
| 3.13 | .4991 | .9991 | .0009 | .0030 |
| 3.14 | .4992 | .9992 | .0008 | .0029 |

APPENDIX B (*Continued*)

| (1)<br>z<br>Standard<br>Score | (2)<br><br>Area from<br>Mean to x/σ | (3)<br>Area in<br>Larger<br>Portion | (4)<br>Area in<br>Smaller<br>Portion | (5)<br>y<br><br>Ordinate<br>at x/σ |
|---|---|---|---|---|
| 3.15 | .4992 | .9992 | .0008 | .0028 |
| 3.16 | .4992 | .9992 | .0008 | .0027 |
| 3.17 | .4992 | .9992 | .0008 | .0026 |
| 3.18 | .4993 | .9993 | .0007 | .0025 |
| 3.19 | .4993 | .9993 | .0007 | .0025 |
| 3.20 | .4993 | .9993 | .0007 | .0024 |
| 3.21 | .4993 | .9993 | .0007 | .0023 |
| 3.22 | .4994 | .9994 | .0006 | .0022 |
| 3.23 | .4994 | .9994 | .0006 | .0022 |
| 3.24 | .4994 | .9994 | .0006 | .0021 |
| 3.30 | .4995 | .9995 | .0005 | .0017 |
| 3.40 | .4997 | .9997 | .0003 | .0012 |
| 3.50 | .4998 | .9998 | .0002 | .0009 |
| 3.60 | .4998 | .9998 | .0002 | .0006 |
| 3.70 | .4999 | .9999 | .0001 | .0004 |

# Appendix C
## Distribution of t Probability

| df | .1 | .05 | .01 | .001 |
|---|---|---|---|---|
| 1 | 6.314 | 12.706 | 63.657 | 636.619 |
| 2 | 2.920 | 4.303 | 9.925 | 31.598 |
| 3 | 2.353 | 3.182 | 5.841 | 12.941 |
| 4 | 2.132 | 2.776 | 4.604 | 8.610 |
| 5 | 2.015 | 2.571 | 4.032 | 6.859 |
| 6 | 1.943 | 2.447 | 3.707 | 5.959 |
| 7 | 1.895 | 2.365 | 3.499 | 5.405 |
| 8 | 1.860 | 2.306 | 3.355 | 5.041 |
| 9 | 1.833 | 2.262 | 3.250 | 4.781 |
| 10 | 1.812 | 2.228 | 3.169 | 4.587 |
| 11 | 1.796 | 2.201 | 3.106 | 4.437 |
| 12 | 1.782 | 2.179 | 3.055 | 4.318 |
| 13 | 1.771 | 2.160 | 3.012 | 4.221 |
| 14 | 1.761 | 2.145 | 2.977 | 4.140 |
| 15 | 1.753 | 2.131 | 2.947 | 4.073 |
| 16 | 1.746 | 2.120 | 2.921 | 4.015 |
| 17 | 1.740 | 2.110 | 2.898 | 3.965 |
| 18 | 1.734 | 2.101 | 2.878 | 3.922 |
| 19 | 1.729 | 2.093 | 2.861 | 3.883 |
| 20 | 1.725 | 2.086 | 2.845 | 3.850 |
| 21 | 1.721 | 2.080 | 2.831 | 3.819 |
| 22 | 1.717 | 2.074 | 2.819 | 3.792 |
| 23 | 1.714 | 2.069 | 2.807 | 3.767 |
| 24 | 1.711 | 2.064 | 2.797 | 3.745 |
| 25 | 1.708 | 2.060 | 2.787 | 3.725 |
| 26 | 1.706 | 2.056 | 2.779 | 3.707 |
| 27 | 1.703 | 2.052 | 2.771 | 3.690 |
| 28 | 1.701 | 2.048 | 2.763 | 3.674 |
| 29 | 1.699 | 2.045 | 2.756 | 3.659 |
| 30 | 1.697 | 2.042 | 2.750 | 3.646 |
| 40 | 1.684 | 2.021 | 2.704 | 3.551 |
| 60 | 1.671 | 2.000 | 2.660 | 3.460 |
| 120 | 1.658 | 1.980 | 2.617 | 3.373 |
| ∞ | 1.645 | 1.960 | 2.576 | 3.291 |

SOURCE: Appendix C is abridged from Table III of R. A. Fisher and F. Yates. *Statistical Tables for Biological, Agricultural, and Medical Research*, published by Oliver and Boyd Ltd., Edinburgh. Abridged with permission of the authors and publisher.

# Appendix D
## Distribution of $\chi^2$

| df | .99 | .98 | .95 | .90 | .80 | .70 | .50 | .30 | .20 | .10 | .05 | .02 | .01 | .001 |
|----|-----|-----|-----|-----|-----|-----|-----|-----|-----|-----|-----|-----|-----|------|
| 1 | $.0^3157$ | $.0^3628$ | .00393 | .0158 | .0642 | .148 | .455 | 1.074 | 1.642 | 2.706 | 3.841 | 5.412 | 6.635 | 10.827 |
| 2 | .0201 | .0404 | .103 | .211 | .446 | .713 | 1.386 | 2.408 | 3.219 | 4.605 | 5.991 | 7.824 | 9.210 | 13.815 |
| 3 | .115 | .185 | .352 | .584 | 1.005 | 1.424 | 2.366 | 3.665 | 4.642 | 6.251 | 7.815 | 9.837 | 11.345 | 16.268 |
| 4 | .297 | .429 | .711 | 1.064 | 1.649 | 2.195 | 3.357 | 4.878 | 5.989 | 7.779 | 9.488 | 11.668 | 13.277 | 18.465 |
| 5 | .554 | .752 | 1.145 | 1.610 | 2.343 | 3.000 | 4.351 | 6.064 | 7.289 | 9.236 | 11.070 | 13.388 | 15.086 | 20.517 |
| 6 | .872 | 1.134 | 1.635 | 2.204 | 3.070 | 3.828 | 5.348 | 7.231 | 8.558 | 10.645 | 12.592 | 15.033 | 16.812 | 22.457 |
| 7 | 1.239 | 1.564 | 2.167 | 2.833 | 3.822 | 4.671 | 6.346 | 8.383 | 9.803 | 12.017 | 14.067 | 16.622 | 18.475 | 24.322 |
| 8 | 1.646 | 2.032 | 2.733 | 3.490 | 4.594 | 5.527 | 7.344 | 9.524 | 11.030 | 13.362 | 15.507 | 18.168 | 20.090 | 26.125 |
| 9 | 2.088 | 2.532 | 3.325 | 4.168 | 5.380 | 6.393 | 8.343 | 10.656 | 12.242 | 14.684 | 16.919 | 19.679 | 21.666 | 27.877 |
| 10 | 2.558 | 3.059 | 3.940 | 4.865 | 6.179 | 7.267 | 9.342 | 11.781 | 13.442 | 15.987 | 18.307 | 21.161 | 23.209 | 29.588 |
| 11 | 3.053 | 3.609 | 4.575 | 5.578 | 6.989 | 8.148 | 10.341 | 12.899 | 14.631 | 17.275 | 19.675 | 22.618 | 24.725 | 31.264 |
| 12 | 3.571 | 4.178 | 5.226 | 6.304 | 7.807 | 9.034 | 11.340 | 14.011 | 15.812 | 18.549 | 21.026 | 24.054 | 26.217 | 32.909 |
| 13 | 4.107 | 4.765 | 5.892 | 7.042 | 8.634 | 9.926 | 12.340 | 15.119 | 16.985 | 19.812 | 22.362 | 25.472 | 27.688 | 34.528 |
| 14 | 4.660 | 5.368 | 6.571 | 7.790 | 9.467 | 10.821 | 13.339 | 16.222 | 18.151 | 21.064 | 23.685 | 26.873 | 29.141 | 36.123 |
| 15 | 5.229 | 5.985 | 7.261 | 8.547 | 10.307 | 11.721 | 14.339 | 17.322 | 19.311 | 22.307 | 24.996 | 28.259 | 30.578 | 37.697 |
| 16 | 5.812 | 6.614 | 7.962 | 9.312 | 11.152 | 12.624 | 15.338 | 18.418 | 20.465 | 23.542 | 26.296 | 29.633 | 32.000 | 39.252 |
| 17 | 6.408 | 7.255 | 8.672 | 10.085 | 12.002 | 13.531 | 16.338 | 19.511 | 21.615 | 24.769 | 27.587 | 30.995 | 33.409 | 40.790 |
| 18 | 7.015 | 7.906 | 9.390 | 10.865 | 12.857 | 14.440 | 17.338 | 20.601 | 22.760 | 25.989 | 28.869 | 32.346 | 34.805 | 42.312 |
| 19 | 7.633 | 8.567 | 10.117 | 11.651 | 13.716 | 15.352 | 18.338 | 21.689 | 23.900 | 27.204 | 30.144 | 33.687 | 36.191 | 43.820 |
| 20 | 8.260 | 9.237 | 10.851 | 12.443 | 14.578 | 16.266 | 19.337 | 22.775 | 25.038 | 28.412 | 31.410 | 35.020 | 37.566 | 45.315 |
| 21 | 8.897 | 9.915 | 11.591 | 13.240 | 15.445 | 17.182 | 20.337 | 23.858 | 26.171 | 29.615 | 32.671 | 36.343 | 38.932 | 46.797 |
| 22 | 9.542 | 10.600 | 12.338 | 14.041 | 16.314 | 18.101 | 21.337 | 24.939 | 27.301 | 30.813 | 33.924 | 37.659 | 40.289 | 48.268 |
| 23 | 10.196 | 11.293 | 13.091 | 14.848 | 17.187 | 19.021 | 22.337 | 26.018 | 28.429 | 32.007 | 35.172 | 38.968 | 41.638 | 49.728 |
| 24 | 10.856 | 11.992 | 13.848 | 15.659 | 18.062 | 19.943 | 23.337 | 27.096 | 29.553 | 33.196 | 36.415 | 40.270 | 42.980 | 51.179 |
| 25 | 11.524 | 12.697 | 14.611 | 16.473 | 18.940 | 20.867 | 24.337 | 28.172 | 30.675 | 34.382 | 37.652 | 41.566 | 44.314 | 52.620 |
| 26 | 12.198 | 13.409 | 15.379 | 17.292 | 19.820 | 21.792 | 25.336 | 29.246 | 31.795 | 35.563 | 38.885 | 42.856 | 45.642 | 54.052 |
| 27 | 12.879 | 14.125 | 16.151 | 18.114 | 20.703 | 22.719 | 26.336 | 30.319 | 32.912 | 36.741 | 40.113 | 44.140 | 46.963 | 55.476 |
| 28 | 13.565 | 14.847 | 16.928 | 18.939 | 21.588 | 23.647 | 27.336 | 31.391 | 34.027 | 37.916 | 41.337 | 45.419 | 48.278 | 56.893 |
| 29 | 14.256 | 15.574 | 17.708 | 19.768 | 22.475 | 24.577 | 28.336 | 32.461 | 35.139 | 39.087 | 42.557 | 46.693 | 49.588 | 58.302 |
| 30 | 14.953 | 16.306 | 18.493 | 20.599 | 23.364 | 25.508 | 29.336 | 33.530 | 36.250 | 40.256 | 43.773 | 47.962 | 50.892 | 59.703 |

SOURCE: Appendix D is reprinted from Table IV of R. A. Fisher and F. Yates. *Statistical Tables for Biological, Agricultural, and Medical Research*, published by Oliver and Boyd Ltd., Edinburgh. Reprinted with permission of the authors and publishers.

# Appendix E
## 5 Percent (Lightface Type) and 1 Percent (Boldface Type) Points for the Distribution of $F$

| $f_2$ | 1 | 2 | 3 | 4 | 5 | 6 | 7 | 8 | 9 | 10 | 11 | 12 |
|---|---|---|---|---|---|---|---|---|---|---|---|---|
| | | | | | | | | $f_1$ Degrees of Freedom (for | | | | |
| 1 | 161 | 200 | 216 | 225 | 230 | 234 | 237 | 239 | 241 | 242 | 243 | 244 |
| | **4,052** | **4,999** | **5,403** | **5,625** | **5,764** | **5,859** | **5,928** | **5,981** | **6,022** | **6,056** | **6,082** | **6,106** |
| 2 | 18.51 | 19.00 | 19.16 | 19.25 | 19.30 | 19.33 | 19.36 | 19.37 | 19.38 | 19.39 | 19.40 | 19.41 |
| | **98.49** | **99.00** | **99.17** | **99.25** | **99.30** | **99.33** | **99.34** | **99.36** | **99.38** | **99.40** | **99.41** | **99.42** |
| 3 | 10.13 | 9.55 | 9.28 | 9.12 | 9.01 | 8.94 | 8.88 | 8.84 | 8.81 | 8.78 | 8.76 | 8.74 |
| | **34.12** | **30.82** | **29.46** | **28.71** | **28.24** | **27.91** | **29.67** | **27.49** | **27.34** | **27.23** | **27.13** | **27.05** |
| 4 | 7.71 | 6.94 | 6.59 | 6.39 | 6.26 | 6.16 | 6.09 | 6.04 | 6.00 | 5.96 | 5.93 | 5.91 |
| | **21.20** | **18.00** | **16.69** | **15.98** | **15.52** | **15.21** | **14.98** | **14.80** | **14.66** | **14.54** | **14.45** | **14.37** |
| 5 | 6.61 | 5.79 | 5.41 | 5.19 | 5.05 | 4.95 | 4.88 | 4.82 | 4.78 | 4.74 | 4.70 | 4.68 |
| | **16.26** | **13.27** | **12.06** | **11.39** | **10.97** | **10.67** | **10.45** | **10.27** | **10.15** | **10.05** | **9,96** | **9.89** |
| 6 | 5.99 | 5.14 | 4.76 | 4.53 | 4.39 | 4.28 | 4.21 | 4.15 | 4.10 | 4.06 | 4.03 | 4.00 |
| | **13.74** | **10.92** | **9.78** | **9.15** | **8.75** | **8.47** | **8.26** | **8.10** | **7.98** | **7.87** | **7.79** | **7.72** |
| 7 | 5.59 | 4.74 | 4.35 | 4.12 | 3.97 | 3.87 | 3.79 | 3.73 | 3.68 | 3.63 | 3.60 | 3.57 |
| | **12.25** | **9.55** | **8.45** | **7.85** | **7.46** | **7.19** | **7.00** | **6.84** | **6.71** | **6.62** | **6.54** | **6.47** |
| 8 | 5.32 | 4.46 | 4.07 | 3.84 | 3.69 | 3.58 | 3.50 | 3.44 | 3.39 | 3.34 | 3.31 | 3.28 |
| | **11.26** | **8.65** | **7.59** | **7.01** | **6.63** | **6.37** | **6.19** | **6.03** | **5.91** | **5.82** | **5.74** | **5.67** |
| 9 | 5.12 | 4.26 | 3.86 | 3.63 | 3.48 | 3.37 | 3.29 | 3.23 | 3.18 | 3.13 | 3.10 | 3.07 |
| | **10.56** | **8.02** | **6.99** | **6.42** | **6.06** | **5.80** | **5.62** | **5.47** | **5.35** | **5.26** | **5.18** | **5.11** |
| 10 | 4.96 | 4.10 | 3.71 | 3.48 | 3.33 | 3.22 | 3.14 | 3.07 | 3.02 | 2.97 | 2.94 | 2.91 |
| | **10.04** | **7.56** | **6.55** | **5.99** | **5.64** | **5.39** | **5.21** | **5.06** | **4.95** | **4.85** | **4.78** | **4.71** |
| 11 | 4.84 | 3.98 | 3.59 | 3.36 | 3.20 | 3.09 | 3.01 | 2.95 | 2.90 | 2.86 | 2.82 | 2.79 |
| | **9.65** | **7.20** | **6.22** | **5.67** | **5.32** | **5.07** | **4.88** | **4.74** | **4.63** | **4.54** | **4.46** | **4.40** |
| 12 | 4.75 | 3.88 | 3.49 | 3.26 | 3.11 | 3.00 | 2.92 | 2.85 | 2.80 | 2.76 | 2.72 | 2.69 |
| | **9.33** | **6.93** | **5.95** | **5.41** | **5.06** | **4.82** | **4.65** | **4.50** | **4.39** | **4.30** | **4.22** | **4.16** |
| 13 | 4.67 | 3.80 | 3.41 | 3.18 | 3.02 | 2.92 | 2.84 | 2.72 | 2.77 | 2.63 | 2.63 | 2.60 |
| | **9.07** | **6.70** | **5.74** | **5.20** | **4.86** | **4.62** | **4.44** | **4.30** | **4.19** | **4.10** | **4.02** | **3.96** |
| 14 | 4.60 | 3.74 | 3.34 | 3.11 | 2.96 | 2.85 | 2.77 | 2.70 | 2.65 | 2.60 | 2.56 | 2.53 |
| | **8.86** | **6.51** | **5.56** | **5.03** | **4.69** | **4.46** | **4.28** | **4.14** | **4.03** | **3.94** | **3.86** | **3.80** |
| 15 | 4.54 | 3.68 | 3.29 | 3.06 | 2.90 | 2.79 | 2.70 | 2.64 | 2.59 | 2.55 | 2.51 | 2.48 |
| | **8.68** | **6.36** | **5.42** | **4.89** | **4.56** | **4.32** | **4.14** | **4.00** | **3.89** | **3.80** | **3.73** | **3.67** |
| 16 | 4.49 | 3.63 | 3.24 | 3.01 | 2.85 | 2.74 | 2.66 | 2.59 | 2.54 | 2.49 | 2.45 | 2.42 |
| | **8.53** | **6.23** | **5.29** | **4.77** | **4.44** | **4.20** | **4.03** | **3.89** | **3.78** | **3.69** | **3.61** | **3.55** |
| 17 | 4.45 | 3.59 | 3.20 | 2.96 | 2.81 | 2.70 | 2.62 | 2.55 | 2.50 | 2.45 | 2.41 | 2.38 |
| | **8.40** | **6.11** | **5.18** | **4.67** | **4.34** | **4.10** | **3.93** | **3.79** | **3.68** | **3.59** | **3.52** | **3.45** |
| 18 | 4.41 | 3.55 | 3.16 | 2.93 | 2.77 | 2.66 | 2.58 | 2.51 | 2.46 | 2.41 | 2.37 | 2.34 |
| | **8.28** | **6.01** | **5.09** | **4.58** | **4.25** | **4.01** | **3.85** | **3.71** | **3.60** | **3.51** | **3.44** | **3.37** |

SOURCE: G. W. Snedecor. *Statistical Methods.* Ames, Iowa: Iowa State College Press, 1956. Reprinted by permission of the author and publisher.

APPENDIX E (*Continued*)

Greater Mean Square)

| 14 | 16 | 20 | 24 | 30 | 40 | 50 | 75 | 100 | 200 | 500 | ∞ | $f_2$ |
|---|---|---|---|---|---|---|---|---|---|---|---|---|
| 245 | 246 | 248 | 249 | 250 | 251 | 252 | 253 | 253 | 254 | 254 | 254 | 1 |
| 6,142 | 6,169 | 6,208 | 6,234 | 6,258 | 6,286 | 6,302 | 6,323 | 6,334 | 6,352 | 6,361 | 6,366 | |
| 19.42 | 19.43 | 19.44 | 19.45 | 19.46 | 19.47 | 19.47 | 19.48 | 19.49 | 19.49 | 19.50 | 19.50 | 2 |
| 99.43 | 99.44 | 99.45 | 99.46 | 99.47 | 99.48 | 99.48 | 99.49 | 99.49 | 99.49 | 99.50 | 99.50 | |
| 8.71 | 8.69 | 8.66 | 8.64 | 8.62 | 8.60 | 8.58 | 8.57 | 8.56 | 8.54 | 8.54 | 8.53 | 3 |
| 26.92 | 26.83 | 26.69 | 26.60 | 26.50 | 26.41 | 26.35 | 26.27 | 26.23 | 26.18 | 26.14 | 26.12 | |
| 5.87 | 5.84 | 5.80 | 5.77 | 5.74 | 5.71 | 5.70 | 5.68 | 5.66 | 5.65 | 5.64 | 5.63 | 4 |
| 14.24 | 14.15 | 14.02 | 13.93 | 13.83 | 13.74 | 13.69 | 13.61 | 13.57 | 13.52 | 13.48 | 13.46 | |
| 4.64 | 4.60 | 4.56 | 4.53 | 4.50 | 4.46 | 4.44 | 4.42 | 4.40 | 4.38 | 4.37 | 4.36 | 5 |
| 9.77 | 9.68 | 9.55 | 9.47 | 9.38 | 9.29 | 9.24 | 9.17 | 9.13 | 9.07 | 9.04 | 9.02 | |
| 3.96 | 3.92 | 3.87 | 3.84 | 3.81 | 3.77 | 3.75 | 3.72 | 3.71 | 3.69 | 3.68 | 3.67 | 6 |
| 7.60 | 7.52 | 7.39 | 7.31 | 7.23 | 7.14 | 7.09 | 7.02 | 6.99 | 6.94 | 6.90 | 6.88 | |
| 3.52 | 3.49 | 3.44 | 3.41 | 3.38 | 3.34 | 3.32 | 3.29 | 3.28 | 3.25 | 3.24 | 3.23 | 7 |
| 6.35 | 6.27 | 6.15 | 6.07 | 5.98 | 5.90 | 5.85 | 5.78 | 5.75 | 5.70 | 5.67 | 5.65 | |
| 3.23 | 3.20 | 3.15 | 3.12 | 3.08 | 3.05 | 3.03 | 3.00 | 2.98 | 2.96 | 2.94 | 2.93 | 8 |
| 5.56 | 5.48 | 5.36 | 5.28 | 5.20 | 5.11 | 5.06 | 5.00 | 4.96 | 4.91 | 4.88 | 4.86 | |
| 3.02 | 2.98 | 2.93 | 2.90 | 2.86 | 2.82 | 2.80 | 2.77 | 2.76 | 2.73 | 2.72 | 2.71 | 9 |
| 5.00 | 4.92 | 4.80 | 4.73 | 4.64 | 4.56 | 4.51 | 4.45 | 4.41 | 4.36 | 4.33 | 4.31 | |
| 2.86 | 2.82 | 2.77 | 2.74 | 2.70 | 2.67 | 2.64 | 2.61 | 2.59 | 2.56 | 2.55 | 2.54 | 10 |
| 4.60 | 4.52 | 4.41 | 4.33 | 4.25 | 4.17 | 4.12 | 4.05 | 4.01 | 3.96 | 3.93 | 3.91 | |
| 2.74 | 2.70 | 2.65 | 2.61 | 2.57 | 2.53 | 2.50 | 2.47 | 2.45 | 2.42 | 2.41 | 2.40 | 11 |
| 4.29 | 4.21 | 4.10 | 4.02 | 3.94 | 3.86 | 3.80 | 3.74 | 3.70 | 3.66 | 3.62 | 3.60 | |
| 2.64 | 2.60 | 2.54 | 2.50 | 2.46 | 2.42 | 2.40 | 2.36 | 2.35 | 2.32 | 2.31 | 2.30 | 12 |
| 4.05 | 3.98 | 3.86 | 3.78 | 3.70 | 3.61 | 3.56 | 3.49 | 3.46 | 3.41 | 3.38 | 3.36 | |
| 2.55 | 2.51 | 2.46 | 2.42 | 2.38 | 2.34 | 2.32 | 2.28 | 2.26 | 2.24 | 2.22 | 2.21 | 13 |
| 3.85 | 3.78 | 3.67 | 3.59 | 3.51 | 3.42 | 3.37 | 3.30 | 3.27 | 3.21 | 3.18 | 3.16 | |
| 2.48 | 2.44 | 2.39 | 2.35 | 2.31 | 2.27 | 2.24 | 2.21 | 2.19 | 2.16 | 2.14 | 2.13 | 14 |
| 3.70 | 3.62 | 3.51 | 3.43 | 3.34 | 3.26 | 3.21 | 3.14 | 3.11 | 3.06 | 3.02 | 3.00 | |
| 2.43 | 2.39 | 2.33 | 2.29 | 2.25 | 2.21 | 2.18 | 2.15 | 2.12 | 2.10 | 2.08 | 2.07 | 15 |
| 3.56 | 3.48 | 3.36 | 3.29 | 3.20 | 3.12 | 3.07 | 3.00 | 2.97 | 2.92 | 2.89 | 2.87 | |
| 2.37 | 2.33 | 2.28 | 2.24 | 2.20 | 2.16 | 2.13 | 2.09 | 2.07 | 2.04 | 2.02 | 2.01 | 16 |
| 3.45 | 3.37 | 3.25 | 3.18 | 3.10 | 3.01 | 2.96 | 2.89 | 2.86 | 2.80 | 2.77 | 2.75 | |
| 2.33 | 2.29 | 2.23 | 2.19 | 2.15 | 2.11 | 2.08 | 2.04 | 2.02 | 1.99 | 1.97 | 1.96 | 17 |
| 3.35 | 3.27 | 3.16 | 3.08 | 3.00 | 2.92 | 2.86 | 2.79 | 2.76 | 2.70 | 2.67 | 2.65 | |
| 2.29 | 2.25 | 2.19 | 2.15 | 2.11 | 2.07 | 2.04 | 2.00 | 1.98 | 1.95 | 1.93 | 1.92 | 18 |
| 3.27 | 3.19 | 3.07 | 3.00 | 2.91 | 2.83 | 2.78 | 2.71 | 2.68 | 2.62 | 2.59 | 2.57 | |

APPENDIX E (*Continued*)

| $f_2$ | 1 | 2 | 3 | 4 | 5 | 6 | 7 | 8 | 9 | 10 | 11 | 12 |
|-------|---|---|---|---|---|---|---|---|---|----|----|----|
| | | | | | | | | | $f_1$ Degrees of Freedom (for | | | |
| 19 | 4.38 | 3.52 | 3.13 | 2.90 | 2.74 | 2.63 | 2.55 | 2.48 | 2.43 | 2.38 | 2.34 | 2.31 |
| | 8.18 | 5.93 | 5.01 | 4.50 | 4.17 | 3.94 | 3.77 | 3.63 | 3.52 | 3.43 | 3.36 | 3.30 |
| 20 | 4.35 | 3.49 | 3.10 | 2.87 | 2.71 | 2.60 | 2.52 | 2.45 | 2.40 | 2.35 | 2.31 | 2.28 |
| | 8.10 | 5.85 | 4.94 | 4.43 | 4.10 | 3.87 | 3.71 | 3.56 | 3.45 | 3.37 | 3.30 | 3.23 |
| 21 | 4.32 | 3.47 | 3.07 | 2.84 | 2.68 | 2.57 | 2.49 | 2.42 | 2.37 | 2.32 | 2.28 | 2.25 |
| | 8.02 | 5.78 | 4.87 | 4.37 | 4.04 | 3.81 | 3.65 | 3.51 | 3.40 | 3.31 | 3.24 | 3.17 |
| 22 | 4.30 | 3.44 | 3.05 | 2.82 | 2.66 | 2.55 | 2.47 | 2.40 | 2.35 | 2.30 | 2.26 | 2.23 |
| | 7.94 | 5.72 | 4.82 | 4.31 | 3.99 | 3.76 | 3.59 | 3.45 | 3.35 | 3.26 | 3.18 | 3.12 |
| 23 | 4.28 | 3.42 | 3.03 | 2.80 | 2.64 | 2.53 | 2.45 | 2.38 | 2.32 | 2.28 | 2.24 | 2.20 |
| | 7.88 | 5.66 | 4.76 | 4.26 | 3.94 | 3.71 | 3.54 | 3.41 | 3.30 | 3.21 | 3.14 | 3.07 |
| 24 | 4.26 | 3.40 | 3.01 | 2.78 | 2.62 | 2.51 | 2.43 | 2.36 | 2.30 | 2.26 | 2.22 | 2.18 |
| | 7.82 | 5.61 | 4.72 | 4.22 | 3.90 | 3.67 | 3.50 | 3.36 | 3.25 | 3.17 | 3.09 | 3.03 |
| 25 | 4.24 | 3.38 | 2.99 | 2.76 | 2.60 | 2.49 | 2.41 | 2.34 | 2.28 | 2.24 | 2.20 | 2.16 |
| | 7.77 | 5.57 | 4.68 | 4.18 | 3.86 | 3.63 | 3.46 | 3.32 | 3.21 | 3.13 | 3.05 | 2.99 |
| 26 | 4.22 | 3.37 | 2.98 | 2.74 | 2.59 | 2.47 | 2.39 | 2.32 | 2.27 | 2.22 | 2.18 | 2.15 |
| | 7.72 | 5.53 | 4.64 | 4.14 | 3.82 | 3.59 | 3.42 | 3.29 | 3.17 | 3.09 | 3.02 | 2.96 |
| 27 | 4.21 | 3.35 | 2.96 | 2.73 | 2.57 | 2.46 | 2.37 | 2.30 | 2.25 | 2.20 | 2.16 | 2.13 |
| | 7.68 | 5.49 | 4.60 | 4.11 | 3.79 | 3.56 | 3.39 | 3.26 | 3.14 | 3.06 | 2.98 | 2.93 |
| 28 | 4.20 | 3.34 | 2.95 | 2.71 | 2.56 | 2.44 | 2.36 | 2.29 | 2.24 | 2.19 | 2.15 | 2.12 |
| | 7.64 | 5.45 | 4.57 | 4.07 | 3.76 | 3.53 | 3.36 | 3.23 | 3.11 | 3.03 | 2.95 | 2.90 |
| 29 | 4.18 | 3.33 | 2.93 | 2.70 | 2.54 | 2.43 | 2.35 | 2.28 | 2.22 | 2.18 | 2.14 | 2.10 |
| | 7.60 | 5.42 | 4.54 | 4.04 | 3.73 | 3.50 | 3.33 | 3.20 | 3.08 | 3.00 | 2.92 | 2.87 |
| 30 | 4.17 | 3.32 | 2.92 | 2.69 | 2.53 | 2.42 | 2.34 | 2.27 | 2.21 | 2.16 | 2.12 | 2.09 |
| | 7.56 | 5.39 | 4.51 | 4.02 | 3.70 | 3.47 | 3.30 | 3.17 | 3.06 | 2.98 | 2.90 | 2.84 |
| 32 | 4.15 | 3.30 | 2.90 | 2.67 | 2.51 | 2.40 | 2.32 | 2.25 | 2.19 | 2.14 | 2.10 | 2.07 |
| | 7.50 | 5.34 | 4.46 | 3.97 | 3.66 | 3.42 | 3.25 | 3.12 | 3.01 | 2.94 | 2.86 | 2.80 |
| 34 | 4.13 | 3.28 | 2.88 | 2.65 | 2.49 | 2.38 | 2.30 | 2.23 | 2.17 | 2.12 | 2.08 | 2.05 |
| | 7.44 | 5.29 | 4.42 | 3.93 | 3.61 | 3.38 | 3.21 | 3.08 | 2.97 | 2.89 | 2.82 | 2.76 |
| 36 | 4.11 | 3.26 | 2.86 | 2.63 | 2.48 | 2.36 | 2.28 | 2.21 | 2.15 | 2.10 | 2.06 | 2.03 |
| | 7.39 | 5.25 | 4.38 | 3.89 | 3.58 | 3.35 | 3.18 | 3.04 | 2.94 | 2.86 | 2.78 | 2.72 |
| 38 | 4.10 | 3.25 | 2.85 | 2.62 | 2.46 | 2.35 | 2.26 | 2.19 | 2.14 | 2.09 | 2.05 | 2.02 |
| | 7.35 | 5.21 | 4.34 | 3.86 | 3.54 | 3.32 | 3.15 | 3.02 | 2.91 | 2.82 | 2.75 | 2.69 |
| 40 | 4.08 | 3.23 | 2.84 | 2.61 | 2.45 | 2.34 | 2.25 | 2.18 | 2.12 | 2.07 | 2.04 | 2.00 |
| | 7.31 | 5.18 | 4.31 | 3.83 | 3.51 | 3.29 | 3.12 | 2.99 | 2.88 | 2.80 | 2.73 | 2.66 |
| 42 | 4.07 | 3.22 | 2.83 | 2.59 | 2.44 | 2.32 | 2.24 | 2.17 | 2.11 | 2.06 | 2.02 | 1.99 |
| | 7.27 | 5.15 | 4.29 | 3.80 | 3.49 | 3.26 | 3.10 | 2.96 | 2.86 | 2.77 | 2.70 | 2.64 |
| 44 | 4.06 | 3.21 | 2.82 | 2.58 | 2.43 | 2.31 | 2.23 | 2.16 | 2.10 | 2.05 | 2.01 | 1.98 |
| | 7.24 | 5.12 | 4.26 | 3.78 | 3.46 | 3.24 | 3.07 | 2.94 | 2.84 | 2.75 | 2.68 | 2.62 |

APPENDIX E (*Continued*)

Greater Mean Square)

| 14 | 16 | 20 | 24 | 30 | 40 | 50 | 75 | 100 | 200 | 500 | ∞ | $f_2$ |
|---|---|---|---|---|---|---|---|---|---|---|---|---|
| 2.26 | 2.21 | 2.15 | 2.11 | 2.07 | 2.02 | 2.00 | 1.96 | 1.94 | 1.91 | 1.90 | 1.88 | 19 |
| **3.19** | **3.12** | **3.00** | **2.92** | **2.84** | **2.76** | **2.70** | **2.63** | **2.60** | **2.54** | **2.51** | **2.49** | |
| 2.23 | 2.18 | 2.12 | 2.08 | 2.04 | 1.99 | 1.96 | 1.92 | 1.90 | 1.87 | 1.85 | 1.84 | 20 |
| **3.13** | **3.05** | **2.94** | **2.86** | **2.77** | **2.69** | **2.63** | **2.56** | **2.53** | **2.47** | **2.44** | **2.42** | |
| 2.20 | 2.15 | 2.09 | 2.05 | 2.00 | 1.96 | 1.93 | 1.89 | 1.87 | 1.84 | 1.82 | 1.81 | 21 |
| **3.07** | **2.99** | **2.88** | **2.80** | **2.72** | **2.63** | **2.58** | **2.51** | **2.47** | **2.42** | **2.38** | **2.36** | |
| 2.18 | 2.13 | 2.07 | 2.03 | 1.98 | 1.93 | 1.91 | 1.87 | 1.84 | 1.81 | 1.80 | 1.78 | 22 |
| **3.02** | **2.94** | **2.83** | **2.75** | **2.67** | **2.58** | **2.53** | **2.46** | **2.42** | **2.37** | **2.33** | **2.31** | |
| 2.14 | 2.10 | 2.04 | 2.00 | 1.96 | 1.91 | 1.88 | 1.84 | 1.82 | 1.79 | 1.77 | 1.76 | 23 |
| **2.97** | **2.89** | **2.78** | **2.70** | **2.62** | **2.53** | **2.48** | **2.41** | **2.37** | **2.32** | **2.28** | **2.26** | |
| 2.13 | 2.09 | 2.02 | 1.98 | 1.94 | 1.89 | 1.86 | 1.82 | 1.80 | 1.76 | 1.74 | 1.73 | 24 |
| **2.93** | **2.85** | **2.74** | **2.66** | **2.58** | **2.49** | **2.44** | **2.36** | **2.33** | **2.27** | **2.23** | **2.21** | |
| 2.11 | 2.06 | 2.00 | 1.96 | 1.92 | 1.87 | 1.84 | 1.80 | 1.77 | 1.74 | 1.72 | 1.71 | 25 |
| **2.89** | **2.81** | **2.70** | **2.62** | **2.54** | **2.45** | **2.40** | **2.32** | **2.29** | **2.23** | **2.19** | **2.17** | |
| 2.10 | 2.05 | 1.99 | 1.95 | 1.90 | 1.85 | 1.82 | 1.78 | 1.76 | 1.72 | 1.70 | 1.69 | 26 |
| **2.86** | **2.77** | **2.66** | **2.58** | **2.50** | **2.41** | **2.36** | **2.28** | **2.25** | **2.19** | **2.15** | **2.13** | |
| 2.08 | 2.03 | 1.97 | 1.93 | 1.88 | 1.84 | 1.80 | 1.76 | 1.74 | 1.71 | 1.68 | 1.67 | 27 |
| **2.83** | **2.74** | **2.63** | **2.55** | **2.47** | **2.38** | **2.33** | **2.25** | **2.21** | **2.16** | **2.12** | **2.10** | |
| 2.06 | 2.02 | 1.96 | 1.91 | 1.87 | 1.81 | 1.78 | 1.75 | 1.72 | 1.69 | 1.67 | 1.65 | 28 |
| **2.80** | **2.71** | **2.60** | **2.52** | **2.44** | **2.35** | **2.30** | **2.22** | **2.18** | **2.13** | **2.09** | **2.06** | |
| 2.05 | 2.00 | 1.94 | 1.90 | 1.85 | 1.80 | 1.77 | 1.73 | 1.71 | 1.68 | 1.65 | 1.64 | 29 |
| **2.77** | **2.68** | **2.57** | **2.49** | **2.41** | **2.32** | **2.27** | **2.19** | **2.15** | **2.10** | **2.06** | **2.03** | |
| 2.04 | 1.99 | 1.93 | 1.89 | 1.84 | 1.79 | 1.76 | 1.72 | 1.69 | 1.66 | 1.64 | 1.62 | 30 |
| **2.74** | **2.66** | **2.55** | **2.47** | **2.38** | **2.29** | **2.24** | **2.16** | **2.13** | **2.07** | **2.03** | **2.01** | |
| 2.02 | 1.97 | 1.91 | 1.86 | 1.82 | 1.76 | 1.74 | 1.69 | 1.67 | 1.64 | 1.61 | 1.59 | 32 |
| **2.70** | **2.62** | **2.51** | **2.42** | **2.34** | **2.25** | **2.20** | **2.12** | **2.08** | **2.02** | **1.98** | **1.96** | |
| 2.00 | 1.95 | 1.89 | 1.84 | 1.80 | 1.74 | 1.71 | 1.67 | 1.64 | 1.61 | 1.59 | 1.57 | 34 |
| **2.66** | **2.58** | **2.47** | **2.38** | **2.30** | **2.21** | **2.15** | **2.08** | **2.04** | **1.98** | **1.94** | **1.91** | |
| 1.98 | 1.93 | 1.87 | 1.82 | 1.78 | 1.72 | 1.69 | 1.65 | 1.62 | 1.59 | 1.56 | 1.55 | 36 |
| **2.62** | **2.54** | **2.43** | **2.35** | **2.26** | **2.17** | **2.12** | **2.04** | **2.00** | **1.94** | **1.90** | **1.87** | |
| 1.96 | 1.92 | 1.85 | 1.80 | 1.76 | 1.71 | 1.67 | 1.63 | 1.60 | 1.57 | 1.54 | 1.53 | 38 |
| **2.59** | **2.51** | **2.40** | **2.32** | **2.22** | **2.14** | **2.08** | **2.00** | **1.97** | **1.90** | **1.86** | **1.84** | |
| 1.95 | 1.90 | 1.84 | 1.79 | 1.74 | 1.69 | 1.66 | 1.61 | 1.59 | 1.55 | 1.53 | 1.51 | 40 |
| **2.56** | **2.49** | **2.37** | **2.29** | **2.20** | **2.11** | **2.05** | **1.97** | **1.94** | **1.88** | **1.84** | **1.81** | |
| 1.94 | 1.89 | 1.82 | 1.78 | 1.73 | 1.68 | 1.64 | 1.60 | 1.57 | 1.54 | 1.51 | 1.49 | 42 |
| **2.54** | **2.46** | **2.35** | **2.26** | **2.17** | **2.08** | **2.02** | **1.94** | **1.91** | **1.85** | **1.80** | **1.78** | |
| 1.92 | 1.88 | 1.81 | 1.76 | 1.72 | 1.66 | 1.63 | 1.58 | 1.56 | 1.52 | 1.50 | 1.48 | 44 |
| **2.52** | **2.44** | **2.32** | **2.24** | **2.15** | **2.06** | **2.00** | **1.92** | **1.88** | **1.82** | **1.78** | **1.75** | |

APPENDIX E (*Continued*)

| $f_2$ | 1 | 2 | 3 | 4 | 5 | 6 | 7 | 8 | 9 | 10 | 11 | 12 |
|-------|---|---|---|---|---|---|---|---|---|----|----|----|
| | | | | | | | | | $f_1$ Degrees of Freedom (for | | | |
| 46 | 4.05 | 3.20 | 2.81 | 2.57 | 2.42 | 2.30 | 2.22 | 2.14 | 2.09 | 2.04 | 2.00 | 1.97 |
| | **7.21** | **5.10** | **4.24** | **3.76** | **3.44** | **3.22** | **3.05** | **2.92** | **2.82** | **2.73** | **2.66** | **2.60** |
| 48 | 4.04 | 3.19 | 2.80 | 2.56 | 2.41 | 2.30 | 2.21 | 2.14 | 2.08 | 2.03 | 1.99 | 1.96 |
| | **7.19** | **5.08** | **4.22** | **3.74** | **3.42** | **3.20** | **3.04** | **2.90** | **2.80** | **2.71** | **2.64** | **2.58** |
| 50 | 4.03 | 3.18 | 2.79 | 2.56 | 2.40 | 2.29 | 2.20 | 2.13 | 2.07 | 2.02 | 1.98 | 1.95 |
| | **7.17** | **5.06** | **4.20** | **3.72** | **3.41** | **3.18** | **3.02** | **2.88** | **2.78** | **2.70** | **2.62** | **2.56** |
| 55 | 4.02 | 3.17 | 2.78 | 2.54 | 2.38 | 2.27 | 2.18 | 2.11 | 2.05 | 2.00 | 1.97 | 1.93 |
| | **7.12** | **5.01** | **4.16** | **3.68** | **3.37** | **3.15** | **2.98** | **2.85** | **2.75** | **2.66** | **2.59** | **2.53** |
| 60 | 4.00 | 3.15 | 2.76 | 2.52 | 2.37 | 2.25 | 2.17 | 2.10 | 2.04 | 1.99 | 1.95 | 1.92 |
| | **7.08** | **4.98** | **4.13** | **3.65** | **3.34** | **3.12** | **2.95** | **2.82** | **2.72** | **2.63** | **2.56** | **2.50** |
| 65 | 3.99 | 3.14 | 2.75 | 2.51 | 2.36 | 2.24 | 2.15 | 2.08 | 2.02 | 1.98 | 1.94 | 1.90 |
| | **7.04** | **4.95** | **4.10** | **3.62** | **3.31** | **3.09** | **2.93** | **2.79** | **2.70** | **2.61** | **2.54** | **2.47** |
| 70 | 3.98 | 3.13 | 2.74 | 2.50 | 2.35 | 2.23 | 2.14 | 2.07 | 2.01 | 1.97 | 1.93 | 1.89 |
| | **7.01** | **4.92** | **4.08** | **3.60** | **3.29** | **3.07** | **2.91** | **2.77** | **2.67** | **2.59** | **2.51** | **2.45** |
| 80 | 3.96 | 3.11 | 2.72 | 2.48 | 2.33 | 2.21 | 2.12 | 2.05 | 1.99 | 1.95 | 1.91 | 1.88 |
| | **6.96** | **4.88** | **4.04** | **3.56** | **3.25** | **3.04** | **2.87** | **2.74** | **2.64** | **2.55** | **2.48** | **2.41** |
| 100 | 3.94 | 3.09 | 2.70 | 2.46 | 2.30 | 2.19 | 2.10 | 2.03 | 1.97 | 1.92 | 1.88 | 1.85 |
| | **6.90** | **4.82** | **3.98** | **3.51** | **3.20** | **2.99** | **2.82** | **2.69** | **2.59** | **2.51** | **2.43** | **2.36** |
| 125 | 3.92 | 3.07 | 2.68 | 2.44 | 2.29 | 2.17 | 2.08 | 2.01 | 1.95 | 1.90 | 1.86 | 1.83 |
| | **6.84** | **4.78** | **3.94** | **3.47** | **3.17** | **2.95** | **2.79** | **2.65** | **2.56** | **2.47** | **2.40** | **2.33** |
| 150 | 3.91 | 3.06 | 2.67 | 2.43 | 2.27 | 2.16 | 2.07 | 2.00 | 1.94 | 1.89 | 1.85 | 1.82 |
| | **6.81** | **4.75** | **3.91** | **3.44** | **3.14** | **2.92** | **2.76** | **2.62** | **2.53** | **2.44** | **2.37** | **2.30** |
| 200 | 3.89 | 3.04 | 2.65 | 2.41 | 2.26 | 2.14 | 2.05 | 1.98 | 1.92 | 1.87 | 1.83 | 1.80 |
| | **6.76** | **4.71** | **3.88** | **3.41** | **3.11** | **2.90** | **2.73** | **2.60** | **2.50** | **2.41** | **2.34** | **2.28** |
| 400 | 3.86 | 3.02 | 2.62 | 2.39 | 2.23 | 2.12 | 2.03 | 1.96 | 1.90 | 1.85 | 1.81 | 1.78 |
| | **6.70** | **4.66** | **3.83** | **3.36** | **3.06** | **2.85** | **2.69** | **2.55** | **2.46** | **2.37** | **2.29** | **2.23** |
| 1000 | 3.85 | 3.00 | 2.61 | 2.38 | 2.22 | 2.10 | 2.02 | 1.95 | 1.89 | 1.84 | 1.80 | 1.76 |
| | **6.66** | **4.62** | **3.80** | **3.34** | **3.04** | **2.82** | **2.66** | **2.53** | **2.43** | **2.34** | **2.26** | **2.20** |
| ∞ | 3.84 | 2.99 | 2.60 | 2.37 | 2.21 | 2.09 | 2.01 | 1.94 | 1.88 | 1.83 | 1.79 | 1.75 |
| | **6.64** | **4.60** | **3.78** | **3.32** | **3.02** | **2.80** | **2.64** | **2.51** | **2.41** | **2.32** | **2.24** | **2.18** |

APPENDIX E (*Continued*)

Greater Mean Square)

| 14 | 16 | 20 | 24 | 30 | 40 | 50 | 75 | 100 | 200 | 500 | ∞ | $f_2$ |
|---|---|---|---|---|---|---|---|---|---|---|---|---|
| 1.91 | 1.87 | 1.80 | 1.75 | 1.71 | 1.65 | 1.62 | 1.57 | 1.54 | 1.51 | 1.48 | 1.46 | 46 |
| **2.50** | **2.42** | **2.30** | **2.22** | **2.13** | **2.04** | **1.98** | **1.90** | **1.86** | **1.80** | **1.76** | **1.72** | |
| 1.90 | 1.86 | 1.79 | 1.74 | 1.70 | 1.64 | 1.61 | 1.56 | 1.53 | 1.50 | 1.47 | 1.45 | 48 |
| **2.48** | **2.40** | **2.28** | **2.20** | **2.11** | **2.02** | **1.96** | **1.88** | **1.84** | **1.78** | **1.73** | **1.70** | |
| 1.90 | 1.85 | 1.78 | 1.74 | 1.69 | 1.63 | 1.60 | 1.55 | 1.52 | 1.48 | 1.46 | 1.44 | 50 |
| **2.46** | **2.39** | **2.26** | **2.18** | **2.10** | **2.00** | **1.94** | **1.86** | **1.82** | **1.76** | **1.71** | **1.68** | |
| 1.88 | 1.83 | 1.76 | 1.72 | 1.67 | 1.61 | 1.58 | 1.52 | 1.50 | 1.46 | 1.43 | 1.41 | 55 |
| **2.43** | **2.35** | **2.23** | **2.15** | **2.06** | **1.96** | **1.90** | **1.82** | **1.78** | **1.71** | **1.66** | **1.64** | |
| 1.86 | 1.81 | 1.75 | 1.70 | 1.65 | 1.59 | 1.56 | 1.50 | 1.48 | 1.44 | 1.41 | 1.39 | 60 |
| **2.40** | **2.32** | **2.20** | **2.12** | **2.03** | **1.93** | **1.87** | **1.79** | **1.74** | **1.68** | **1.63** | **1.60** | |
| 1.85 | 1.80 | 1.73 | 1.68 | 1.63 | 1.57 | 1.54 | 1.49 | 1.46 | 1.42 | 1.39 | 1.37 | 65 |
| **2.37** | **2.30** | **2.18** | **2.09** | **2.00** | **1.90** | **1.84** | **1.76** | **1.71** | **1.64** | **1.60** | **1.56** | |
| 1.84 | 1.79 | 1.72 | 1.67 | 1.62 | 1.56 | 1.53 | 1.47 | 1.45 | 1.40 | 1.37 | 1.35 | 70 |
| **2.35** | **2.28** | **2.15** | **2.07** | **1.98** | **1.88** | **1.82** | **1.74** | **1.69** | **1.62** | **1.56** | **1.53** | |
| 1.82 | 1.77 | 1.70 | 1.65 | 1.60 | 1.54 | 1.51 | 1.45 | 1.42 | 1.38 | 1.35 | 1.32 | 80 |
| **2.32** | **2.24** | **2.11** | **2.03** | **1.94** | **1.84** | **1.78** | **1.70** | **1.65** | **1.57** | **1.52** | **1.49** | |
| 1.79 | 1.75 | 1.68 | 1.63 | 1.57 | 1.51 | 1.48 | 1.42 | 1.39 | 1.34 | 1.30 | 1.28 | 100 |
| **2.26** | **2.19** | **2.06** | **1.98** | **1.89** | **1.79** | **1.73** | **1.64** | **1.59** | **1.51** | **1.46** | **1.43** | |
| 1.77 | 1.72 | 1.65 | 1.60 | 1.55 | 1.49 | 1.45 | 1.39 | 1.36 | 1.31 | 1.27 | 1.25 | 125 |
| **2.23** | **2.15** | **2.03** | **1.94** | **1.85** | **1.75** | **1.68** | **1.59** | **1.54** | **1.46** | **1.40** | **1.37** | |
| 1.76 | 1.71 | 1.64 | 1.59 | 1.54 | 1.47 | 1.44 | 1.37 | 1.34 | 1.29 | 1.25 | 1.22 | 150 |
| **2.20** | **2.12** | **2.00** | **1.91** | **1.83** | **1.72** | **1.66** | **1.56** | **1.51** | **1.43** | **1.37** | **1.33** | |
| 1.74 | 1.69 | 1.62 | 1.57 | 1.52 | 1.45 | 1.42 | 1.35 | 1.32 | 1.26 | 1.22 | 1.19 | 200 |
| **2.17** | **2.09** | **1.97** | **1.88** | **1.79** | **1.69** | **1.62** | **1.53** | **1.48** | **1.39** | **1.33** | **1.28** | |
| 1.72 | 1.67 | 1.60 | 1.54 | 1.49 | 1.42 | 1.38 | 1.32 | 1.28 | 1.22 | 1.16 | 1.13 | 400 |
| **2.12** | **2.04** | **1.92** | **1.84** | **1.74** | **1.64** | **1.57** | **1.47** | **1.42** | **1.32** | **1.24** | **1.19** | |
| 1.70 | 1.65 | 1.58 | 1.53 | 1.47 | 1.41 | 1.36 | 1.30 | 1.26 | 1.19 | 1.13 | 1.08 | 1000 |
| **2.09** | **2.01** | **1.89** | **1.81** | **1.71** | **1.61** | **1.54** | **1.44** | **1.38** | **1.28** | **1.19** | **1.11** | |
| 1.69 | 1.64 | 1.57 | 1.52 | 1.46 | 1.40 | 1.35 | 1.28 | 1.24 | 1.17 | 1.11 | 1.00 | ∞ |
| **2.07** | **1.99** | **1.87** | **1.79** | **1.69** | **1.59** | **1.52** | **1.41** | **1.36** | **1.25** | **1.15** | **1.00** | |

# Appendix F
## Values of r for Different Levels of Significance

| df | .1 | .05 | .02 | .01 | .001 |
|----|------|------|------|------|------|
| 1 | .98769 | .99692 | .999507 | .999877 | .9999988 |
| 2 | .90000 | .95000 | .98000 | .990000 | .99900 |
| 3 | .8054 | .8783 | .93433 | .95873 | .99116 |
| 4 | .7293 | .8114 | .8822 | .91720 | .97406 |
| 5 | .6694 | .7545 | .8329 | .8745 | .95074 |
| 6 | .6215 | .7067 | .7887 | .8343 | .92493 |
| 7 | .5822 | .6664 | .7498 | .7977 | .8982 |
| 8 | .5494 | .6319 | .7155 | .7646 | .8721 |
| 9 | .5214 | .6021 | .6851 | .7348 | .8471 |
| 10 | .4973 | .5760 | .6581 | .7079 | .8233 |
| 11 | .4762 | .5529 | .6339 | .6835 | .8010 |
| 12 | .4575 | .5324 | .6120 | .6614 | .7800 |
| 13 | .4409 | .5139 | .5923 | .6411 | .7603 |
| 14 | .4259 | .4973 | .5742 | .6226 | .7420 |
| 15 | .4124 | .4821 | .5577 | .6055 | .7246 |
| 16 | .4000 | .4683 | .5425 | .5897 | .7084 |
| 17 | .3887 | .4555 | .5285 | .5751 | .6932 |
| 18 | .3783 | .4438 | .5155 | .5614 | .6787 |
| 19 | .3687 | .4329 | .5034 | .5487 | .6652 |
| 20 | .3598 | .4227 | .4921 | .5368 | .6524 |
| 25 | .3233 | .3809 | .4451 | .4869 | .5974 |
| 30 | .2960 | .3494 | .4093 | .4487 | .5541 |
| 35 | .2746 | .3246 | .3810 | .4182 | .5189 |
| 40 | .2573 | .3044 | .3578 | .3932 | .4896 |
| 45 | .2428 | .2875 | .3384 | .3721 | .4648 |
| 50 | .2306 | .2732 | .3218 | .3541 | .4433 |
| 60 | .2108 | .2500 | .2948 | .3248 | .4078 |
| 70 | .1954 | .2319 | .2737 | .3017 | .3799 |
| 80 | .1829 | .2172 | .2565 | .2830 | .3568 |
| 90 | .1726 | .2050 | .2422 | .2673 | .3375 |
| 100 | .1638 | .1946 | .2301 | .2540 | .3211 |

SOURCE: Appendix F is reprinted from Table VI of R. A. Fisher and F. Yates. *Statistical Tables for Biological, Agricultural, and Medical Research*, published by Oliver and Boyd Ltd., Edinburgh. Reprinted with permission of the authors and publishers.

# Appendix G
## Table of z Values for $r$[a]

| r | z | r | z | r | z | r | z | r | z |
|---|---|---|---|---|---|---|---|---|---|
| .000 | .000 | .200 | .203 | .400 | .424 | .600 | .693 | .800 | 1.099 |
| .005 | .005 | .205 | .208 | .405 | .430 | .605 | .701 | .805 | 1.113 |
| .010 | .010 | .210 | .213 | .410 | .436 | .610 | .709 | .810 | 1.127 |
| .015 | .015 | .215 | .218 | .415 | .442 | .615 | .717 | .815 | 1.142 |
| .020 | .020 | .220 | .224 | .420 | .448 | .620 | .725 | .820 | 1.157 |
| .025 | .025 | .225 | .229 | .425 | .454 | .625 | .733 | .825 | 1.172 |
| .030 | .030 | .230 | .234 | .430 | .460 | .630 | .741 | .830 | 1.188 |
| .035 | .035 | .235 | .239 | .435 | .466 | .635 | .750 | .835 | 1.204 |
| .040 | .040 | .240 | .245 | .440 | .472 | .640 | .758 | .840 | 1.221 |
| .045 | .045 | .245 | .250 | .445 | .478 | .645 | .767 | .845 | 1.238 |
| .050 | .050 | .250 | .255 | .450 | .485 | .650 | .775 | .850 | 1.256 |
| .055 | .055 | .255 | .261 | .455 | .491 | .655 | .784 | .855 | 1.274 |
| .060 | .060 | .260 | .266 | .460 | .497 | .660 | .793 | .860 | 1.293 |
| .065 | .065 | .265 | .271 | .465 | .504 | .665 | .802 | .865 | 1.313 |
| .070 | .070 | .270 | .277 | .470 | .510 | .670 | .811 | .870 | 1.333 |
| .075 | .075 | .275 | .282 | .475 | .517 | .675 | .820 | .875 | 1.354 |
| .080 | .080 | .280 | .288 | .480 | .523 | .680 | .829 | .880 | 1.376 |
| .085 | .085 | .285 | .293 | .485 | .530 | .685 | .838 | .885 | 1.398 |
| .090 | .090 | .290 | .299 | .490 | .536 | .690 | .848 | .890 | 1.422 |
| .095 | .095 | .295 | .304 | .495 | .543 | .695 | .858 | .895 | 1.447 |
| .100 | .100 | .300 | .310 | .500 | .549 | .700 | .867 | .900 | 1.472 |
| .105 | .105 | .305 | .315 | .505 | .556 | .705 | .877 | .905 | 1.499 |
| .110 | .110 | .310 | .321 | .510 | .563 | .710 | .887 | .910 | 1.528 |
| .115 | .116 | .315 | .326 | .515 | .570 | .715 | .897 | .915 | 1.557 |
| .120 | .121 | .320 | .332 | .520 | .576 | .720 | .908 | .920 | 1.589 |
| .125 | .126 | .325 | .337 | .525 | .583 | .725 | .918 | .925 | 1.623 |
| .130 | .131 | .330 | .343 | .530 | .590 | .730 | .929 | .930 | 1.658 |
| .135 | .136 | .335 | .348 | .535 | .597 | .735 | .940 | .935 | 1.697 |
| .140 | .141 | .340 | .354 | .540 | .604 | .740 | .950 | .940 | 1.738 |
| .145 | .146 | .345 | .360 | .545 | .611 | .745 | .962 | .945 | 1.783 |
| .150 | .151 | .350 | .365 | .550 | .618 | .750 | .973 | .950 | 1.832 |
| .155 | .156 | .355 | .371 | .555 | .626 | .755 | .984 | .955 | 1.886 |
| .160 | .161 | .360 | .377 | .560 | .633 | .760 | .996 | .960 | 1.946 |
| .165 | .167 | .365 | .383 | .565 | .640 | .765 | 1.008 | .965 | 2.014 |
| .170 | .172 | .370 | .388 | .570 | .648 | .770 | 1.020 | .970 | 2.092 |
| .175 | .177 | .375 | .394 | .575 | .655 | .775 | 1.033 | .975 | 2.185 |
| .180 | .182 | .380 | .400 | .580 | .662 | .780 | 1.045 | .980 | 2.298 |
| .185 | .187 | .385 | .406 | .585 | .670 | .785 | 1.058 | .985 | 2.443 |
| .190 | .192 | .390 | .412 | .590 | .678 | .790 | 1.071 | .990 | 2.647 |
| .195 | .198 | .395 | .418 | .595 | .685 | .795 | 1.085 | .995 | 2.994 |

[a] Appendix G was constructed by F. P. Kilpatrick and D. A. Buchanan from the formula
$$z = \tfrac{1}{2}[\log_e (1 + r) - \log_e (1 - r)]$$
SOURCE: A. L. Edwards. *Statistical Methods for the Behavioral Sciences.* New York: Holt, Rinehart and Winston, 1954. Reprinted by permission of the publisher.

# Appendix H
## Estimates of $r_{tet}$ for Various Values of $ad/bc$

| $r_{tet}$ | $ad/bc$ | $r_{tet}$ | $ad/bc$ | $r_{tet}$ | $ad/bc$ |
|---|---|---|---|---|---|
| .00 | 0–1.00 | .35 | 2.49–2.55 | .70 | 8.50–8.90 |
| .01 | 1.01–1.03 | .36 | 2.56–2.63 | .71 | 8.91–9.35 |
| .02 | 1.04–1.06 | .37 | 2.64–2.71 | .72 | 9.36–9.82 |
| .03 | 1.07–1.08 | .38 | 2.72–2.79 | .73 | 9.83–10.33 |
| .04 | 1.09–1.11 | .39 | 2.80–2.87 | .74 | 10.34–10.90 |
| .05 | 1.12–1.14 | .40 | 2.88–2.96 | .75 | 10.91–11.51 |
| .06 | 1.15–1.17 | .41 | 2.97–3.05 | .76 | 11.52–12.16 |
| .07 | 1.18–1.20 | .42 | 3.06–3.14 | .77 | 12.17–12.89 |
| .08 | 1.21–1.23 | .43 | 3.15–3.24 | .78 | 12.90–13.70 |
| .09 | 1.24–1.27 | .44 | 3.25–3.34 | .79 | 13.71–14.58 |
| .10 | 1.28–1.30 | .45 | 3.35–3.45 | .80 | 14.59–15.57 |
| .11 | 1.31–1.33 | .46 | 3.46–3.56 | .81 | 15.58–16.65 |
| .12 | 1.34–1.37 | .47 | 3.57–3.68 | .82 | 16.66–17.88 |
| .13 | 1.38–1.40 | .48 | 3.69–3.80 | .83 | 17.89–19.28 |
| .14 | 1.41–1.44 | .49 | 3.81–3.92 | .84 | 19.29–20.85 |
| .15 | 1.45–1.48 | .50 | 3.93–4.06 | .85 | 20.86–22.68 |
| .16 | 1.49–1.52 | .51 | 4.07–4.20 | .86 | 22.69–24.76 |
| .17 | 1.53–1.56 | .52 | 4.21–4.34 | .87 | 24.77–27.22 |
| .18 | 1.57–1.60 | .53 | 4.35–4.49 | .88 | 27.23–30.09 |
| .19 | 1.61–1.64 | .54 | 4.50–4.66 | .89 | 30.10–33.60 |
| .20 | 1.65–1.69 | .55 | 4.67–4.82 | .90 | 33.61–37.79 |
| .21 | 1.70–1.73 | .56 | 4.83–4.99 | .91 | 37.80–43.06 |
| .22 | 1.74–1.78 | .57 | 5.00–5.18 | .92 | 43.07–49.83 |
| .23 | 1.79–1.83 | .58 | 5.19–5.38 | .93 | 49.84–58.79 |
| .24 | 1.84–1.88 | .59 | 5.39–5.59 | .94 | 58.80–70.95 |
| .25 | 1.89–1.93 | .60 | 5.60–5.80 | .95 | 70.96–89.01 |
| .26 | 1.94–1.98 | .61 | 5.81–6.03 | .96 | 89.02–117.54 |
| .27 | 1.99–2.04 | .62 | 6.04–6.28 | .97 | 117.55–169.67 |
| .28 | 2.05–2.10 | .63 | 6.29–6.54 | .98 | 169.68–293.12 |
| .29 | 2.11–2.15 | .64 | 6.55–6.81 | .99 | 293.13–923.97 |
| .30 | 2.16–2.22 | .65 | 6.82–7.10 | 1.00 | 923.98 . . . |
| .31 | 2.23–2.28 | .66 | 7.11–7.42 | | |
| .32 | 2.29–2.34 | .67 | 7.43–7.75 | | |
| .33 | 2.35–2.41 | .68 | 7.76–8.11 | | |
| .34 | 2.42–2.48 | .69 | 8.12–8.49 | | |

SOURCE: M. D. Davidoff and H. W. Goheen. "A table for the rapid determination of the tetrachoric correlation coefficient." *Psychometrika*, 1953, *18*, 115–121. Reprinted with the permission of the authors and publisher.

# Appendix I
## Table of Critical Values of *T* in the Wilcoxon Matched-Pairs Signed-Ranks Test

| | Level of Significance for One-Tailed Test | | |
|---|---|---|---|
| | .025 | .01 | .005 |
| *N* | Level of Significance for Two-Tailed Test | | |
| | .05 | .02 | .01 |
| 6 | 0 | — | — |
| 7 | 2 | 0 | — |
| 8 | 4 | 2 | 0 |
| 9 | 6 | 3 | 2 |
| 10 | 8 | 5 | 3 |
| 11 | 11 | 7 | 5 |
| 12 | 14 | 10 | 7 |
| 13 | 17 | 13 | 10 |
| 14 | 21 | 16 | 13 |
| 15 | 25 | 20 | 16 |
| 16 | 30 | 24 | 20 |
| 17 | 35 | 28 | 23 |
| 18 | 40 | 33 | 28 |
| 19 | 46 | 38 | 32 |
| 20 | 52 | 43 | 38 |
| 21 | 59 | 49 | 43 |
| 22 | 66 | 56 | 49 |
| 23 | 73 | 62 | 55 |
| 24 | 81 | 69 | 61 |
| 25 | 89 | 77 | 68 |

SOURCE: Adapted from Table I of F. Wilcoxon. *Some Rapid Approximate Statistical Procedures.* New York: American Cyanamid Company, 1949, p. 13. Reproduced from S. Seigel. *Nonparametric Statistics for the Behavioral Sciences.* New York: McGraw-Hill, 1956. Reprinted by permission of the author, American Cyanamid Company, and McGraw-Hill Book Company.

# Appendix J
## Table of Critical Values of $U$ in the Mann-Whitney Test

(a) Critical Values of $U$ for a One-Tailed Test at .001 or for a Two-Tailed Test at .002

| $n_1$ \ $n_2$ | 9 | 10 | 11 | 12 | 13 | 14 | 15 | 16 | 17 | 18 | 19 | 20 |
|---|---|---|---|---|---|---|---|---|---|---|---|---|
| 1 | | | | | | | | | | | | |
| 2 | | | | | | | | | | | | |
| 3 | | | | | | | | | 0 | 0 | 0 | 0 |
| 4 | | 0 | 0 | 0 | 1 | 1 | 1 | 2 | 2 | 3 | 3 | 3 |
| 5 | 1 | 1 | 2 | 2 | 3 | 3 | 4 | 5 | 5 | 6 | 7 | 7 |
| 6 | 2 | 3 | 4 | 4 | 5 | 6 | 7 | 8 | 9 | 10 | 11 | 12 |
| 7 | 3 | 5 | 6 | 7 | 8 | 9 | 10 | 11 | 13 | 14 | 15 | 16 |
| 8 | 5 | 6 | 8 | 9 | 11 | 12 | 14 | 15 | 17 | 18 | 20 | 21 |
| 9 | 7 | 8 | 10 | 12 | 14 | 15 | 17 | 19 | 21 | 23 | 25 | 26 |
| 10 | 8 | 10 | 12 | 14 | 17 | 19 | 21 | 23 | 25 | 27 | 29 | 32 |
| 11 | 10 | 12 | 15 | 17 | 20 | 22 | 24 | 27 | 29 | 32 | 34 | 37 |
| 12 | 12 | 14 | 17 | 20 | 23 | 25 | 28 | 31 | 34 | 37 | 40 | 42 |
| 13 | 14 | 17 | 20 | 23 | 26 | 29 | 32 | 35 | 38 | 42 | 45 | 48 |
| 14 | 15 | 19 | 22 | 25 | 29 | 32 | 36 | 39 | 43 | 46 | 50 | 54 |
| 15 | 17 | 21 | 24 | 28 | 32 | 36 | 40 | 43 | 47 | 51 | 55 | 59 |
| 16 | 19 | 23 | 27 | 31 | 35 | 39 | 43 | 48 | 52 | 56 | 60 | 65 |
| 17 | 21 | 25 | 29 | 34 | 38 | 43 | 47 | 52 | 57 | 61 | 66 | 70 |
| 18 | 23 | 27 | 32 | 37 | 42 | 46 | 51 | 56 | 61 | 66 | 71 | 76 |
| 19 | 25 | 29 | 34 | 40 | 45 | 50 | 55 | 60 | 66 | 71 | 77 | 82 |
| 20 | 26 | 32 | 37 | 42 | 48 | 54 | 59 | 65 | 70 | 76 | 82 | 88 |

(b) Critical Values of $U$ for a One-Tailed Test at .01 or for a Two-Tailed Test at .02

| $n_1$ \ $n_2$ | 9 | 10 | 11 | 12 | 13 | 14 | 15 | 16 | 17 | 18 | 19 | 20 |
|---|---|---|---|---|---|---|---|---|---|---|---|---|
| 1 | | | | | | | | | | | | |
| 2 | | | | | 0 | 0 | 0 | 0 | 0 | 0 | 1 | 1 |
| 3 | 1 | 1 | 1 | 2 | 2 | 2 | 3 | 3 | 4 | 4 | 4 | 5 |
| 4 | 3 | 3 | 4 | 5 | 5 | 6 | 7 | 7 | 8 | 9 | 9 | 10 |
| 5 | 5 | 6 | 7 | 8 | 9 | 10 | 11 | 12 | 13 | 14 | 15 | 16 |
| 6 | 7 | 8 | 9 | 11 | 12 | 13 | 15 | 16 | 18 | 19 | 20 | 22 |
| 7 | 9 | 11 | 12 | 14 | 16 | 17 | 19 | 21 | 23 | 24 | 26 | 28 |
| 8 | 11 | 13 | 15 | 17 | 20 | 22 | 24 | 26 | 28 | 30 | 32 | 34 |
| 9 | 14 | 16 | 18 | 21 | 23 | 26 | 28 | 31 | 33 | 36 | 38 | 40 |
| 10 | 16 | 19 | 22 | 24 | 27 | 30 | 33 | 36 | 38 | 41 | 44 | 47 |
| 11 | 18 | 22 | 25 | 28 | 31 | 34 | 37 | 41 | 44 | 47 | 50 | 53 |
| 12 | 21 | 24 | 28 | 31 | 35 | 38 | 42 | 46 | 49 | 53 | 56 | 60 |
| 13 | 23 | 27 | 31 | 35 | 39 | 43 | 47 | 51 | 55 | 59 | 63 | 67 |
| 14 | 26 | 30 | 34 | 38 | 43 | 47 | 51 | 56 | 60 | 65 | 69 | 73 |
| 15 | 28 | 33 | 37 | 42 | 47 | 51 | 56 | 61 | 66 | 70 | 75 | 80 |
| 16 | 31 | 36 | 41 | 46 | 51 | 56 | 61 | 66 | 71 | 76 | 82 | 87 |
| 17 | 33 | 38 | 44 | 49 | 55 | 60 | 66 | 71 | 77 | 82 | 88 | 93 |
| 18 | 36 | 41 | 47 | 53 | 59 | 65 | 70 | 76 | 82 | 88 | 94 | 100 |
| 19 | 38 | 44 | 50 | 56 | 63 | 69 | 75 | 82 | 88 | 94 | 101 | 107 |
| 20 | 40 | 47 | 53 | 60 | 67 | 73 | 80 | 87 | 93 | 100 | 107 | 114 |

SOURCE: Adapted and abridged from Tables 1, 3, 5, and 7 of D. Auble. "Extended tables for the Mann-Whitney statistic." *Bulletin of the Institute of Educational Research at Indiana University*, 1953, *1*, No. 2. Reproduced from S. Siegel. *Nonparametric Statistics for the Behavioral Sciences*. New York: McGraw-Hill, 1956. Reprinted by permission of the author, Institute of Educational Research, and McGraw-Hill Book Company.

APPENDIX J (*Continued*)

(*c*) Critical Values of *U* for a One-Tailed Test at .025 or for a Two-Tailed Test at .05

| $n_1$ \ $n_2$ | 9 | 10 | 11 | 12 | 13 | 14 | 15 | 16 | 17 | 18 | 19 | 20 |
|---|---|---|---|---|---|---|---|---|---|---|---|---|
| 1 | | | | | | | | | | | | |
| 2 | 0 | 0 | 0 | 1 | 1 | 1 | 1 | 1 | 2 | 2 | 2 | 2 |
| 3 | 2 | 3 | 3 | 4 | 4 | 5 | 5 | 6 | 6 | 7 | 7 | 8 |
| 4 | 4 | 5 | 6 | 7 | 8 | 9 | 10 | 11 | 11 | 12 | 13 | 13 |
| 5 | 7 | 8 | 9 | 11 | 12 | 13 | 14 | 15 | 17 | 18 | 19 | 20 |
| 6 | 10 | 11 | 13 | 14 | 16 | 17 | 19 | 21 | 22 | 24 | 25 | 27 |
| 7 | 12 | 14 | 16 | 18 | 20 | 22 | 24 | 26 | 28 | 30 | 32 | 34 |
| 8 | 15 | 17 | 19 | 22 | 24 | 26 | 29 | 31 | 34 | 36 | 38 | 41 |
| 9 | 17 | 20 | 23 | 26 | 28 | 31 | 34 | 37 | 39 | 42 | 45 | 48 |
| 10 | 20 | 23 | 26 | 29 | 33 | 36 | 39 | 42 | 45 | 48 | 52 | 55 |
| 11 | 23 | 26 | 30 | 33 | 37 | 40 | 44 | 47 | 51 | 55 | 58 | 62 |
| 12 | 26 | 29 | 33 | 37 | 41 | 45 | 49 | 53 | 57 | 61 | 65 | 69 |
| 13 | 28 | 33 | 37 | 41 | 45 | 50 | 54 | 59 | 63 | 67 | 72 | 76 |
| 14 | 31 | 36 | 40 | 45 | 50 | 55 | 59 | 64 | 67 | 74 | 78 | 83 |
| 15 | 34 | 39 | 44 | 49 | 54 | 59 | 64 | 70 | 75 | 80 | 85 | 90 |
| 16 | 37 | 42 | 47 | 53 | 59 | 64 | 70 | 75 | 81 | 86 | 92 | 98 |
| 17 | 39 | 45 | 51 | 57 | 63 | 67 | 75 | 81 | 87 | 93 | 99 | 105 |
| 18 | 42 | 48 | 55 | 61 | 67 | 74 | 80 | 86 | 93 | 99 | 106 | 112 |
| 19 | 45 | 52 | 58 | 65 | 72 | 78 | 85 | 92 | 99 | 106 | 113 | 119 |
| 20 | 48 | 55 | 62 | 69 | 76 | 83 | 90 | 98 | 105 | 112 | 119 | 127 |

(*d*) Critical Values of *U* for a One-Tailed Test at .05 or for a Two-Tailed Test at .10

| $n_2$ \ $n_1$ | 9 | 10 | 11 | 12 | 13 | 14 | 15 | 16 | 17 | 18 | 19 | 20 |
|---|---|---|---|---|---|---|---|---|---|---|---|---|
| 1 | | | | | | | | | | | 0 | 0 |
| 2 | 1 | 1 | 1 | 2 | 2 | 2 | 3 | 3 | 3 | 4 | 4 | 4 |
| 3 | 3 | 4 | 5 | 5 | 6 | 7 | 7 | 8 | 9 | 9 | 10 | 11 |
| 4 | 6 | 7 | 8 | 9 | 10 | 11 | 12 | 14 | 15 | 16 | 17 | 18 |
| 5 | 9 | 11 | 12 | 13 | 15 | 16 | 18 | 19 | 20 | 22 | 23 | 25 |
| 6 | 12 | 14 | 16 | 17 | 19 | 21 | 23 | 25 | 26 | 28 | 30 | 32 |
| 7 | 15 | 17 | 19 | 21 | 24 | 26 | 28 | 30 | 33 | 35 | 37 | 39 |
| 8 | 18 | 20 | 23 | 26 | 28 | 31 | 33 | 36 | 39 | 41 | 44 | 47 |
| 9 | 21 | 24 | 27 | 30 | 33 | 36 | 39 | 42 | 45 | 48 | 51 | 54 |
| 10 | 24 | 27 | 31 | 34 | 37 | 41 | 44 | 48 | 51 | 55 | 58 | 62 |
| 11 | 27 | 31 | 34 | 38 | 42 | 46 | 50 | 54 | 57 | 61 | 65 | 69 |
| 12 | 30 | 34 | 38 | 42 | 47 | 51 | 55 | 60 | 64 | 68 | 72 | 77 |
| 13 | 33 | 37 | 42 | 47 | 51 | 56 | 61 | 65 | 70 | 75 | 80 | 84 |
| 14 | 36 | 41 | 46 | 51 | 56 | 61 | 66 | 71 | 77 | 82 | 87 | 92 |
| 15 | 39 | 44 | 50 | 55 | 61 | 66 | 72 | 77 | 83 | 88 | 94 | 100 |
| 16 | 42 | 48 | 54 | 60 | 65 | 71 | 77 | 83 | 89 | 95 | 101 | 107 |
| 17 | 45 | 51 | 57 | 64 | 70 | 77 | 83 | 89 | 96 | 102 | 109 | 115 |
| 18 | 48 | 55 | 61 | 68 | 75 | 82 | 88 | 95 | 102 | 109 | 116 | 123 |
| 19 | 51 | 58 | 65 | 72 | 80 | 87 | 94 | 101 | 109 | 116 | 123 | 130 |
| 20 | 54 | 62 | 69 | 77 | 84 | 92 | 100 | 107 | 115 | 123 | 130 | 138 |

# Appendix K
## Table of Critical Values
## of *r* in the Runs Test

In the bodies of Appendix K (*a*) and (*b*) are various critical values of *r* for various values of $n_1$ and $n_2$. For the one-sample runs test, any value of *r* which is equal to or smaller than that shown in Appendix K (*a*) or equal to or larger than that shown in Appendix K (*b*) is significant at the .05 level. For the Wald-Wolwitz two-sample runs test, any value of *r* which is equal to or smaller than that shown in Appendix K (*a*) is significant at the .05 level.

### (a)

| $n_1$ \ $n_2$ | 2 | 3 | 4 | 5 | 6 | 7 | 8 | 9 | 10 | 11 | 12 | 13 | 14 | 15 | 16 | 17 | 18 | 19 | 20 |
|---|---|---|---|---|---|---|---|---|---|---|---|---|---|---|---|---|---|---|---|
| 2 | | | | | | | | | | | 2 | 2 | 2 | 2 | 2 | 2 | 2 | 2 | 2 |
| 3 | | | | | 2 | 2 | 2 | 2 | 2 | 2 | 2 | 2 | 2 | 3 | 3 | 3 | 3 | 3 | 3 |
| 4 | | | | 2 | 2 | 3 | 3 | 3 | 3 | 3 | 3 | 3 | 3 | 3 | 4 | 4 | 4 | 4 | 4 |
| 5 | | | 2 | 2 | 3 | 3 | 3 | 3 | 3 | 4 | 4 | 4 | 4 | 4 | 4 | 4 | 5 | 5 | 5 |
| 6 | | 2 | 2 | 3 | 3 | 3 | 3 | 4 | 4 | 4 | 4 | 5 | 5 | 5 | 5 | 5 | 5 | 6 | 6 |
| 7 | | 2 | 2 | 3 | 3 | 3 | 4 | 4 | 5 | 5 | 5 | 5 | 5 | 6 | 6 | 6 | 6 | 6 | 6 |
| 8 | | 2 | 3 | 3 | 3 | 4 | 4 | 5 | 5 | 5 | 6 | 6 | 6 | 6 | 6 | 7 | 7 | 7 | 7 |
| 9 | | 2 | 3 | 3 | 4 | 4 | 5 | 5 | 5 | 6 | 6 | 6 | 7 | 7 | 7 | 7 | 8 | 8 | 8 |
| 10 | | 2 | 3 | 3 | 4 | 5 | 5 | 5 | 6 | 6 | 7 | 7 | 7 | 7 | 8 | 8 | 8 | 8 | 9 |
| 11 | | 2 | 3 | 4 | 4 | 5 | 5 | 6 | 6 | 7 | 7 | 7 | 8 | 8 | 8 | 9 | 9 | 9 | 9 |
| 12 | 2 | 2 | 3 | 4 | 4 | 5 | 6 | 6 | 7 | 7 | 8 | 8 | 8 | 9 | 9 | 9 | 10 | 10 | 10 |
| 13 | 2 | 2 | 3 | 4 | 5 | 5 | 6 | 6 | 7 | 8 | 8 | 9 | 9 | 9 | 10 | 10 | 10 | 10 | 10 |
| 14 | 2 | 2 | 3 | 4 | 5 | 5 | 6 | 7 | 7 | 8 | 8 | 9 | 9 | 9 | 10 | 10 | 10 | 11 | 11 |
| 15 | 2 | 3 | 3 | 4 | 5 | 6 | 6 | 7 | 7 | 8 | 8 | 9 | 9 | 10 | 10 | 11 | 11 | 11 | 12 |
| 16 | 2 | 3 | 4 | 4 | 5 | 6 | 6 | 7 | 8 | 8 | 9 | 10 | 10 | 11 | 11 | 11 | 12 | 12 | 12 |
| 17 | 2 | 3 | 4 | 4 | 5 | 6 | 7 | 7 | 8 | 9 | 9 | 10 | 10 | 11 | 11 | 11 | 12 | 12 | 13 |
| 18 | 2 | 3 | 4 | 5 | 5 | 6 | 7 | 8 | 8 | 9 | 9 | 10 | 10 | 11 | 11 | 12 | 12 | 13 | 13 |
| 19 | 2 | 3 | 4 | 5 | 6 | 6 | 7 | 8 | 8 | 9 | 10 | 10 | 11 | 11 | 12 | 12 | 13 | 13 | 13 |
| 20 | 2 | 3 | 4 | 5 | 6 | 6 | 7 | 8 | 9 | 9 | 10 | 10 | 11 | 12 | 12 | 13 | 13 | 13 | 14 |

### (b)

| $n_1$ \ $n_2$ | 2 | 3 | 4 | 5 | 6 | 7 | 8 | 9 | 10 | 11 | 12 | 13 | 14 | 15 | 16 | 17 | 18 | 19 | 20 |
|---|---|---|---|---|---|---|---|---|---|---|---|---|---|---|---|---|---|---|---|
| 2 | | | | | | | | | | | | | | | | | | | |
| 3 | | | | | | | | | | | | | | | | | | | |
| 4 | | | | 9 | 9 | | | | | | | | | | | | | | |
| 5 | | | 9 | 10 | 10 | 11 | 11 | | | | | | | | | | | | |
| 6 | | | 9 | 10 | 11 | 12 | 12 | 13 | 13 | 13 | 13 | | | | | | | | |
| 7 | | | | 11 | 12 | 13 | 13 | 14 | 14 | 14 | 14 | 15 | 15 | 15 | | | | | |
| 8 | | | | | 11 | 12 | 13 | 14 | 14 | 15 | 15 | 16 | 16 | 16 | 17 | 17 | 17 | 17 | 17 |
| 9 | | | | | 13 | 14 | 14 | 15 | 16 | 16 | 16 | 17 | 17 | 18 | 18 | 18 | 18 | 18 | 18 |
| 10 | | | | | 13 | 14 | 15 | 16 | 16 | 17 | 17 | 18 | 18 | 18 | 19 | 19 | 19 | 20 | 20 |
| 11 | | | | | 13 | 14 | 15 | 16 | 17 | 17 | 18 | 19 | 19 | 19 | 20 | 20 | 20 | 21 | 21 |
| 12 | | | | | 13 | 14 | 16 | 16 | 17 | 18 | 19 | 19 | 20 | 20 | 21 | 21 | 21 | 22 | 22 |
| 13 | | | | | | 15 | 16 | 17 | 18 | 19 | 19 | 20 | 20 | 21 | 21 | 22 | 22 | 23 | 23 |
| 14 | | | | | | 15 | 16 | 17 | 18 | 19 | 20 | 20 | 21 | 22 | 22 | 23 | 23 | 23 | 24 |
| 15 | | | | | | 15 | 16 | 18 | 18 | 19 | 20 | 21 | 22 | 22 | 23 | 23 | 24 | 24 | 25 |
| 16 | | | | | | | 17 | 18 | 19 | 20 | 21 | 22 | 22 | 23 | 23 | 24 | 25 | 25 | 26 |
| 17 | | | | | | | 17 | 18 | 19 | 20 | 21 | 22 | 23 | 23 | 24 | 25 | 25 | 26 | 26 |
| 18 | | | | | | | 17 | 18 | 19 | 20 | 21 | 22 | 23 | 24 | 25 | 25 | 26 | 26 | 27 |
| 19 | | | | | | | 17 | 18 | 20 | 21 | 22 | 23 | 23 | 24 | 25 | 26 | 26 | 27 | 27 |
| 20 | | | | | | | 17 | 18 | 20 | 21 | 22 | 23 | 24 | 25 | 25 | 26 | 27 | 27 | 28 |

SOURCE: Adapted from Frieda S. Swed and C. Eisenhart. "Tables for testing randomness of grouping in a sequence of alternatives." *Annals of Mathematical Statistics*, 1943, *14*, 83–86. Reproduced from S. Siegel. *Nonparametric Statistics for the Behavioral Sciences.* New York: McGraw-Hill, 1956. Reprinted by permission of the authors, Institute of Mathematical Statistics, and McGraw-Hill Book Company.

# Appendix L
## Values of *H* for Three Samples Significant
## at the 10, 5, and 1 Percent Levels

| Sample Sizes | | | Level | | |
|---|---|---|---|---|---|
| $N_1$ | $N_2$ | $N_3$ | .10 | .05 | .01 |
| 2 | 2 | 2 | 4.57 | | |
| 3 | 2 | 1 | 4.29 | | |
| 3 | 2 | 2 | 4.50 | 4.71 | |
| 3 | 3 | 1 | 4.57 | 5.14 | |
| 3 | 3 | 2 | 4.56 | 5.36 | 6.25 |
| 3 | 3 | 3 | 4.62 | 5.60 | 6.49 |
| 4 | 2 | 1 | 4.50 | | |
| 4 | 2 | 2 | 4.46 | 5.33 | |
| 4 | 3 | 1 | 4.06 | 5.21 | |
| 4 | 3 | 2 | 4.51 | 5.44 | 6.30 |
| 4 | 3 | 3 | 4.70 | 5.73 | 6.75 |
| 4 | 4 | 1 | 4.17 | 4.79 | 6.67 |
| 4 | 4 | 2 | 4.55 | 5.45 | 6.87 |
| 4 | 4 | 3 | 4.55 | 5.60 | 7.14 |
| 4 | 4 | 4 | 4.65 | 5.69 | 7.54 |
| 5 | 2 | 1 | 4.20 | 5.00 | |
| 5 | 2 | 2 | 4.37 | 5.16 | 6.53 |
| 5 | 3 | 1 | 4.02 | 4.96 | |
| 5 | 3 | 2 | 4.49 | 5.25 | 6.82 |
| 5 | 3 | 3 | 4.53 | 5.44 | 6.98 |
| 5 | 4 | 1 | 3.99 | 4.99 | 6.84 |
| 5 | 4 | 2 | 4.52 | 5.27 | 7.12 |
| 5 | 4 | 3 | 4.55 | 5.63 | 7.40 |
| 5 | 4 | 4 | 4.62 | 5.62 | 7.74 |
| 5 | 5 | 1 | 4.11 | 5.13 | 6.84 |
| 5 | 5 | 2 | 4.51 | 5.25 | 7.27 |
| 5 | 5 | 3 | 4.55 | 5.63 | 7.54 |
| 5 | 5 | 4 | 4.52 | 5.64 | 7.79 |
| 5 | 5 | 5 | 4.56 | 5.66 | 7.98 |

SOURCE: Abridged from Table 6.1 of W. H Kruskal and W. A. Wallis. "Use of ranks on one-criterion variance analysis." *Journal of the American Statistical Association*, 1952, *47*, 584–621. Reproduced from M. W. Tate and R. C. Clelland. *Nonparametric and Shortcut Statistics*. Danville, Ill.: The Interstate Printers and Publishers, 1957. Reprinted by permission of the authors, *Journal of the American Statistical Association*, and The Interstate Printers and Publishers.

# Appendix M
## Values of the Coefficient of Concordance $W$ Significant at the 20, 10, 5, and 1 Percent Levels

| $m$ | $a$ | 3 | 4 | $n$ 5 | 6 | 7 | 8 | 9 | 10 |
|---|---|---|---|---|---|---|---|---|---|
|   | .20 | .78 | .60 | .53 | .49 | .47 | .46 | .45 | .44 |
|   | .10 |     | .73 | .62 | .58 | .55 | .53 | .52 | .51 |
| 3 | .05 | 1.00 | .82 | .71 | .65 | .62 | .60 | .58 | .56 |
|   | .01 |     | .96 | .84 | .77 | .73 | .70 | .67 | .65 |
|   | .20 | .56 | .40 | .38 | .37 | .36 | .35 | .34 | .33 |
|   | .10 | .75 | .52 | .47 | .44 | .42 | .41 | .40 | .39 |
| 4 | .05 | .81 | .65 | .54 | .51 | .48 | .46 | .45 | .44 |
|   | .01 | 1.00 | .80 | .67 | .62 | .59 | .56 | .54 | .52 |
|   | .20 | .36 | .34 | .30 | .29 | .28 | .28 | .27 | .27 |
|   | .10 | .52 | .42 | .38 | .36 | .34 | .33 | .32 | .31 |
| 5 | .05 | .64 | .52 | .44 | .41 | .39 | .38 | .36 | .35 |
|   | .01 | .84 | .66 | .56 | .52 | .49 | .46 | .44 | .43 |
|   | .20 | .33 | .27 | .25 | .24 | .24 | .23 | .23 | .23 |
|   | .10 | .44 | .36 | .32 | .30 | .29 | .28 | .27 | .26 |
| 6 | .05 | .58 | .42 | .37 | .35 | .33 | .32 | .31 | .30 |
|   | .01 | .75 | .56 | .49 | .45 | .42 | .40 | .38 | .37 |
|   | .20 | .27 | .23 | .22 | .21 | .20 | .20 | .20 | .19 |
|   | .10 | .39 | .30 | .27 | .26 | .25 | .24 | .23 | .23 |
| 7 | .05 | .51 | .36 | .32 | .30 | .29 | .27 | .26 | .26 |
|   | .01 | .63 | .48 | .43 | .39 | .36 | .34 | .33 | .32 |
|   | .20 | .25 | .20 | .19 | .18 | .18 | .17 | .17 | .17 |
|   | .10 | .33 | .26 | .24 | .23 | .22 | .21 | .20 | .20 |
| 8 | .05 | .39 | .32 | .29 | .27 | .25 | .24 | .23 | .23 |
|   | .01 | .56 | .43 | .38 | .35 | .32 | .31 | .29 | .28 |
|   | .20 | .20 | .18 | .17 | .16 | .16 | .16 | .15 | .15 |
|   | .10 | .31 | .23 | .21 | .20 | .19 | .19 | .18 | .18 |
| 9 | .05 | .35 | .28 | .26 | .24 | .23 | .22 | .21 | .20 |
|   | .01 | .48 | .38 | .34 | .31 | .29 | .27 | .26 | .25 |
|   | .20 | .19 | .16 | .15 | .15 | .14 | .14 | .14 | .13 |
|   | .10 | .25 | .21 | .19 | .18 | .17 | .17 | .16 | .16 |
| 10 | .05 | .31 | .25 | .23 | .21 | .20 | .20 | .19 | .18 |
|   | .01 | .48 | .35 | .31 | .28 | .26 | .25 | .24 | .23 |
|   | .20 | .14 | .13 | .13 | .12 | .12 | .12 | .11 | .11 |
|   | .10 | .19 | .17 | .16 | .15 | .15 | .14 | .14 | .13 |
| 12 | .05 | .25 | .21 | .19 | .18 | .17 | .16 | .16 | .15 |
|   | .01 | .36 | .30 | .26 | .24 | .22 | .21 | .20 | .19 |
|   | .20 | .12 | .11 | .11 | .10 | .10 | .10 | .10 | .10 |
|   | .10 | .17 | .15 | .14 | .13 | .13 | .12 | .12 | .12 |
| 14 | .05 | .21 | .18 | .17 | .16 | .15 | .14 | .14 | .13 |
|   | .01 | .31 | .26 | .23 | .21 | .19 | .18 | .17 | .17 |

Source: M. W. Tate and R. C. Clelland. *Nonparametric and Shortcut Statistics*. Danville, Ill.: The Interstate Printers and Publishers, 1957. Reprinted by permission.

APPENDIX M (*Continued*)

| m | a | 3 | 4 | $n$ 5 | 6 | 7 | 8 | 9 | 10 |
|---|---|---|---|---|---|---|---|---|----|
|    | .20 | .10 | .10 | .09 | .09 | .09 | .09 | .09 | .08 |
|    | .10 | .15 | .13 | .12 | .12 | .11 | .11 | .10 | .10 |
| 16 | .05 | .19 | .16 | .15 | .14 | .13 | .12 | .12 | .12 |
|    | .01 | .28 | .23 | .20 | .18 | .17 | .16 | .15 | .15 |
|    | .20 | .09 | .09 | .08 | .08 | .08 | .08 | .08 | .07 |
|    | .10 | .13 | .12 | .11 | .10 | .10 | .09 | .09 | .09 |
| 18 | .05 | .17 | .14 | .13 | .12 | .11 | .11 | .11 | .10 |
|    | .01 | .25 | .20 | .18 | .16 | .15 | .14 | .14 | .13 |
|    | .20 | .08 | .08 | .07 | .07 | .07 | .07 | .07 | .07 |
|    | .10 | .11 | .10 | .10 | .09 | .09 | .08 | .08 | .08 |
| 20 | .05 | .15 | .13 | .12 | .11 | .10 | .10 | .10 | .09 |
|    | .01 | .22 | .18 | .16 | .15 | .14 | .13 | .12 | .11 |
|    | .20 | .07 | .06 | .06 | .06 | .06 | .06 | .05 | .05 |
|    | .10 | .09 | .08 | .08 | .07 | .07 | .07 | .07 | .06 |
| 25 | .05 | .12 | .10 | .09 | .09 | .08 | .08 | .08 | .07 |
|    | .01 | .18 | .15 | .13 | .12 | .11 | .10 | .10 | .09 |
|    | .20 | .05 | .05 | .05 | .05 | .05 | .05 | .05 | .04 |
|    | .10 | .08 | .07 | .06 | .06 | .06 | .06 | .06 | .05 |
| 30 | .05 | .10 | .09 | .08 | .07 | .07 | .07 | .07 | .06 |
|    | .01 | .15 | .12 | .11 | .10 | .09 | .09 | .08 | .08 |

| Row | 1 | 2 | 3 | 4 | 5 | 6 | 7 | 8 | 9 | 10 | 11 | 12 | 13 | 14 | 15 | 16 | 17 | 18 | 19 |
|---|---|---|---|---|---|---|---|---|---|---|---|---|---|---|---|---|---|---|---|
| | | | | | | | | Column Number | | | | | | | | | | | |
| 1 | 9 | 8 | 9 | 6 | 9 | 9 | 0 | 9 | 6 | 3 | 2 | 3 | 3 | 8 | 6 | 8 | 4 | 4 | 2 |
| 2 | 3 | 5 | 6 | 1 | 7 | 4 | 1 | 3 | 2 | 6 | 8 | 6 | 0 | 4 | 7 | 5 | 2 | 0 | 3 |
| 3 | 4 | 0 | 6 | 1 | 6 | 9 | 6 | 1 | 5 | 9 | 5 | 4 | 5 | 4 | 8 | 6 | 7 | 4 | 0 |
| 4 | 6 | 5 | 6 | 3 | 1 | 6 | 8 | 6 | 7 | 2 | 0 | 7 | 2 | 3 | 2 | 1 | 5 | 0 | 9 |
| 5 | 2 | 4 | 9 | 7 | 9 | 1 | 0 | 3 | 9 | 6 | 7 | 4 | 1 | 5 | 4 | 9 | 6 | 9 | 8 |
| 6 | 7 | 6 | 1 | 2 | 7 | 5 | 6 | 9 | 4 | 8 | 4 | 2 | 8 | 5 | 2 | 4 | 1 | 8 | 0 |
| 7 | 8 | 2 | 1 | 3 | 4 | 7 | 4 | 6 | 3 | 0 | 7 | 5 | 0 | 9 | 2 | 9 | 0 | 6 | 1 |
| 8 | 6 | 9 | 5 | 6 | 5 | 6 | 0 | 9 | 0 | 7 | 7 | 1 | 4 | 1 | 8 | 3 | 1 | 9 | 3 |
| 9 | 7 | 2 | 1 | 9 | 9 | 8 | 0 | 1 | 6 | 1 | 6 | 2 | 3 | 6 | 9 | 5 | 5 | 8 | 4 |
| 10 | 2 | 9 | 0 | 7 | 3 | 0 | 8 | 9 | 6 | 3 | 3 | 8 | 5 | 5 | 6 | 5 | 2 | 0 | 9 |
| 11 | 9 | 3 | 5 | 4 | 5 | 7 | 4 | 0 | 3 | 0 | 1 | 0 | 4 | 3 | 3 | 9 | 5 | 3 | 2 |
| 12 | 9 | 7 | 5 | 7 | 9 | 4 | 8 | 6 | 8 | 7 | 6 | 1 | 6 | 8 | 2 | 5 | 5 | 5 | 3 |
| 13 | 4 | 1 | 7 | 8 | 6 | 8 | 1 | 0 | 5 | 8 | 8 | 6 | 1 | 6 | 8 | 2 | 9 | 0 | 4 |
| 14 | 5 | 0 | 8 | 3 | 3 | 4 | 5 | 4 | 4 | 2 | 5 | 3 | 0 | 4 | 9 | 6 | 1 | 2 | 3 |
| 15 | 3 | 5 | 0 | 2 | 9 | 4 | 1 | 0 | 0 | 3 | 9 | 0 | 5 | 8 | 6 | 0 | 9 | 9 | 6 |
| 16 | 0 | 3 | 8 | 2 | 3 | 5 | 1 | 0 | 1 | 0 | 6 | 8 | 5 | 2 | 4 | 8 | 0 | 3 | 8 |
| 17 | 1 | 7 | 2 | 9 | 1 | 2 | 7 | 8 | 4 | 7 | 0 | 3 | 3 | 1 | 5 | 8 | 2 | 7 | 3 |
| 18 | 5 | 0 | 5 | 7 | 9 | 5 | 8 | 7 | 8 | 9 | 3 | 5 | 3 | 4 | 4 | 6 | 1 | 1 | 3 |
| 19 | 7 | 7 | 3 | 3 | 5 | 3 | 6 | 1 | 3 | 2 | 8 | 5 | 4 | 1 | 4 | 8 | 3 | 9 | 0 |
| 20 | 1 | 0 | 9 | 1 | 3 | 8 | 2 | 5 | 3 | 0 | 3 | 8 | 0 | 9 | 3 | 3 | 0 | 4 | 5 |
| 21 | 1 | 3 | 8 | 5 | 1 | 8 | 5 | 9 | 4 | 1 | 9 | 3 | 9 | 3 | 6 | 5 | 5 | 9 | 4 |
| 22 | 8 | 6 | 4 | 7 | 8 | 7 | 5 | 9 | 4 | 1 | 9 | 3 | 9 | 3 | 6 | 5 | 5 | 9 | 4 |
| 23 | 0 | 6 | 6 | 9 | 6 | 5 | 1 | 0 | 3 | 2 | 6 | 7 | 7 | 4 | 9 | 6 | 0 | 3 | 4 |
| 24 | 7 | 6 | 7 | 4 | 7 | 0 | 8 | 3 | 8 | 7 | 3 | 2 | 5 | 1 | 2 | 4 | 2 | 9 | 7 |
| 25 | 3 | 2 | 3 | 8 | 1 | 3 | 1 | 8 | 7 | 4 | 5 | 9 | 0 | 0 | 2 | 4 | 1 | 2 | 1 |
| 26 | 9 | 2 | 1 | 6 | 4 | 2 | 3 | 8 | 7 | 6 | 2 | 6 | 2 | 6 | 4 | 8 | 1 | 0 | 1 |
| 27 | 3 | 7 | 4 | 2 | 2 | 8 | 1 | 7 | 8 | 0 | 6 | 0 | 0 | 0 | 3 | 2 | 2 | 9 | 7 |
| 28 | 0 | 7 | 7 | 8 | 0 | 8 | 5 | 1 | 5 | 0 | 2 | 6 | 5 | 8 | 7 | 5 | 3 | 0 | 6 |
| 29 | 7 | 4 | 2 | 3 | 3 | 2 | 6 | 0 | 0 | 6 | 5 | 2 | 2 | 3 | 6 | 3 | 9 | 0 | 4 |
| 30 | 1 | 8 | 2 | 7 | 5 | 9 | 5 | 3 | 0 | 6 | 5 | 2 | 9 | 9 | 1 | 1 | 7 | 3 | 3 |
| 31 | 4 | 3 | 1 | 8 | 7 | 0 | 6 | 0 | 8 | 6 | 5 | 0 | 1 | 0 | 4 | 0 | 6 | 1 | 5 |
| 32 | 8 | 5 | 8 | 0 | 6 | 1 | 4 | 1 | 2 | 0 | 4 | 4 | 1 | 4 | 7 | 6 | 3 | 5 | 1 |
| 33 | 4 | 5 | 8 | 5 | 0 | 4 | 5 | 8 | 3 | 9 | 2 | 8 | 7 | 8 | 9 | 0 | 8 | 4 | 3 |
| 34 | 5 | 0 | 2 | 5 | 4 | 9 | 2 | 2 | 1 | 1 | 0 | 0 | 5 | 4 | 8 | 7 | 6 | 4 | 0 |
| 35 | 0 | 8 | 1 | 7 | 0 | 6 | 3 | 3 | 4 | 7 | 6 | 2 | 6 | 8 | 9 | 3 | 4 | 1 | 4 |
| 36 | 2 | 5 | 9 | 3 | 4 | 6 | 0 | 7 | 5 | 2 | 0 | 0 | 9 | 6 | 0 | 8 | 2 | 2 | 5 |
| 37 | 2 | 1 | 3 | 1 | 3 | 7 | 8 | 9 | 8 | 4 | 9 | 3 | 8 | 0 | 2 | 2 | 1 | 8 | 1 |
| 38 | 3 | 8 | 8 | 6 | 8 | 5 | 1 | 3 | 3 | 4 | 6 | 7 | 2 | 6 | 3 | 4 | 8 | 6 | 7 |
| 39 | 0 | 9 | 9 | 8 | 5 | 9 | 8 | 4 | 4 | 2 | 2 | 1 | 1 | 0 | 1 | 7 | 6 | 1 | 3 |
| 40 | 2 | 2 | 3 | 5 | 3 | 9 | 7 | 4 | 4 | 2 | 1 | 4 | 0 | 5 | 8 | 2 | 3 | 0 | 8 |

APPENDIX N (*Continued*)

| 20 | 21 | 22 | 23 | 24 | 25 | 26 | 27 | Column Number 28 | 29 | 30 | 31 | 32 | 33 | 34 | 35 | 36 | 37 | 38 | 39 | 40 | Row |
|----|----|----|----|----|----|----|----|----|----|----|----|----|----|----|----|----|----|----|----|----|-----|
| 0 | 9 | 7 | 1 | 1 | 9 | 1 | 2 | 7 | 3 | 5 | 1 | 8 | 4 | 0 | 4 | 1 | 0 | 6 | 0 | 3 | 1 |
| 8 | 3 | 7 | 7 | 9 | 1 | 4 | 9 | 9 | 5 | 9 | 2 | 0 | 1 | 6 | 1 | 2 | 6 | 6 | 7 | 0 | 2 |
| 2 | 5 | 6 | 3 | 7 | 8 | 3 | 3 | 8 | 4 | 3 | 9 | 3 | 9 | 0 | 0 | 9 | 8 | 3 | 5 | 2 | 3 |
| 4 | 7 | 0 | 8 | 6 | 6 | 5 | 9 | 6 | 2 | 7 | 3 | 5 | 9 | 0 | 1 | 8 | 0 | 9 | 6 | 9 | 4 |
| 0 | 9 | 8 | 7 | 3 | 5 | 6 | 8 | 8 | 1 | 2 | 0 | 2 | 3 | 2 | 6 | 4 | 3 | 1 | 9 | 7 | 5 |
| 5 | 1 | 8 | 8 | 4 | 7 | 0 | 1 | 7 | 6 | 8 | 2 | 1 | 6 | 3 | 2 | 1 | 8 | 1 | 8 | 3 | 6 |
| 1 | 3 | 7 | 8 | 6 | 9 | 5 | 4 | 1 | 7 | 3 | 8 | 7 | 1 | 5 | 6 | 5 | 6 | 4 | 3 | 6 | 7 |
| 5 | 9 | 0 | 1 | 5 | 2 | 8 | 6 | 5 | 5 | 7 | 8 | 1 | 8 | 7 | 1 | 2 | 4 | 0 | 4 | 1 | 8 |
| 2 | 2 | 5 | 5 | 2 | 1 | 8 | 6 | 9 | 8 | 9 | 8 | 0 | 5 | 8 | 9 | 9 | 4 | 1 | 3 | 4 | 9 |
| 1 | 3 | 4 | 2 | 8 | 5 | 0 | 7 | 9 | 8 | 4 | 3 | 5 | 8 | 0 | 9 | 4 | 6 | 6 | 0 | 5 | 10 |
| 2 | 6 | 8 | 6 | 6 | 4 | 7 | 1 | 5 | 1 | 6 | 4 | 6 | 7 | 6 | 0 | 8 | 7 | 3 | 5 | 2 | 11 |
| 8 | 6 | 0 | 1 | 4 | 2 | 9 | 8 | 6 | 8 | 0 | 7 | 6 | 5 | 1 | 9 | 1 | 3 | 7 | 0 | 3 | 12 |
| 9 | 5 | 7 | 0 | 9 | 8 | 7 | 6 | 9 | 0 | 6 | 5 | 4 | 0 | 3 | 6 | 5 | 6 | 3 | 5 | 0 | 13 |
| 2 | 2 | 3 | 4 | 7 | 8 | 0 | 2 | 0 | 8 | 0 | 3 | 4 | 9 | 2 | 5 | 7 | 7 | 8 | 6 | 4 | 14 |
| 2 | 4 | 6 | 1 | 0 | 5 | 0 | 6 | 1 | 4 | 9 | 4 | 7 | 3 | 9 | 1 | 7 | 6 | 4 | 5 | 8 | 15 |
| 6 | 3 | 4 | 8 | 1 | 6 | 9 | 5 | 6 | 2 | 0 | 4 | 6 | 1 | 6 | 8 | 1 | 9 | 9 | 1 | 0 | 16 |
| 9 | 0 | 5 | 1 | 3 | 6 | 1 | 9 | 5 | 4 | 1 | 2 | 5 | 4 | 2 | 9 | 5 | 6 | 2 | 4 | 0 | 17 |
| 3 | 6 | 7 | 0 | 3 | 5 | 3 | 7 | 4 | 1 | 7 | 5 | 4 | 8 | 3 | 7 | 4 | 8 | 5 | 7 | 2 | 18 |
| 4 | 3 | 6 | 6 | 3 | 6 | 3 | 0 | 0 | 9 | 4 | 2 | 2 | 5 | 1 | 8 | 9 | 5 | 1 | 9 | 7 | 19 |
| 1 | 0 | 6 | 9 | 0 | 2 | 7 | 3 | 9 | 8 | 4 | 0 | 6 | 9 | 8 | 2 | 3 | 2 | 8 | 0 | 4 | 20 |
| 9 | 1 | 3 | 5 | 7 | 9 | 6 | 2 | 4 | 3 | 4 | 6 | 4 | 9 | 1 | 3 | 1 | 7 | 5 | 2 | 2 | 21 |
| 6 | 4 | 2 | 2 | 2 | 1 | 4 | 5 | 2 | 2 | 8 | 3 | 2 | 1 | 2 | 6 | 6 | 0 | 1 | 8 | 9 | 22 |
| 7 | 2 | 6 | 9 | 0 | 7 | 5 | 3 | 2 | 5 | 6 | 2 | 7 | 6 | 3 | 8 | 1 | 4 | 1 | 5 | 1 | 23 |
| 8 | 2 | 8 | 2 | 4 | 4 | 2 | 9 | 1 | 9 | 8 | 3 | 4 | 4 | 1 | 0 | 4 | 6 | 9 | 6 | 0 | 24 |
| 7 | 3 | 1 | 4 | 3 | 0 | 4 | 7 | 1 | 3 | 7 | 4 | 8 | 6 | 7 | 3 | 2 | 6 | 6 | 2 | 0 | 25 |
| 0 | 6 | 4 | 5 | 8 | 3 | 1 | 4 | 8 | 1 | 8 | 3 | 1 | 6 | 4 | 3 | 0 | 2 | 8 | 7 | 3 | 26 |
| 4 | 2 | 2 | 8 | 3 | 2 | 1 | 9 | 3 | 0 | 1 | 7 | 5 | 9 | 0 | 9 | 1 | 2 | 5 | 8 | 2 | 27 |
| 2 | 9 | 8 | 7 | 2 | 0 | 6 | 4 | 0 | 2 | 7 | 1 | 3 | 1 | 6 | 8 | 7 | 0 | 9 | 2 | 5 | 28 |
| 0 | 8 | 0 | 5 | 6 | 8 | 2 | 4 | 3 | 6 | 1 | 3 | 5 | 2 | 3 | 5 | 9 | 8 | 6 | 2 | 1 | 29 |
| 0 | 1 | 7 | 6 | 1 | 5 | 7 | 9 | 0 | 3 | 5 | 3 | 4 | 2 | 4 | 8 | 5 | 6 | 4 | 0 | 6 | 30 |
| 5 | 1 | 9 | 8 | 5 | 2 | 4 | 5 | 1 | 7 | 5 | 3 | 2 | 4 | 6 | 7 | 9 | 9 | 6 | 7 | 2 | 31 |
| 0 | 3 | 6 | 6 | 3 | 7 | 8 | 6 | 9 | 7 | 2 | 8 | 9 | 0 | 7 | 2 | 9 | 4 | 0 | 8 | 6 | 32 |
| 5 | 0 | 0 | 0 | 2 | 0 | 8 | 9 | 0 | 1 | 0 | 6 | 2 | 0 | 4 | 6 | 9 | 6 | 5 | 4 | 9 | 33 |
| 1 | 9 | 4 | 4 | 2 | 6 | 4 | 2 | 4 | 1 | 0 | 2 | 7 | 9 | 6 | 8 | 7 | 5 | 6 | 9 | 3 | 34 |
| 0 | 0 | 5 | 3 | 8 | 3 | 2 | 7 | 5 | 0 | 4 | 7 | 6 | 4 | 6 | 3 | 0 | 4 | 7 | 5 | 3 | 35 |
| 6 | 2 | 6 | 2 | 0 | 6 | 0 | 1 | 4 | 8 | 9 | 6 | 5 | 9 | 7 | 3 | 6 | 7 | 6 | 5 | 4 | 36 |
| 6 | 3 | 9 | 0 | 3 | 5 | 0 | 6 | 1 | 2 | 0 | 5 | 9 | 7 | 3 | 2 | 5 | 9 | 3 | 0 | 2 | 37 |
| 9 | 7 | 3 | 3 | 5 | 4 | 0 | 6 | 4 | 9 | 4 | 7 | 9 | 1 | 4 | 3 | 9 | 7 | 7 | 1 | 8 | 38 |
| 1 | 9 | 6 | 2 | 9 | 4 | 2 | 9 | 7 | 0 | 3 | 8 | 9 | 5 | 7 | 0 | 6 | 9 | 7 | 2 | 5 | 39 |
| 5 | 9 | 4 | 5 | 8 | 6 | 2 | 3 | 0 | 6 | 2 | 9 | 8 | 6 | 3 | 0 | 4 | 1 | 0 | 7 | 6 | 40 |

# Appendix O
## Formulas

**Formula number** page

**Formula number** page

**Formula number** page

**Formula number**                                                        **page**

**Formula number**                                                      **page**

## Formula number page

# Appendix P
Answers to Exercises

## Chapter 2

1. a. 110      e. 2.61
   b. −75.4      f. 20
   c. .000049      g. −1.1
   d. −.056      h. 7000

2. a. 2.08      i. 0
   b. .62      j. 625
   c. 1.62      k. −61
   d. .47      l. $x^2y$
   e. −.21      m. .56
   f. 1.17      n. $x^2 + 2xy + y^2$
   g. .53      o. $fx^2$
   h. 1120      p. $x^4y^2$

3. a. 87.9      f. .0084
   b. 278      g. .013
   c. 2.78      h. 2.65
   d. 386      i. .35
   e. 38.6      j. 276.5

4. a. 13.5      d. .1
   b. 42.2      e. 29.2
   c. 17.9      f. 29.8

5. a. 5      d. 2
   b. 4      e. 5
   c. 2

6. a. .047      e. 1.708
   b. .375      f. .002
   c. .454      g. .318
   d. −.045

## Chapter 3

| | | Exact Limits | Midpoint | Interval Size |
|---|---|---|---|---|
| 1. | a. | 2.5–5.5 | 4 | 3 |
| | b. | 24.5–29.5 | 27 | 5 |
| | c. | 19.5–29.5 | 24.5 | 10 |
| | d. | (−5.5)–5.5 | 0 | 11 |
| | e. | (−9.5)–(−4.5) | −7 | 5 |
| | f. | .45–1.45 | .95 | 1 |
| | g. | .245–.495 | .37 | .25 |

2.  a.  1
    b.  3, 4, 5
    c.  10—easiest to use

    d.  2
    e.  .3, .4, or .5

5.  a.  28.5    e.  44.9
    b.  55.3    f.  37.2
    c.  35.1    g.  50.5
    d.  51.7    h.  49.4

6.  a.  6    d.  4.3
    b.  5.5    e.  6
    c.  7.9

8.  a.  2.3    c.  32
    b.  13    d.  98

## Chapter 4

1.  $X = 7.9$
2.  $X = 55.4$; Mdn $= 55.5$; Mo $= 55$

|   | X | Mdn | Mo |
|---|---|---|---|
| 3. a. | 15.0 | 15.0 | — |
| b. | 11.4 | 10.0 | 8 |
| c. | 7.8 | 7.2 | 7 |
| d. | 18.1 | 18.1 | 18 |

4.  $X = 23.9$; Mdn $= 23.9$; Mo $= 24$
5.  a.  Mdn $= 1.8$
    b.  $Q_1 = .9$; $Q_3 = 3.1$
6.  a.  Mdn $= 13512$
    b.  $Q_1 = 11524.5$; $Q_3 = 14502.62$
7.  a.  44.2
    b.  50.9
8.  49.3 mph
9.  48.2 mph

## Chapter 5

1.  (1) $s = 3.1$    (3) $s =$ a: 2.9; b: 3.7; c: 2.1; d: 1.6
    (2) $s = 9.9$    (4) $s = 2.2$
2.  $X = 29.6$; $s = 8.2$
3.  $X = 531.8$; $s = 10$
4.  $X = 85.2$; $s = 13.1$
5.  5: $Q = 1.1$; 6: $Q = 1489.06$
6.  $Q = 2.2$
7.  $g_1 = .59$; $g_2 = -.95$

**Chapter 6**

1. a. 2.5; 1.0; 0; −1.33; −2
   b. 75; 60; 50; 37; 30
   c. 3; 79; 250; 454; 489
   d. 99; 84; 50; 9; 2
2. a. 38
   b. 475/10000 or 5/100
   c. 125
   d. $p = .1558$
   e. 724
3. a. 47, 38, 69
   b. 37%; 11%; 97%
4. A, 67 and above; B, 58–66; C, 43–57; D, 34–42; F, 33 and below
6. a. 48%; 22%; 46%
   b. 43.5%, 57 cases; 12.1%, 16 cases; 3.58%, 5 cases
   c. 25; 22

**Chapter 7**

3. $r = -.86$
4. c. $r = .005$
5. a. $r = .93$
   b. Mental ability: $X = 43.8$; $s = 6.1$
      English: $X = 175.6$; $s = 19.4$
6. $r = .83$
7.

| | WAIS* | J | C | ML | P & H |
|---|---|---|---|---|---|
| WAIS | — | −.55 | .54 | −.74 | −.50 |
| J | | — | −.86 | .82 | .68 |
| C | | | — | −.87 | −.90 |
| ML | | | | — | .73 |
| P & H | | | | | — |

* All $r$'s computed by raw-score formula.

**Chapter 8**

1. .45
2. .54
3. .71
4. .40
5. .59
6. .31
7. .10
8. .92

9. .82

10. a. .44
    b. −.006

11. $R = .54$

12.

| | Minn.Cl. Nos. | Minn.Cl. Names | MPFB | Otis |
|---|---|---|---|---|
| Minn.Cl. Nos. | $x$ | .76 | .60 | .35 |
| Minn.Cl. Names | | $x$ | .34 | .41 |
| MPFB | | | $x$ | .72 |
| Otis | | | | $x$ |

## Chapter 9

1. a. 7
   b. 24
   c. −7
   d. 50.88

2. $Y' = 5.3$

3. .19

4. a. $Y' = 3.8 + .35X$
   b. $s_{yx} = 2.1$

5. b. $b_{yx} = .70; a_{yx} = 7.0$
      $b_{xy} = 1.2; a_{xy} = 3.1$
   d. $s_{yx} = 9.7; s_{xy} = 12.8$
   e. 90

6. a. $Y' = 5.0$
   b. $Y' = 4.2$
   c. $s_{yx} = .33$

7. a. 0    c. 1.25
   b. .625    d. −.375

## Chapter 10

1. a. .0769    c. .25
   b. .0385    d. .019
              e. .3077

2. a. .0039
   b. .1445
   c. .1094
   d. .0078

3. a. .50
   b. .00038

 c. .00001
 d. 3 to 1; 51 to 1

4. b. .1667
 c. .1667
 d. .1944
 e. .50

5. a. .001  d. .064
 b. .001  e. .002
 c. .216  f. .729

6. a. .21   c. .0625
 b. .09

7. .0087

8. $p = .19$; No

9. a. 8
 b. .084
 c. .046

10. Coca-Cola: $p = .08$; Pepsi-Cola: $p = .22$; No

11. a. .0918
 b. .0336
 c. .1836

## Chapter 11

4. .0091

5. .1762

6. 95.9–104.1

7. 95%: 79.06–80.94
 99%: 78.76–81.24

8. .19–.38

## Chapter 12

1. $z = 1.83$

2. $z = 2.50$

3. $t = 1.16; F = 1.14$

4. $t = .76$

5. $t = 3.36$

6. 11.9–23.1

7. $t = 4.81$

8. $z = 4.56$

9. $t = 1.35$

**Chapter 13**

1. a. .53
   b. .61
   c. .35
2. $t = 4.17$
3. $t = 2.15$
4. $t = 1.91$
5. $t = 3.37$
6. a. 1.59
   b. 1.94
7. 95% : .54–.70
   99% : .52–.72

**Chapter 14**

1. 5; yes, at 5% level
2. 1.2; no
3. 4, yes; 1.58, no
4. 8.01; yes, at 1% level
5. 1.76; no
6. 7.99; corrected 7.93
7. 17.5; yes at 1% level
8. 3.61; no
9. 3.56; no
10. 19.73; yes, at .1% level
11. 20.46; yes, at .1% level
12. .95; no
13. a. .31
    b. .23

**Chapter 15**

1. a. $F = 11.37$
   b. $t = 3.37$
2. a. $F = 10.80$
   b. A differs from C at 1% level; A differs from B at the 5% level
3. $F = .167$
4. a. $F = 4.81$
   b. A differs from B at the 5% level

5. a. $F = 23.43$
   b. $F = 2.63$
   c. $r = .96$

6. Between classes: $F = 11.96$
   Between fraternity–non-fraternity: $F = .77$
   Interaction: $F = 1.26$

## Chapter 16

1. $t = 1.52$

2. a. Significant at 2% level    c. Not significant
   b. Not significant    d. Significant at 1% level

3. $z = .83$

4. $z = -8.57$

5. .79

6. a. Not significant
   b. Significant at 5% level
   c. Not significant

7. $t = 2.90$

8. 

| Item | Significance |
|------|--------------|
| 1 | 1% level |
| 2 | 5% level |
| 3 | Not significant |
| 4 | 1% level |
| 5 | 1% level |
| 6 | Not significant |

9. $F = 4.69$, significant beyond 1% point

10. a. Significant at 2% level    c. Significant at 1% level
    b. Significant at 5% level    d. Significant at .1% level

11. a. Significant at 5% level
    b. Not significant

## Chapter 17

1. a. .70    c. .33
   b. .76    d. 1.6 using the K-R 20 coefficient

2. a. .91
   b. .82
   c. 100

3. a. 9
   c. 12.3

5. .50

8. a. .20
   b. .24

Chapter 18

1. $p = .0390$; two-tailed test
2. $z = 1.81$
3. Reject $H_0$ at 1% level; one-tailed test
4. $X^2 = .81$
5. $X^2 = 3.17$
6. $U = 31$
7. $U = 42.5$
8. $z = 1.86$
9. Difference not significant
10. $H = 3.28$
11. Median test; $X^2 = 9.94$; $H = 11.84$

# References

American Psychological Association. *Standards for Educational and Psychological Tests and Manuals.* Washington, D.C.: American Psychological Association, 1966.

Bartlett, M. S. "Some Examples of Statistical Research in Agriculture and Applied Biology." *Journal of the Royal Statistical Society,* 1937, *4,* 137–170.

Boneau, C. A. "The Effects of Violating of Assumptions Underlying the *t*-Test." *Psychological Bulletin,* 1960, *57,* 49–64.

Bradley, J. V. *Distribution-Free Statistical Tests.* Englewood Cliffs, N.J.: Prentice-Hall, 1968.

Cochran, W. G. "Some Consequences When the Assumptions Underlying the Analysis of Variance Have Not Been Met." *Biometrics,* 1947, *3,* 22–28.

Cochran, W. G. and G. M. Cox. *Experimental Design,* 2nd ed., New York: Wiley, 1957.

Curtis, E. W. "Predictive Value Applied to Predictive Validity." *American Psychologist,* 1971, *26,* 908–914.

Downie, N. M. *Fundamentals of Measurement,* 2nd ed. New York: Oxford University Press, 1967.

Dudycha, A. L. and L. W. Dudycha. "Behavioral Statistics: An Historical Perspective," in Kirk, R. E., *Statistical Issues. A Reader for the Behavioral Sciences.* Monterey, Calif.: Brooks/Cole, 1972.

Ebel, R. L. "Estimation of the Reliability of Ratings." *Psychometrika,* 1951, *16,* 407–424.

Ebel, R. L. *Essentials of Educational Measurement,* 2nd ed. Englewood Cliffs, N.J.: Prentice Hall, 1972.

Edwards, A. L. *Experimental Design in Psychological Research,* 3rd ed. New York: Holt, Rinehart & Winston, 1968.

Edwards, A. L. *Statistical Methods,* 2nd ed. New York: Holt, Rinehart & Winston, 1967.

Ferguson, G. A. *Statistical Analysis in Psychology and Education*, 2nd ed. New York: McGraw-Hill, 1966.

Fienberg, S. E. "Randomization and Social Affairs: The 1970 Draft Lottery." *Science*, 1971, *171*, 255–261.

Guilford, J. P. *Psychometric Methods*, rev. ed. New York: McGraw-Hill, 1954.

Guilford, J. P. *Fundamental Statistics in Psychology and Education*, 4th ed. New York: McGraw-Hill, 1965.

Gulliksen, H. *Theory of Mental Tests*. New York: Wiley, 1950.

Hays, W. L. *Statistics*. New York: Holt, Rinehart & Winston, 1965.

Henryssen, S. "Gathering, Analyzing, and Using Data on Test Items," in Thorndike, R. L., *Educational Measurement*, 2nd ed. Washington, D.C.: American Council on Education, 1971, pp. 130–159.

Hicks, C. R. *Fundamental Concepts in the Design of Experiments*. New York: Holt, Rinehart & Winston, 1964.

Kraft, C. H. and VanEeden, C. *A Nonparametric Introduction to Statistics*. New York: Macmillan, 1968.

Lord, F. M. "Nomograph for Computing Multiple Correlation Coefficients." *Journal of the American Statistical Association*, 1955, *50*, 1073–1077.

Lord, F. M. "Tests of the Same Length Do Have the Same Standard Error." *Educational and Psychological Measurement*, 1959, *19*, 233–239.

Magnusson, D. *Test Theory*. Reading. Mass.: Addison-Wesley, 1967.

McNemar, Q. *Psychological Statistics*, 4th ed. New York: Wiley, 1969.

Moses, E. L. "Non-Parametric Statistics for Psychological Research." *Psychological Bulletin*, 1952, *49*, 122–143.

Nunnally, J. *Psychometric Theory*. New York: McGraw-Hill, 1967.

Pierce, A. *Fundamentals of Nonparametric Statistics*. Belmont, Calif.: Dickenson, 1970.

Pirie, W. R. and Hamden, M. A., "Some Revised Continuity Corrections for Discrete Data." *Biometrics*, 1972, *28*, 693–701.

Plackett, R. L. "The Continuity Correction on 2 × 2 Tables." *Biometrika*, 1964, *51*, 327–337.

Rosenblatt, J. R. and Filliben, J. J. "Randomization and the Draft Lottery." *Science*, 1971, *171*, 306–308.

Scheffé, H. *The Analysis of Variance*. New York: Wiley, 1957.

Selltiz, C. et al. *Research Methods in Social Relations*. New York: Holt, Rinehart & Winston, 1959.

Siegel, S. *Nonparametric Statistics for the Behavioral Sciences*. New York: McGraw-Hill, 1956.

Snedecor, G. W. and Cochran, W. G. *Statistical Methods*, 6th ed. Ames: Iowa State College Press, 1967.

Stanley, J. C. "Reliability," in Thorndike, R. L. *Educational Measurement*, 2nd ed. Washington, D.C.: American Council on Education, 1971, pp. 356–442.

Stevens, S. S. "Measurement, Statistics, and the Schemapiric View." *Science*, 1968, *161*, 845–856.

Tate, M. W. and Clelland, R. C. *Non-Parametric and Short-Cut Statistics*. Danville, Ill.: Interstate Printers and Publishers, 1957.
Thorndike, R. L. *Educational Measurement*, 2nd ed. Washington, D.C.: American Council on Education, 1971.
Tukey, J. W. "Comparing Individual Means in the Analysis of Variance." *Biometrics*, 1949, *5*, 99–114.

Walker, H. M. *Studies in the History of Statistical Method*. Baltimore, Md.: Williams and Wilkins, 1929.
Walker, H. M. and Lev, J. *Statistical Inference*. New York: Holt, Rinehart & Winston, 1953.
Winer, B. J. *Statistical Principles in Experimental Design*, 2nd ed. New York: McGraw-Hill, 1971.

# Index

74 75 76   9 8 7 6 5 4 3